JULIAN HUXLEY

Biologist and Statesman of Science

Julian Huxley

Rice Institute, ca. 1915

Julian Huxley

Biologist and Statesman of Science

PROCEEDINGS OF A CONFERENCE
HELD AT RICE UNIVERSITY
25–27 SEPTEMBER 1987
EDITED BY

C. Kenneth Waters
Albert Van Helden

RICE UNIVERSITY PRESS

Houston, Texas

Copyright 1992 by C. Kenneth Waters and Albert Van Helden
Printed in the United States of America
First Edition
Requests for permission to reproduce material from this work
should be addressed to

Rice University Press
Post Office Box 1892
Houston, Texas 77251

Book design by Patricia D. Crowder

LIBRARY OF CONGRESS CATALOGING-IN-PUBLICATION DATA

Julian Huxley: biologist and statesman of science / edited by C. Kenneth Waters,
Albert Van Helden. — 1st ed.
p. cm.
Based on a conference held at Rice University, Sept. 25–27, 1987.
Includes bibliographical references and index.
ISBN 0–89263–314-X : $32.50
1. Huxley, Julian, 1887–1975—Congresses. 2. Biologists—England—
Biography—Congresses. I. Waters, C. Kenneth, 1956–   . II. Van Helden, Albert.
QH31.H88J85   1992
574'.092—dc20
[B]   92-50136
                                                                              CIP

Preface

Julian Huxley (1887–1975) taught at the Rice Institute for three years, from 1913 to 1916. At that time the Rice Institute had just opened its doors to its first class of students, and Huxley was at the beginning of his career. Huxley was one member of an outstanding faculty that Rice's first president, Edgar Odell Lovett, had recruited by traveling all over the United States and Europe. During his three years at the Rice Institute, Huxley founded the Biology Department and recruited a faculty, among whom was Hermann Muller. Huxley went on to a brilliant career as a biologist, statesman, and intellectual, but he never forgot his sojourn in Texas.

After Sir Julian's death in 1975, Lady Huxley made it possible for Rice (now a university) to obtain his papers. The papers were deposited in the Woodson Research Center of the Rice University library in 1980 and, after organization and cataloging, were opened to scholars in 1984. Lady Huxley was present at the opening ceremony.

In 1987 Rice held a symposium to mark the centennial of the founder of its Biology Department. At this event, which took place from 25 to 27 September, papers on various aspects of Huxley's career were delivered by scholars from the United States and the United Kingdom (see list of participants). This volume stems from the papers delivered on this occasion.

Julian Huxley was a man of many talents and enormous energy, whose many activities and prodigious intellectual output could not possibly be covered completely in a three-day meeting. We chose, therefore, to concentrate on his research in biology and his statesmanship of science. The resulting collection of papers provides the first comprehensive account of Huxley's contributions to field and laboratory biology. In addition, they are the first in-depth examinations of his efforts to popularize science and to advance the human species through eugenics. We think these essays should generate greater interest in Huxley and encourage scholars to examine important facets of his thinking and career that are not represented or only touched upon in this volume. In particular, we hope this volume will spur further research into Huxley's views on religion and his work as secretary of the London Zoological Society and as the first director-general of UNESCO. What this book is intended to offer, then, is a substantial beginning for serious scholarship on Huxley that might lead to further critical examinations of his career and perhaps a complete intellectual biography.

In organizing the symposium and editing the proceedings, we have incurred a number of debts. First of all, we thank Mr. and Mrs. C. M. Hudspeth, whose abiding interest in this university and generous and continuing contributions made this project possible. We also thank Mrs. Hardin Craig, Jr., Mr. and Mrs. David Hannah, Mr. and Mrs. John F. Heard, and the M. D. Anderson Foundation for their financial support. British Caledonian Airways (now British Airways) underwrote the cost of transatlantic travel. Martin Hime, Her Britannic Majesty's Consul General in Houston, encouraged our effort and gave invaluable advice. Mary G. Winkler helped symposium participants in locating manuscript sources in the Huxley Papers. Nancy Boothe made the collection available to participants, mounted an exhibition, and entertained the symposium participants in the Woodson Research Center. Franz R. Brotzen, Don H. Johnson, Robert L. Patten, Ronald L. Sass, and William L. Taylor, Jr. helped plan the symposium. Roger Chadwick, Kenneth A. DeVille, Don H. Johnson, Lea R. Strichartz, and Charles L. Zelden made certain that the event ran smoothly. Ronald L. Sass was a most entertaining after-dinner speaker. Stephen J. Gould delivered a public lecture. Last but not least, Susan Stewart was involved in the planning and execution of the symposium as well as the preparation of the manuscript. To her energy and intelligence we owe a large part of the success of this project.

C. Kenneth Waters
Albert Van Helden
August 1992

Contents

Conference Participants

Garland E. Allen
Department of Biology / Washington University

Carl Jay Bajema
Department of Biology / Grand Valley State College

Elazar Barkan
Division of the Humanities and Social Sciences /
California Institute of Technology

John Beatty
Department of Ecology and Behavioral Biology / University of Minnesota

Ludy T. Benjamin
Department of Psychology / Texas A&M University

Richard W. Burkhardt, Jr.
Department of History / University of Illinois

Elof A. Carlson
Department of Biochemistry /
State University of New York at Stony Brook

Frederick B. Churchill
Department of History and Philosophy of Science / Indiana University

Colin Divall
Department of Interdisciplinary Studies / Manchester Polytechnic

K. R. Dronamraju
Genetics Center / Houston, Texas

John R. Durant
Science Museum / London

Stephen Jay Gould
Museum of Comparative Zoology / Harvard University

John C. Greene
Department of History / University of Connecticut

Daniel J. Kevles
Division of Humanities and Social Sciences /
California Institute of Technology

Otto Klineberg
Department of Psychology / Columbia University

D. L. LeMahieu
Department of History / Lake Forest College

Francis L. Loewenheim
Department of History / Rice University

Everett Mendelsohn
Department of the History of Science / Harvard University

Robert Olby
Division of History and Philosophy of Science / University of Leeds

Robert L. Patten
Department of English / Rice University

Diane B. Paul
Department of Political Science / University of Massachusetts at Boston

William B. Provine
Section of Ecology and Systematics / Cornell University

Michael Ruse
Departments of History and Philosophy / University of Guelph

Peter Stansky
Department of History / Stanford University

Martin J. Wiener
Department of History / Rice University

J. A. Witkowski
Banbury Center / Cold Spring Harbor Laboratory

Solly Zuckerman
University of East Anglia

C. KENNETH WATERS

Introduction:
Revising Our Picture of Julian Huxley

Julian Huxley did not fit the traditional mold of a research scientist working in isolation on narrow and esoteric problems. Some have suggested that he was more a statesman of science than a research scientist. In previous decades, this might have made him an unlikely candidate for historical attention, but now that science historians are taking greater interest in the wider dimensions of science, Huxley makes an ideal and timely subject for study. The primary aim of this volume is to gain a better understanding of Julian Huxley as both biologist and statesman of science.

As a biologist, Huxley covered an extraordinary range of topics in his research, though in general his laboratory work was related to the development of individual organisms and his field work concerned evolution—in particular the evolution of ritualized behavior exhibited by birds. In addition to dozens of substantial articles, Huxley wrote three major scientific books in which he attempted to synthesize a broad range of biological findings.[1]

Huxley's ambitions extended beyond the confines of professional biology, and he wrote a number of essays and books relating biological theories to areas of general humanistic concern, including religion and ethics. Like his famous grandfather, T. H. Huxley (better known as "Darwin's bulldog"), he was a popularizer. He wrote hundreds of articles and nearly two dozen books on popular science. He also directed the London zoo for several years as fulltime secretary of the London Zoological Society. His statesmanship was not limited to popularization, and he had a keen interest in the possibility of using scientific knowledge for the improvement of the human condition. This interest motivated his work in eugenics and eventually led him well beyond science when he became the first director-general of UNESCO.

The essays that follow challenge standing views about Huxley's contributions to the growth of twentieth-century experimental biology, the founding of ethology, the synthesis of evolutionary theory, the reform of eugenics, and the popularization of science. The essays also have something to say about these developments themselves. Hence, this volume's second aim is to help refine our understanding of central themes in twentieth-century biological thought by examining them through the contributions of Huxley.

This introduction includes a biographical sketch followed by an overview of the volume's essays and a brief discussion of how these essays will revise our understanding of Huxley as biologist and statesman of science.

Biographical Sketch

Julian Huxley's life consisted of a chain of intense periods—stints of brilliant laboratory and field research, episodes of self-doubt and clinical depression, intervals of writing prodigious synthetic works, and periods of public administration and global politicking. While the focus of Huxley's professional activities was never fixed for long, the motivation and interests underlying his work remained fairly constant. He sought to develop grand syntheses in biology, to create a religion of evolutionary humanism based on biology, and to bring these efforts to fruition through popularization and liberal political action. He was also motivated by personal ambition. He wished to live up to his intellectual heritage by establishing himself among the British intellectual elite, and he clearly enjoyed his reigns of power and influence in public administration. This summation of Huxley's motivations, however, fails to do justice to the complexity of his personality and the resulting tension in his works. Although it is true that Huxley cared immensely about science popularization, he also championed views tinged by elitism and declined to support science education in third-world countries as director-general of UNESCO. While his political philosophy was one of liberalism, his views on eugenics and birth control betrayed remnants of racism. And despite the time and effort he devoted to acquiring public offices, his failure to work well with governing boards forced his early departures. This is the story of a man whose extraordinary talents and hard work often propelled him to the heights of success, but whose personality and insecurities periodically marred his achievements and literally drove him to the depths of despair.

Julian Huxley came in with the fireworks and festivities of Queen Victoria's Jubilee, on 22 June 1887. He was, as one writer has put it, part of a "remarkable intellectual crop." His paternal grandfather was the eminent biologist T. H. Huxley, who was catapulted to fame as public defender of Darwin's theory of evolution. The maternal side of his family included his grandfather Thomas Arnold, famous headmaster of Rugby School, his great uncle the poet Matthew Arnold, and his aunt, a novelist writing under the name of Mrs. Humphry Ward. One of his two half-siblings, Andrew Huxley, became a physiologist and Nobel laureate.[2]

Huxley's father, Leonard Huxley, was also an intellect. After taking Firsts in Classical Moderations and in Literae Humaniores, he left Oxford to accept a Junior Mastership at Charterhouse School and to marry Julia Arnold. Upon completion of his father's official biography[3] Leonard left Charterhouse in 1900 to help edit *The Cornhill Magazine*. He eventually became a literary force in his own right as sole editor of the *Cornhill* and intellectual biographer. His wife, Julia Huxley, founded a girls' school, Prior's Field, which she successfully ran until her death in 1908.

When Julia died at the age of forty-seven, she left three sons and a daughter, all of whom were being groomed for intellectual futures. Julian, during the year of his mother's death, won the Newdigate Prize for English Verse and scored a First in Natural Science at Oxford. Trevenen, born in 1889, was on scholarship at Oxford, where he later took a First in mathematics in Moderations, and a Second in Greats. After postgraduate work at Oxford, however, he suffered a bout of severe depression and committed suicide at the age of twenty-five. Aldous, born in 1894, was preparing to go to Oxford, where he took a First in English. Aldous became famous for writing perceptive short stories, intelligent screenplays, and brilliant novels, including the modern classic *Brave New World*.[4] Margaret, born in 1899, was attending Prior's Field at the time of her mother's death and in adulthood founded her own girls' school at Bexhill. It is to the eldest brother, Julian, who took up the scientific legacy of their famous grandfather T. H. Huxley, that our attention now turns.

During his boyhood years, which were spent mostly in the English countryside, Julian Huxley showed a keen interest in observing nature, an interest that was encouraged by his family. In later years, he was especially proud of the letter in which his grandfather, T. H. Huxley, predicted of the precocious five-year-old Julian: "There are some people who see a great deal and some who see very little in the same things. When you grow up I dare say you will be one of the great-deal seers and see things more wonderful than water babies where other folks can see nothing."[5]

Huxley started his formal education at Hillside Preparatory School in 1897. Three years later, he entered Eton College, where his grandfather Huxley had formerly been governor. At Eton, he was influenced by M. D. Hill, master of biology, who Huxley said helped settle his career plans.[6] The writings of Edmund Selous, as reported by J. R. Baker, also exerted a major influence and convinced the ambitious student, who had already practiced birdwatching for four or five years, that there was still much to discover.[7] After finishing at Eton, Huxley went on to Balliol College, Oxford.

In 1908, Huxley won the Naples Biological Scholarship to conduct research at the Naples Stazione Zoologica (which his grandfather had helped rescue from financial difficulty thirty years before). He visited the research station after finishing his degree at Oxford and performed experiments on the development of sponges by separating them into their individual cells and observing the ways in which they reformed and developed. The results of his laboratory work were published,[8] but when he returned to Oxford in 1910 as lecturer at Balliol College and demonstrator in the Department of Zoology and Comparative Anatomy, he directed much of his attention to natural history—and especially to ornithology. During his first vacation from Oxford, Huxley began an important series of studies on the evolution of bird courtship rituals, which he conducted part time over the next decade and a half.[9]

Huxley's stay at Oxford was brief. During his second year as lecturer, he sought another position because, as he explained in a letter of inquiry, "I am informally engaged to be married, and want to become formally so, and this I cannot very well do until I can tell my fiance's father something more definite about my future prospects."[10] He was later offered and accepted a post as assistant professor and founding head of the Department of Biology at Rice Institute in Houston, Texas.

On his way to the inauguration ceremonies of Rice Institute, young Huxley was reportedly greeted by headlines in a New York newspaper proclaiming: "Huxley, Zoologist, Here."[11] His grandfather's much-celebrated lectures in America had not been forgotten. While in New York, he met a number of prominent biologists and visited the famous Morgan fly room. Huxley was favorably impressed with one of Morgan's graduate students, H. J. Muller, whom he hired to be his assistant at Rice Institute. (Muller later won a Nobel Prize for his work in genetics.) After attending the inauguration and looking into the housing situation for himself and his future bride, Huxley traveled to Europe for a year of study in Germany, which was financed by his new employer. But his plans were spoiled when his engagement to Kathleen Fordham was broken and he experienced his first nervous breakdown. Several months later, he recovered from his crisis and returned to Texas, where he stayed, with the exception of one summer in England and another at Woods Hole Marine Biological Laboratory on the coast of Massachusetts, until the outbreak of World War I.

In Texas, Huxley carried out work with Muller on genetics, studied relative growth of the fiddler crab, and pursued his ornithological research by observing an egret colony in Louisiana. He also devoted time to designing and teaching an innovative biology curriculum, which emphasized laboratory and field instruction. Despite his apparent productivity in Texas, Huxley suffered from frequent ailments, which prevented him from carrying out a number of his teaching duties.[12] At the outbreak of World War I, he decided to return home to join the war effort. When Huxley left Rice, he gave a series of public lectures entitled "Biology and Man," presenting his highly controversial views on science and religion, which he referred to as "scientific humanism" (and in later years as "evolutionary humanism"). He further developed these views in *Religion Without Revelation*,[13] where he argued that discoveries in physiology, general biology, and psychology force us to repudiate traditional supernaturalistic religions and that evolutionary biology provides the key for understanding human destiny.

Huxley first supported the British war effort by taking a post in the Censor's office, but was bored and soon enlisted in the Army Service Corps. Shortly thereafter he arranged a transfer to Intelligence and was assigned to a unit at Colchester.[14] It was during a leave from this unit that he became acquainted with Juliette Baillot, a French-Swiss woman ten years his junior. He met Baillot, the governess of Lady Ottoline Morrell's daughter, when he visited his brother Aldous at Lady Ottoline's Garsington Manor. As Juliette recalls, Julian "seemed to climb up to the schoolroom quite often, and I was surprised that he should, wondering what he found in little Julian [Lady Ottoline's ten-year-old daughter] to merit his attention."[15] Huxley was soon sent to the General Headquarters in Italy, where he served as a lieutenant in the Intelligence Corps. Huxley and Bailliot entered a courtship by correspondence and married in the spring of 1920, within three months of Huxley's return to Britain.

After the war Huxley returned to Oxford, this time as fellow of New College and senior demonstrator in the Department of Zoology and Comparative Anatomy. He discovered that his basic knowledge of biology had decayed during his three years with the service and apparently doubted whether he was equal to his appointed tasks at Oxford.[16] He became increasingly depressed and by the end of his first term suffered his second nervous breakdown. His young bride took him to a doctor in Switzerland, where they stayed for several months while he recovered.[17] They returned to Oxford at the end of the summer, and Huxley began a productive period of teaching and research.

During his postwar years at Oxford, Huxley taught a number of promising students, including John Baker, Gavin de Beer, Charles Elton, E. B. Ford, and Alister Hardy. He characterized them as "brilliantly original men who all made their mark."[18] In later years, one of these students, John Baker, said Huxley had "the capacity to stimulate the possessors of eager minds, anxious to forget the war and devote themselves to thought and learning. . . . Beginning straightaway on his arrival back in Oxford, he gave a course of 30 lectures on experimental zoology, and followed it up in the next term with a course of 30 lectures on genetics, and this by another course on animal behaviour."[19] Evidently, Huxley's students thought he was more than competent in his role as senior demonstrator, an assessment at odds with the self-doubts that elicited Huxley's second breakdown.

Huxley's research interests remained diverse during these six years at Oxford (1917–1925). He resumed his ornithological studies, describing the courtship behavior of different bird species and theorizing about the evolutionary origins of their rituals. In addition, he embarked on laboratory studies that continued through the 1920s and well into the 1930s. Most of them focused on the growth and development of individual organisms, and he published a number of articles on dedifferentiation, morphogenesis, the hormonal control of growth and development, and rate genes. Perhaps his most important contribution to laboratory research was his simple allometry formula that related the growth of different parts of an organism.[20]

Huxley's first claim to fame, however, came with a short article in *Nature*,[21] which was misinterpreted in the popular press. The article concerned his laboratory induction of the metamorphosis of the axolotl. Huxley fed thyroid gland of the ox to larval (or "tadpole") forms of this amphibian and observed the resorption of their gills and dorsal fins, the development of their capacity to breathe air, and their movement from water to land. The larval form of this amphibian can reproduce without metamorphosing, and although naturalists had long known that the adult form existed in the laboratory and under special natural conditions,[22] Huxley believed that the adult form had not existed for many thousands of years. Hence, he claimed to have induced the development of a long-lost ancestral form. The popular press was even more extravagant. Huxley was credited with nothing less than discovering the "Elixir of Life."[23]

Huxley's attempt to clarify his work on the axolotl for the public marked his entry into the field of science popularization. During his postwar years at Oxford he wrote a number of articles on popular science, and his first book on biology for public consumption came out in 1923, *Essays of a Biologist*.[24] This was followed by nearly twenty more books of popular science. He also pursued his interests in issues of wider concern, for instance the ramifications of biological knowledge for the humanities and public policy. His early views on these issues can be traced in his popular writings, though most of his major contributions to these areas were yet to come.

It was during their postwar years at Oxford that Julian and Juliette started a family. They had two sons, both of whom developed the family interest in nature. Anthony, born in 1920, became a botanist. Francis, born three years after Anthony, took up social anthropology. The Huxley family seems to have been a reasonably happy one during the Oxford years, and Juliette supported her husband's work in numerous ways, including tending his laboratory experiments while he was away conducting field research.[25]

In 1925, with little prospect of promotion at Oxford, Huxley accepted a chair in zoology at King's College at the University of London. Although his field studies were drawing to an end, he continued his experimental research at his new college. During his second year in the new position, he began to write an encyclopedic work on biology with H. G. Wells and Wells's son, then a zoologist at University College in London. The elder Wells, who had already written a successful encyclopedic work on history, invited Huxley to join the new venture and seems to have played the role of editor and taskmaster. As the following passage from a letter to Huxley makes clear, Wells was extremely demanding of his junior collaborators:

> If a satisfactory mass of material does not show up by Jan. 1st 1928 then I shall cancel the whole thing, return Doran [the American publisher] his money and wash my hands of the project. I can't cluck after you and Gip like an old hen after ducklings. I thought the thing was good enough for both of you to work hard and do your best. I find that holidays, research, the Leeds gathering [a scientific meeting of the British Association], a summer holiday, any little thing of that sort, is sufficient to put off work on the *Science of Life*.[26]

The project became burdensome, and in 1927, after just two years at King's College, Huxley astonished his colleagues by resigning his chair, thus enabling himself to devote greater attention to *The Science of Life*.[27] Huxley did most of the original writing of this three-volume work,[28] which was completed in 1930 and was his greatest popular success.

Huxley maintained a frantic pace during the eight years, beginning with his resignation from King's College in 1927 and ending with his appointment as secretary of the Zoological Society of London in 1935. He continued to use his laboratories at King's College during this period and published at least nine substantive articles and two original books on experimental biology, *Problems of Relative Growth* and *The Elements of Experimental Embryology* (with G. R. de Beer).[29] In addition to these substantive works, which were written for professional biologists, he continued to popularize science and pursued a career in broadcasting. He gave many radio talks[30] and held debates over the air with Hyman Levy on topics concerning science and society.[31] Huxley also supported the call to bring scientific education to the world community and visited East and Central Africa at the invitation of the Colonial Office's Committee on Education.

In her autobiography, Juliette Huxley revealed that her marriage with Julian entered a crisis in 1929 when he announced that he was having an affair. Julian met Miss Weldmeier, a young American about eighteen years of age, on a boat trip to Africa and informed Juliette of their affair in a letter written during the trip. "The letter," Juliette recalled, "stated his wish to pursue this affair, his entire right to do so, while he swore his continued devotion to me from whom he was taking absolutely nothing, but promised an enhanced relationship for ourselves."[32] Juliette was deeply hurt and the Huxleys went their separate ways, searching for a solution. Apparently, plans to marry Waldmeier and resettle in America fell through,[33] and Julian finally wrote:

> You are my wife, I am your husband, we care deeply for each other, we want to make a go of things jointly. It is a very true and very big love, which has survived very much, which is alive, which has shed its larval skin of romantic immaturity without losing its deep core. I have learned my lesson; I know myself and you; and I only say what you already know, that I have no thought or dream of wishing for any other woman as my life companion.[34]

They reunited in 1932, though as Juliette related in her autobiography:

> That he needed me was undeniable. And just as he needed me, so did I need him, and could not imagine life without his vital concern, his intellectual challenge, the ventures whose accomplishment seemed beyond my capacity yet which he so confidently trusted me with. . . . His

somewhat brutal decision against "leading a smaller life" was really based on a sound instinct and wise balancing act. We both needed outside help, the venture into a different kind of love, which he proposed for himself and equally for me.[35]

The Huxleys were to remain together for the rest of Julian's life, but as Juliette makes clear, they pursued outside ventures.

In addition to his laboratory work, lecturing, scientific and popular writing, trips overseas, and broadcasting, the enterprising Huxley joined the film industry. In the early 1930s, he edited "Cosmos, the Story of Evolution" and prepared the Eugenics Society's "Heredity in Man." He also served as general supervisor of biological films for G. B. Instructional Ltd.[36] His greatest achievement in film was undoubtedly his making of "The Private Life of the Gannet" with John Grierson and Ronald M. Lockley. The film documented the nesting and feeding habits of the great white sea birds, capturing their elaborate display behavior as well as their spectacular aerial dives. The film won an Oscar for best documentary of the year and clinched Huxley's candidacy for the head administrative post of the Zoological Society of London.[37]

In 1935, when Huxley was appointed secretary of the Zoological Society, he was a well-known public figure. He had already written several popular books on science, regularly contributed articles to magazines such as *The Spectator*, and appeared frequently on the BBC. The prevalence of his name on lists of "Britain's Five Best Brains" drawn up by readers of *The Spectator* (in 1930) indicates that he had acquired a relatively high public profile—he ranked sixteenth overall, ahead of both Ernest Rutherford and Bertrand Russell.[38] Soon after the outbreak of World War II, Huxley was invited to join a new radio program, "The Brains Trust," that was to be aired by the BBC. Joining Huxley were the philosopher and author C. E. M. Joad and "Commander" Campbell, a former ship's purser who had apparently experienced a wide variety of adventures.[39] The program's host surprised team members with questions submitted by the public. The three then argued about how the questions should be answered. The show was a popular success. Huxley later explained, "The combination of an argumentative philosopher with an equally argumentative biologist, and an endearing buffoon as foil to the two intellectuals, proved irresistible."[40]

In many ways, Huxley was ideally suited to his new position as secretary of the Zoological Society. His experience as a practicing scientist put him in an excellent position to advise the society on how to sponsor pure research, though the society supported little scientific research during his reign. His experience and flair as a science educator and popularizer, however, were fully utilized. He brought about important innovations—the zoo started to sponsor regular public lectures for children (given by Huxley and the curators), present special exhibits designed to illustrate scientific principles, create Pet's Corner (a children's zoo), publish *Zoo Magazine*, produce numerous films, and establish the Studio of Animal Art. In addition, several new exhibits were created and antiquated ones replaced during Huxley's years. The zoo's improvements and Huxley's moves for gaining publicity, which included creating the post of public relations officer, paid off in a dramatic increase in children's attendance.[41] Nevertheless, Huxley's performance must have been less than ideal and he was pushed out of his leadership post while visiting the United States in 1942. It is not clear why the society's fellows forced Huxley to resign.[42] Huxley blamed it on their reluctance to accept his progressive programs, but there is reason to think that his headstrong personality may have been an important contributing factor.[43]

The Zoological Society provided Huxley with a healthy income (which he voluntarily cut in half after the war began), a spacious office, and a comfortable home at the London facility as well as a flat at the country branch of the zoo in Whipsnade. As Huxley's account of his years at the zoo makes clear, he and his family enjoyed a wonderful lifestyle:

> Our greatest pleasure was to spend the week-end on the job at Whipsnade. A part of the grounds was not utilized for animals, and in summer we used to camp there, first in tents and later in a couple of shacks. In winter, we slept comfortably in the flat, and bought food from the restaurant for picnics, looking at the magnificent view across the valley to Ivinghoe Beacon. We bought a couple of Iceland ponies, Thor and Odin, which we and the boys used to ride round the park before the gates were open to the public, or after they were closed. The tigers stalked them with grim if vain purpose, the polar bears scared them, and the ponies were all too anxious to get on. Elsewhere, however, they trotted happily, enjoying the exercise as much as we did.[44]

Huxley also recalled how his sons, Anthony and Francis, provided their own suppers at Whipsnade by cooking rabbits hunted with .22 rifles. A "valuable training in self-sufficiency," the elder Huxley remarked.[45]

Huxley had always been interested in evolution, and the theoretical import of his field work on birds concerned the selective mechanism behind the evolutionary origin of their behavior. But the bulk of Huxley's research publications, if not his popular writings, had thus far stemmed from his work in the laboratory (see Churchill's analysis, this volume). This changed when Huxley went to the zoo. He abandoned his laboratory at King's College and redirected his research attention toward evolutionary theory.

From 1936 to 1941, Huxley wrote several articles on evolutionary biology, coined such key terms as "cline," and wrote the book that gave the name to the consensus emerging in evolutionary biology: *Evolution, the Modern Synthesis*. The central neo-Darwinian theme of this work, which combined elements of the new genetics with Darwin's mechanism of natural selection, was anticipated in several of Huxley's popular writings.[46] But he did not present his synthetic, neo-Darwinian account of evolution to professional biologists until he gave the presidential address to the zoological section of the British Association in 1936.[47] Huxley's work on evolutionary theory, which culminated in *Evolution, the Modern Synthesis*, is typically identified as his most important contribution to biology.

Throughout his lifetime but especially in the 1930s, Huxley pursued a keen interest in eugenics, the highly controversial "science" of "improving" the genetic makeup of the human species. Huxley published a number of articles on eugenics throughout his career, some intended for the public, others for the professional scientist. He was also an active member of the movement's leading professional society, the Eugenics Society, serving as its president from 1959 until 1962.[48] As Allen shows (this volume), he played a key role in the transformation from what has been called "old" eugenics to "new," or "reform" eugenics.[49]

After resigning from the zoo in 1942, Huxley earned his living by giving lectures and talks on the BBC, and kept busy meeting with various groups and committees on higher education and planning. One of these groups was involved with the preliminary plans for forming a United Nations agency concerned with education and culture. Huxley and Joseph Needham led the drive for the inclusion of science as an area of concern and they have been credited with putting the "S" in UNESCO (the United Nations Educational, Scientific and Cultural Organization). In 1944, Huxley traveled to West Africa as a member of the Commission on Higher Education in the British Colonies. By the time he returned to London, during the peak of the blitz, he was suffering from another nervous breakdown. This time, he received electric shock therapy, which, in his own words, "did me good."[50] He recovered in time to travel to Russia, where he and other leading scientists were invited to celebrate the bicentenary of the Academy of Sciences, which happened to coincide with the end of the war. Huxley was appalled at the state of Soviet genetics, which had suffered tremendously at the hands of Lysenko, and called for intellectual freedom in *Soviet Genetics and World Science*.[51]

When he returned from Russia in the spring of 1945, Huxley replaced the ailing Sir Alfred Zimmern as fulltime secretary of the UNESCO Preparatory Commission, which was charged with drawing up a charter and defining the scope of the future United Nations agency. He quickly wrote a pamphlet, "UNESCO, Its Purpose and Philosophy," in which he announced that the future UN organization could not rely on religious doctrines or any of the conflicting systems of academic philosophy. Instead, it was to carry out its work within the framework of "Scientific Humanism," as based on the "established facts of biological adaptation and advance, brought about by means of Darwinian selection, continued into the human sphere by psycho-social pressures, and leading to some kind of advance, even progress, with increased human control and conservation of the environment and of natural forces." The mission of UNESCO, he stated, was to facilitate further advances by supporting the "humanistic ideals of mutual aid, the spread of scientific ideas, and . . . cultural interchange."[52] Huxley's views were attacked as atheism in disguise, and the members of the commission decided not to endorse his document.

UNESCO was formally inaugurated in the fall of the year following Huxley's appointment as secretary of the Preparatory Commission. Huxley was elected its first director-general, though he was offered a term of just two years, rather than the constitutional six.[53]

During his short reign, he traveled the world on behalf of UNESCO, explaining the mission of the new agency to political and academic leaders and exploring ways in which its mission might be furthered around the globe.

As secretary of the Preparatory Commission and then director-general of UNESCO, Huxley advanced many liberal causes, such as the development of national parks for the conservation of sites of natural beauty and historical significance, the establishment of museums around the world to preserve cultural artifacts, and the application of science and technology to improve living conditions in developing countries. He also advanced less liberal causes, such as the selective institution of birth control in developing nations to limit potentially catastrophic increases in human populations. If his views about birth control now appear less than liberal, his stance on education at UNESCO seems downright elitist. Although he was an active popularizer of science at home and had formerly advised his government on colonial education in Africa, he was not a strong advocate of scientific education during his reign at UNESCO.[54] In general, however, Huxley exerted a progressive force and this force was critical in setting the direction of UNESCO during its early years.[55]

After Huxley left UNESCO at the age of sixty-one, he never took another regular position. The remaining twenty-seven years of his life were spent giving lectures, writing, and traveling. During this period, he wrote hundreds of articles and chapters as well as over half a dozen new books. Many of his writings were on general topics, including conditions in various countries, social problems, international organizations, evolutionary ethics, and eugenics.

Huxley never resumed his original scientific research and most of his new scientific writings were popularizations. Yet, despite the fact that he had not held a regular academic position for over twenty years and was no longer conducting original research, he continued to play an important role in the scientific profession and was called upon to give key lectures at scientific meetings, to organize scientific conferences, and to support professional societies, including the Ecological Society, the Societies for the Study of Evolution, and the Society for the Study of Animal Behavior, all of which he helped found. He continued to work for the causes he advanced as director-general of UNESCO and remained active in many of the agency's affiliated international commissions.

Huxley and his wife enjoyed international travel and the later portions of his autobiography read like a travel log, rather than the autobiography of an acclaimed biologist.[56] Nevertheless, the naturalist in him emerges with the occasional mention of an observation of an interesting bird in some distant land. Huxley accepted many prizes and awards in his later years, including the Royal Society's Darwin Medal for extending Darwin's work on evolutionary theory, the Kalinga Prize for the popularization of science, the Lasker Award for his contributions to Planned Parenthood, and a gold medal for his outstanding contribution to scientific research relating to conservation from the International Union for Conservation of Nature and Natural Resources and the World Wildlife Fund.

Despite the awards and the words of praise that went with them, Huxley was still subject to periods of intense self-doubt, and he suffered more nervous breakdowns (in 1951, 1957, and 1966). He was offered a knighthood during his breakdown of 1957 and later recalled that "the sense of unworthiness that goes with depressive neuroses made me hesitate to accept it, but I was finally dubbed in 1958."[57] Huxley spent his final years writing his two-volume autobiography[58] and visiting with old friends. He suffered a stroke in 1973 and ill health afterward. Juliette cared for him until he died on Valentine's day, 14 February 1975.

Overview of Essays

This volume is made up of three parts. The first places Huxley in a broad intellectual context and offers an overview of his contributions to biology as they related to major developments in twentieth-century evolutionary theory. The main areas of Huxley's biological work are investigated more deeply in the second set of essays. The third set examines Huxley as a public scientist and takes a new look at his efforts to bring biology and its potential benefits to the community at large.

HUXLEY AND HIS TIMES: THE INTELLECTUAL CONTEXT. Huxley's contributions to UNESCO suggest that he represented a politics of the future and led the way toward a more modern outlook of the world situation. Colin Divall, however, suggests that "Huxley was a Victorian thinker fated to live in an unsympathetic modern age." He argues that Huxley's scientific humanism, the intellectual base upon which his politics rested, came straight out of an attempt to reconcile the rationalist and ethicist traditions of late-Victorian and Edwardian thought.

Divall interprets Huxley's effort to base morality on a scientifically objective account of evolutionary progress as an attempt to bridge the gap between rationalism, which focused on science as the efficient means to achieve social goals, and ethicism, which critically analyzed what the rationalists seem to have taken for granted—the goals themselves. He also argues, however, that Huxley's conception of progress was tied to the philosophical idealism of ethicism, and hence that the tension in late-Victorian and Edwardian political thought reappeared at the core of Huxley's scientific humanism. Peter Stanksy and Martin Wiener discuss precise connections between Huxley's scientific humanism, politics, and his family and national background in support of Divall's thesis that Huxley attempted "to translate the project of Victorian and Edwardian rationalism into the modern hostile world" (Divall, this volume).

Robert Olby's account of Huxley's place in twentieth-century biology also reinforces Divall's view that Huxley was preoccupied with the late-Victorian and Edwardian project of achieving social progress through the rational method of science. Olby suggests that Huxley's great ambition was centered on the idea that "man could control his own evolution." As Olby sees it, Huxley sought a synthesis in biology that could account for evolutionary progress and help us further the advance of our species. While Olby is quick to point out that Huxley was a naturalist, experimentalist, and theoretician, he maintains that in all of Huxley's roles as biologist, his foremost concern was evolution. Olby suggests, for example, that "genetics was for Huxley not an independent specialism, but a resource, the tools of which could be used to tackle problems raised by the critics of Darwinism, and offering a new foundation upon which to construct sound explanations of the phenomena previously attributed to the recapitulation of phylogeny in ontogeny" (Olby, this volume).

HUXLEY THE BIOLOGIST. The four main essays in the second part of this volume deal respectively with Huxley's laboratory work, his embryological synthesis, his contributions to ethology, and his role in the evolutionary synthesis. Although it is generally assumed that Huxley's zeal for evolution was the impetus behind his biological research, Jan Witkowski claims this zeal did not guide Huxley's eclectic studies in experimental biology. Witkowski presents a general overview of Huxley's laboratory research and examines in detail several of his early papers and his groundbreaking studies on relative growth. In addition, he relates Huxley's laboratory work to several themes in twentieth-century biological research, including the apparent dichotomy between experimental and descriptive biology. Elof Carlson's complementary note discusses factors that encouraged Huxley to devote most of his laboratory research to developmental biology.

Frederick Churchill, in a penetrating analysis of Huxley and de Beer's *Elements of Experimental Embryology*,[59] seeks to deepen our understanding of Huxley as a biologist. According to his analysis, Huxley's laboratory studies represented a sustained program of research that played an important role in his and de Beer's innovative synthesis of diverse elements of embryology and Mendelian genetics. As Churchill sees it, the characteristic feature of Huxley the biologist was "an eagerness to see his own research within the larger framework defined by a number of fields" (this volume).

Huxley's contributions to field biology have been widely celebrated as groundbreaking contributions to the emerging science of ethology. Richard Burkhardt's comprehensive account of Huxley's ornithological papers, however, suggests otherwise. Burkhardt argues that despite what has been written by Huxley and others about his role in the founding of ethology, his works did not represent a turning point in the field and should not be credited with making the study of animal behavior "scientifically respectable." Burkhardt shows that the observations and interpretations of Huxley's key articles were clearly anticipated by Selous, Howard, and other amateur bird-watchers early in the century. What Huxley did, according to Burkhardt, was make the work already carried out by amateurs more accessible to the scientific community by formulating their observations *and* interpretations within a scientific framework and publishing his articles in accessible scientific journals.

Solly Zuckerman and John Durant supplement Burkhardt's revisionist essay. Durant argues that the tension between Huxley's desire to see evolution as progressive on the one hand and purposeless on the other can make several apparently unrelated features of Huxley's ethology more comprehensible. Durant's insightful analysis helps explain Huxley's anthropomorphism, his stress of selection stories based on the good of the species, and his tendency to favor accounts based on natural selection over those involving sexual selection. Zuckerman, a friend and colleague of Huxley's, reports that Huxley was interested in establishing himself as an experimentalist rather than a field biologist. As Zuckerman relates, Huxley's circle of intellectual friends were simply not interested in observations of animal behavior.

Huxley's work on evolutionary theory is typically cited as his greatest contribution to biology. William Provine writes about Huxley's version of the neo-Darwinian theory of evolution and in a section reprinted from an earlier essay[60] discusses the consensus reached by evolutionary biologists during the 1930s and 1940s. Although Huxley named the resulting theory "the evolutionary synthesis," Provine argues that Huxley and his colleagues never offered a truly synthetic account of evolution. Instead of combining a number of different theories of evolution into a synthetic whole, they eliminated many proposed theories from contention and reportedly compiled, rather than synthesized, the remaining theoretical mechanisms. In addition to arguing that Huxley misnamed the evolutionary synthesis, Provine critically examines Huxley's attempt to establish a concept of evolutionary progress. As Provine sees it, Huxley simply imposed his own cultural values to yield a conception of progress from the purposeless process of evolution.

Advocates of the view that the evolutionary synthesis of the 1930s and 1940s was a "constriction" rather than a synthesis have sometimes argued that there was a further constriction, or a "hardening of the synthesis," in the late 1940s, and 1950s. John Beatty investigates Huxley's views on evolution during this later period and finds that Huxley acknowledged the further constriction, but stopped short of endorsing it. Beatty points out that despite the constriction and hardening, there was still much room for disagreement concerning the relative importance of the remaining evolutionary mechanisms.

HUXLEY THE STATESMAN OF SCIENCE. Huxley's work in eugenics provides the clearest expression of his lifelong ambition to advance social progress through science. Garland Allen's essay provides an in-depth examination of Huxley's attempt to develop a science for improving the human species. Allen explains how Huxley's interest in eugenics grew out of his social and political environment and describes Huxley's version of reform eugenics as it matured in the 1930s. Huxley's program for eugenics was based on his evolutionary account of how genetic differences might have arisen along class or ethnic lines. He believed that preselective factors played a role in the initial sorting of people into social classes and that once people were separated into classes, selective forces brought about additional differentiation because different classes lived in different environments.

Unlike older eugenicists, Huxley recognized that many or possibly even all of the differences reported to exist between classes might actually be caused by environmental differences rather than by genetic variations. So his program for eugenics included improving the environment of the lower classes as well as reducing their reproductive rates and increasing the reproductive rates of higher classes. Allen describes the environmental changes that Huxley favored and the traits that Huxley thought should be selected. Allen shows that many features of Huxley's reform program were a reflection of the social and political company he kept and the widespread recognition of the abuses wrought by old-style eugenics.

Allen also uncovers distinctive features of Huxley's eugenical program. One of them was his new strategy for decreasing the reproductive rates of the lower classes. In an interesting twist on his marital state of affairs, Huxley suggested that men and women should "consummate the sexual function with those they love, but to fulfill the reproductive function with those whom on perhaps quite different grounds they admire."[61] Allen states that Huxley introduced "population thinking" (from the modern evolutionary synthesis) to eugenics, and was unique in his stress on the diversity within human populations, a stress that led him to reject the concept of race.

Diane Paul takes issue with Allen's suggestion that population thinking and an emphasis on diversity set Huxley's views apart from those of other eugenic reformers. She points out that one of the most hotly debated questions in evolutionary biology during this period concerned the amount of genetic diversity within populations. Huxley was not alone in stressing diversity. Furthermore, Paul cautions against assuming that reformers who stressed diversity in their eugenics held innocent views. As she reminds us, there was a subgenre of eugenic literature explaining "Why the World Needs Morons," and Huxley himself remarked that it takes all kinds to do the world's work. Elazar Barkan, in his commentary, examines Huxley's early views on race and traces their development as they matured. Barkan's examination of Huxley's early writings on race, including those concerning blacks in Texas, reveals that Huxley harbored a "personal aversion to aliens." Barkan traces this aversion and argues that even long after Huxley denounced his former bigotry, there remained a tension in his eugenics between relics of former prejudices and a moral commitment to liberalism.

Huxley is probably best known for his efforts to make scientific knowledge accessible to the public, and Daniel Kevles writes about Huxley's accomplishments as a science popularizer. Huxley's success is attributed to his skillful use of English and his ability to express his "wonder at the marvelous variety and comprehensible logic of life." The central theme of his popular science, according to Kevles's analysis, concerns the social implications of biology and Kevles discusses Huxley's views on evolutionary progress, ethics, and religion. In his commentary, Robert Patten places Huxley's popular writings in the broader context of British culture. He explains that Huxley's mixture of scientific optimism and cultural humanism was well received in the "twilight" of Comtian positivism and Edwardian rationalism.

D. L. LeMahieu challenges the view that Huxley successfully brought scientific knowledge to the public at large. LeMahieu shows that the readership of the "popular" magazines to which Huxley contributed included a small minority of English readers. Furthermore, LeMahieu cites statistics that suggest that Huxley's BBC broadcasts were probably listened to by less than 2 percent of the broadcast audience. He attributes Huxley's high public profile to his participation in the BBC's Brains Trust program, which attracted 20 to 30 percent of radio listeners. But as LeMahieu explains, the format of the show enabled Huxley to popularize his quick wit, not his scientific knowledge.[62] LeMahieu concludes that Huxley was a successful scientific popularizer only in the narrow sense that he reached an audience that at its widest encompassed the educated middle classes, and at its narrowest included only the cultivated elite.

Revising Our Picture of Huxley

Most of the literature on Huxley has been written by Huxley himself or by former colleagues, students, and sympathizers. These writers have chosen to emphasize those works and activities of Huxley that reinforce the received picture of him, a picture that Huxley was largely responsible for drawing. One part of Huxley's past they chose to ignore was his nearly lifelong work in eugenics and his association with the Eugenics Society. With the exception of David Hubback's brief essay,[63] the current literature includes but a handful of sentences on Huxley's active participation in the eugenics movement.[64] This volume will help close such gaps. The essays by Allen, Paul, and Barkan, for example, offer the first in-depth examination of Huxley's attempt to base the endeavor to "improve" the human species on a sound biological footing. Though several useful essays on Huxley have recently appeared,[65] this volume presents the first in-depth and critical accounts of Huxley's work in the laboratory, field, and public arenas of science.

The importance of this volume extends beyond simply filling gaps in the literature. The essays here challenge the received picture of Huxley and demand at least two fundamental revisions in the way we portray him. The first demand is that we look at Huxley not so much as a modern figure anticipating a late-twentieth-century world outlook, but as a figure from the past stubbornly clinging to the agenda of the late-Victorian and Edwardian period. The second revision concerns our portrayal of how Huxley practiced biology. This requires that we resist the temptation to define his scientific practice purely in terms of his interest in evolutionary biology. Instead, we should base our understanding of how he practiced science largely on his research program in experimental embryology and on the way in which he drew upon his own laboratory findings and those from several related fields to develop an embryological synthesis.

Divall's essay sets the context for viewing Huxley as a thinker from the late-Victorian and Edwardian period. This theme is picked up throughout much of this volume, but especially in the essays on Huxley's evolutionary biology, eugenics, and popularization. Provine's essay on Huxley's contributions to the evolutionary synthesis, for example, stresses the fact that unlike the other main players in the synthesis, Huxley insisted that evolution was progressive. Provine argues that the result of Huxley's search for evolutionary progress was not a scientific finding but an expression of his cultural values, values that might be traced back to the Victorian period.

The essays on Huxley's eugenics reinforce a similar perspective. Although Huxley offered a new program for eugenics and formulated it within the framework consistent with later twentieth-century genetics and evolutionary biology, the essays by Allen, Paul, and Barkan suggest that the motivation behind the program was from an earlier era. In addition, Allen's detailed analysis shows that Huxley's program for birth control was closely tied to his eugenics. Hence, while Huxley's early proposal for population control has been heralded as an enlightened call from the future, the essays here explain how it was essentially a call from the past. Although this volume focuses on Huxley's biology and scientific statesmanship, this revision in our understanding of him should enable historians to shed new light on Huxley's other writings and activities as well, especially on his activities at UNESCO.

The existing literature supports two contrasting pictures of Huxley as a scientist. One portrays him as an important participant in the founding of ethology and a central player in the evolutionary synthesis. Konrad Lorenz, for example, once called Huxley a father of the modern science of ethology,[66] and Ernst Mayr credited him with being one of six persons who first realized that the problems of the origin of species could be solved only by collaboration between geneticists, ecologists, biogeographers, paleontologists, and taxonomists.[67] Such accolades support a picture of Huxley as an important research biologist who advanced our knowledge by conducting groundbreaking field studies and developing innovative syntheses of previously unconnected lines of research.

Other aspects of the public record hint at a less-flattering picture. As Olby points out (this volume), the official honors bestowed upon Huxley usually hail him as public scientist rather than research scientist, as spokesman rather than architect. "There is obvious significance," Olby remarks, "in the award to him of the Kalinga Prize 'for distinguished service in popularizing science and scientific progress,' rather than a Nobel Prize." The deflated picture, never articulated though often implied, is of a generalizer, deft at summarizing the work of others, yet unsuccessful in his attempts to synthesize or contribute original research.

The essays of this volume cast doubt on both pictures. On the one hand, Witkowski's and Churchill's essays on Huxley's laboratory work show that his contributions were more original and his embryological synthesis more genuine than suggested by the view that he was merely a public scientist or spokesman. Burkhardt's and Provine's essays on Huxley's field work and contributions to evolutionary biology, on the other hand, imply that Huxley did not make original contributions to ethology and failed to combine diverse elements of evolutionary theory into a truly synthesized account. In fact, Burkhardt's essay provides a resounding refutation of Huxley's previously unchallenged boasts[68] that his paper on the crested grebe was "a turning point in the study of bird courtship, and indeed of vertebrate ethology in general" and that he "made the field of natural history scientifically respectable."[69]

As a whole, the volume suggests that earlier pictures of Huxley's biological practice have emphasized the wrong contributions.[70] The essays here reveal that Huxley's biological work did not come together in his synthetic account of evolutionary theory. While it is undoubtedly true that Huxley was keenly aware of and interested in the evolutionary implications of his laboratory work, Witkowski's essay shows that evolutionary concerns did not closely motivate or guide his laboratory work. This suggests that instead of viewing Huxley's laboratory research in terms of how it led toward or fit into his evolutionary synthesis, we should place it within the context of his synthesis in embryology. This is exactly what Churchill has done. What emerges from Witkowski's and Churchill's works is a clearer and more accurate picture of how Huxley practiced biology, a picture, I will argue, that can help clarify the nature of his contribution to the evolutionary synthesis and perhaps the nature of the synthesis as well.

The picture of Huxley that emerges from Witkowski's and Churchill's essays is of a research scientist conducting original empirical studies and combining results of his own studies with results obtained by workers in related fields. Churchill's analysis of Huxley's and de Beer's *Elements of Experimental Embryology* reveals that much of Huxley's laboratory work fed into his embryological synthesis. This work was a genuine synthesis, according to Churchill, because it brought together experimental findings and observations from a number of different fields and related them to a general organizing model. Churchill shows that Huxley's laboratory studies were part of a sustained program of research, which was later articulated in *Elements of Experimental Embryology* (not in *Evolution, the Modern Synthesis*).

This picture of Huxley can help us understand part of what he was trying to accomplish in his evolutionary synthesis. I stress the "part of what he was trying to accomplish" because, as the essays in this volume make clear, Huxley's evolutionary synthesis was much motivated and influenced by his secular humanism. What I am interested in focusing on here, however, is why Huxley thought his presentation of evolutionary biology represented a genuine synthesis. Provine (this volume) claims Huxley failed to offer a genuine synthesis because instead of combining theories (or theoretical mechanisms) from various fields, he selectively eliminated most theories and never combined the remaining ones. Olby (this volume) objects to this view and points out that the remaining theories, in particular Darwinism and Mendelian genetics, were brought together in modified form and hence were synthesized. While Provine and Olby disagree about whether Huxley achieved a synthesis in evolutionary biology, they both conceive of a synthesis as the bringing together of theories or theoretical mechanisms from different fields. Churchill, however, uses a different sense of synthesis in explaining Huxley's embryological work. According to Churchill's conception, a synthesis is a gathering of "diverse experimental or observational results around a general organizing model that then serves as an explanation and program for further research." I believe we can gain a better understanding of Huxley's attempt to synthesize evolutionary biology if we use the sense of synthesis that applies so clearly to his work in experimental embryology.

Before I proceed, it will be valuable to clarify the difference between the two concepts of synthesis at issue. Provine's concept entails the bringing together of theoretical explanations. Huxley's account of evolutionary theory does not seem to count as a synthesis in this sense because he rejected most of the theoretical explanations that had been favored in particular fields (for example, the inheritance of acquired characteristics and orthogenesis). An elimination of theories, however, could count as a synthesis under Churchill's conception, provided that experimental results and observations from diverse fields could be related to the remaining body of theory. The chief difference between the two conceptions of synthesis is that Provine's focuses on *bringing together theories* while Churchill's involves *bringing together observations* from different fields.

Huxley did bring together observations from a broad range of fields and he related them to the general neo-Darwinian framework. As Beatty points out (this volume), this framework left much room for a variety of theoretical mechanisms. Nevertheless, the important point is that these mechanisms were now described within a common conceptual framework and that this framework could be used to relate findings from a number of different fields. This, I suggest, captures the sense in which Huxley thought a synthesis was being achieved.

The idea that Huxley conceived of the evolutionary synthesis more as a bringing together of results from diverse fields than as a unification of different theories is supported by Huxley's historical account of evolutionary biology before the synthesis. Instead of stressing the multitude of evolutionary theories as Provine and other historians have, Huxley describes the disunity of *disciplines* whose observations were not related by a common perspective:

> The facts of Mendelism appeared to contradict the facts of paleontology, the theories of the mutationists would not square with the Weismannian views of adaptation, the discoveries of experimental embryology seemed to contradict the classical recapitulatory theories of development. Zoologists who clung to Darwinian views were looked down on by the devotees of the newer disciplines, whether cytology or genetics, *Entwicklungsmechanik* or comparative physiology, as old-fashioned theorizers. . . .[71]

Huxley's description of what happened during the period of synthesis emphasizes the newfound capacity to relate results from different fields:

> Biology in the last twenty years, after a period in which new disciplines were taken up in turn and worked out in comparative isolation, has become a more unified science. It has embarked upon a period of synthesis, until to-day it no longer presents the spectacle of a number of semi-independent and largely contradictory sub-sciences, but is coming to rival the unity of older sciences like physics, in which advance in any one

branch leads almost at once to advance in all other fields, and theory and experiment march hand-in-hand.[72]

It appears that Huxley conceived of the synthesis as a bringing together of previously isolated fields under a common framework such that findings in one field could now be related in a complementary fashion to findings from other fields.

This discussion illustrates how our understanding of central developments in twentieth-century biological thought can be refined by examining them through the contributions of Huxley. For, if this analysis clarifies the way in which Huxley was trying to synthesize evolutionary biology, then it also sheds light on the evolutionary synthesis in general. There are two different senses in which the twentieth-century consensus about evolution might be called a synthesis. It might be considered a synthesis in the sense that a number of different theories were unified. In this sense, the consensus was a narrow synthesis that brought together Darwinism and Mendelian genetics. But the consensus might be regarded as a synthesis in the alternative sense that observations from a broad range of fields were brought together under a common framework. In this sense, the consensus was a broad synthesis that brought together a number of fields, including genetics, ecology, systematics, biogeography, and paleontology. Much of the current disagreement about the consensus might be resolved by carefully sorting out the senses in which "synthesis" has been and is currently being applied.[73]

Our understanding of Huxley's contributions to the evolutionary synthesis and twentieth-century experimental biology, as our understanding of these developments themselves, is far from complete. The essays in this volume, however, significantly advance our understanding of Huxley's contributions to biology and, one hopes, will shed light on developments in biology as well. The same can be said with respect to Huxley's contributions to science popularization and reform eugenics. In general, the essays advance our understanding of Huxley's contributions by placing them in greater historical perspective. While Huxley and his eulogists viewed his contributions in terms of where they led, the essays here critically analyze them in the context from which they emerged. The result is that our understanding of Huxley's most-celebrated contributions (such as those in ethology) as well as our appreciation of his least-celebrated ones (such as those in eugenics) will be fundamentally altered. The essays to follow go a long way toward filling out our picture of Huxley and do so in a way that will force us to look at Julian Huxley as biologist and statesman, differently.

Huxley's Fulltime Positions

1909–1912	Oxford, Lecturer at Balliol College and Demonstrator in the Department of Zoology and Comparative Anatomy
1913	Rice Institute, Research Associate, Department of Biology, study in Germany
1914–1916	Rice Institute, Assistant Professor and Chair of the Department of Biology
1917–1919	Army and Intelligence Corps
1919–1925	Oxford, Fellow of New College and Senior Demonstrator in the Department of Zoology and Comparative Anatomy
1925–1927	London University, Professor of Zoology at King's College
1935–1942	London Zoo, Secretary of the London Zoological Society
1946–1948	UNESCO, Secretary of the Preparatory Commission and Director-General of UNESCO

Huxley and His Times

COLIN DIVALL

From a Victorian to a Modern:
Julian Huxley and the
English Intellectual Climate

How coherent was Julian Huxley's most ambitious intellectual project, "scientific" or (as he later called it) "evolutionary" humanism? The question is important if only because of the central role humanism played in Huxley's life outside the laboratory. The basic ideas were essentially completed in his earliest collection of writings, *Essays of a Biologist*, published in 1923. They still formed the core of his last major collection, *Essays of a Humanist*, in 1964. But as Huxley made clear in the earlier volume, scientific humanism was not simply an intellectual exercise. It was a philosophy of life, a "general conception of the universe," guiding and unifying his many, apparently diverse, practical interests.[1] Huxley's interests in eugenics, education, and internationalism, to name but a few, all bore the mark of this outlook. If we are to understand the man as a statesman of science, we cannot ignore his commitment to scientific humanism.

In this paper the emphasis is on the general relationship between Huxley's politics and the philosophical aspects of his humanism, concentrating on the years before 1950 when he was politically most active. What was central to this relationship? An overwhelming and exclusive belief in the potential of scientific reason for human betterment is the short answer. Yet this was hardly a novel claim, even when Huxley first voiced it in 1923.[2] English intellectuals—and others—had been open to the influence of Comtian positivism prior to the mid-nineteenth century. Positivists followed Comte's mentor,

Saint Simon, in their belief that scientific reason would cast aside
mystified, theological conceptions of humanity. As a result, a new age
would follow of the "scientific administration" of society, bringing
social and human progress. Similarly, Herbert Spencer's somewhat
later (and highly influential) views cast this idea of progress into an
explicitly evolutionary frame, connecting biological and sociological
development in one continuous stream.[3]

The explosive impact of Darwin's theories of biological evolution
from the 1860s—which owed so much to Julian's grandfather and
greatest influence, T. H. Huxley—formed a watershed within this
tradition. Initially, the middle classes in particular saw Darwinian
ideas as a decisive scientific challenge to theology. But as Melvin
Richter has argued, the loss of faith that resulted then took several
further forms, including a concern with social reform.[4] One strand
of late-nineteenth-century secular thought embraced philosophical
materialism and an objective, utilitarian image of science. Another
stressed philosophical idealism and a secularized account of morality
as the central defining characteristic of social life.[5] Julian Huxley was
clearly sympathetic to the first tradition, that of rationalism. But he
drew also upon the second tradition, of ethicism. In part, I shall
argue, this eclecticism is to be understood in terms of Huxley's family
background and education. But it was also a product of a philosophical
difficulty buried deep within Victorian rationalism, a flaw that only
became fully apparent in the vastly changed intellectual, social, and
economic climate after World War I. Huxley was a Victorian thinker
fated to live in an unsympathetic modern age.

It is useful to first appreciate in a little more detail the political and
philosophical background to the late-Victorian and Edwardian rationalism that Huxley imbibed from his father, Leonard. Rationalists
shared many, but not all, of the beliefs of the earlier Comtian positivists. Philosophically, they were more clearly materialist, denying
theology and religion any autonomous intellectual sphere. But anti-
theism was not their exclusive concern. They strongly emphasized
the Saint Simonian concern for social reform. Science, applied to
society, would generate enormous social gains through increasing the
efficiency of government, the economy, and a myriad of other social
functions.[6]

Rationalists such as H. G. Wells and Sidney Webb distinguished

sharply between their scientific analyses of social malaise and contemporary "irrational" politics. The Rationalist Press Association,
for instance, founded in 1899 as a source of literature for the movement, stood aside from party politics. Yet the majority of its members
were liberals or Fabian socialists, wishing for moderate reform of a
society that by and large afforded them a comfortable existence. They
also had a common interest in education as an important means to
reform. Scientific popularization was seen as a crucial element in this
process. Popularizers had appeared from the 1870s, in response to the
increasing specialization in science that made it practically impossible
for even an educated individual to keep abreast of all new developments. The rationalists concentrated their efforts on the influential
middle classes. Where the middle classes led, so it was believed, the
working classes would follow. This strong belief in social hierarchy
and the importance of experts set rationalists firmly apart from
working-class anti-theists, such as those in the National Secular
Society.[7]

Victorian and Edwardian rationalism faced an implicit tension between its inevitably political attempts at social reform and the philosophical image of science rhetorically supporting these efforts. In the
first place, the Saint Simonian ideal of social control demanded that
science be objective. It had to show social processes "as they really
are." In so doing, it indicated the points at which intervention could
be made to change the outcome of events. This image of science also
tended to be analytical and reductionist in tone. If "control" were the
ultimate meaning and use of science, then the social world must be
capable of being split up into manageable, comprehensible parcels.
One could not control the whole of society at once. Together, these
requirements encouraged an image of science as a highly efficient
tool, a tool that could be used for whatever social ends humanity desired. In short, science proposed but humanity (politically) disposed.

The problem for rationalists was precisely that this philosophical
model left the political choice of goals outside the purview of science.
This was scarcely adequate for a creed that claimed "science" as the
only possible kind of rationality. But historically, the concern of
Edwardian rationalists to have "science" accepted as an "effective"
method of government and reform overshadowed discussion about
which ends should be so efficiently promoted, or whether these ends

could be "scientifically" defined. As Martin Wiener has argued, few Edwardians considered the issues of "efficiency" and the "goals of life" in relation to one another. While rationalists took the former topic to their hearts—or, rather, minds—the various branches of ethical idealism emphasized the latter. Edwardian society was intellectually schizophrenic.[8]

It is this intellectual split that must color our understanding of Huxley's political activity and philosophy. In many respects his concerns were those of Edwardian rationalism. For instance, the mammoth and successful *Science of Life* (1931), written with H. G. Wells and his son Gip, stood firmly in the rationalist tradition of scientific popularization. Given Huxley's own acknowledgment of his grandfather's influence on him, we need look no further for an explanation of this particular, lifelong interest.[9] Nor need we look further than the younger Huxley's class and family background for an explanation of his politics. His views fell well within the spectrum of opinion acceptable to the English liberal intellectual elite. In the 1920s, 1930s, and 1940s, Huxley's politics were closely allied with the outlook of rationalists, such as Sir Richard Gregory (editor of *Nature*), and individuals active within professional and popularizing scientific organizations like the British Association for the Advancement of Science.[10]

A good example of the "scientific" approach to politics was *Nature's* rhetoric throughout the inter-war period. This formed a series of variations on the theme of the potential scientific contribution to "national efficiency," including eugenic issues.[11] Until the early 1930s, Huxley considered that the social sciences were incapable, at their present level of development, of contributing to the solution of political problems. But this did not prevent him from seeing other ways in which "science" could intervene in social affairs. Up to 1935 he shared *Nature's* enthusiastic eugenical stance, campaigning on these grounds for birth control and the "voluntary" sterilization of sections of the working class. For Huxley, eugenics was not seen as a social science. Rather, it was a natural science and the "racial" and social issues with which it dealt were conceived as fundamentally biological problems.[12]

The British economic crisis of 1931 and the perceived threat of political revolution encouraged Huxley to reappraise the utilitarian potential of the social sciences. Economics, sociology, and psychology now became essential to his enthusiasm for centralized social and economic planning, an enthusiasm shared with a considerable number of other intellectuals and politicians such as the Tory backbencher Harold Macmillan. The appeal of planning was that "the application of scientific method to human affairs" offered the "possibility of intelligent control." In Comtian mood, Huxley proclaimed scientific planning as nothing less than the next "great step in human history."[13]

Clearly associated with this enthusiasm for planning was the Edwardian rationalists' philosophical image of science (including now the fledgling social disciplines) as morally unproblematic. Scientific knowledge was a means, "a tool, which like other tools can be used for whatever ends its possessor sees fit."[14] At a time of high unemployment, perceived by many as technologically induced, such rhetoric enabled Huxley to shift the blame for current problems onto the "irrationalities" of politics, thereby preserving the supposedly innocent "science" for more felicitous times. In some respects, the perspective was the same as that employed by several of the younger Marxist scientists of the 1930s such as J. D. Bernal and J. B. S. Haldane (an old friend of Huxley's). They saw science under capitalism as a distortion from its true potential, which would only be realized under socialism.[15]

But what of Huxley's vision? To which goals was science ideally the means? He was no socialist, as he made explicit in *If I Were Dictator*, published in 1934 as a manifesto for the planned society. There were, it is true, signs of convergence between the centrist politics for which he was by now a valued propagandist and the socialist scientists. Both groups were, for instance, keen industrial reformers, defenders of production against finance capital.[16] Here, surely, was an instance of the irrepressible modernizer, sweeping away the restrictions of the old Victorian laissez-faire economy in the search for greater efficiency. Yet there was a strand within Huxley's thought that tempered unrestrained technocratic expansion. His views accorded with the anti-industrial values of much conservative as well as left-leaning ideology between the wars.[17] If industry were to continue as the powerhouse of the new England, the economic materialism that had engendered it in the past was nonetheless to be kept firmly at bay. The fruits of industry were to be directed by benevolent planners toward the fulfillment of higher human needs. Planned society would be an organic, ordered whole: individual bound to class, and class to class, by shared moral values. Huxley's ideal was a pacified capitalism.

This theme continued in his wartime writings. Books such as *Democracy Marches* developed the notion that economic values "must lose their primacy, and become subordinated to social values."[18] Huxley's own conception of UNESCO's role told a similar story writ large. But what precisely were such "social values"? And what was their relation to scientific rationality? Did they really stand outside its scope, as Huxley's comments of 1931 on science as a tool suggested? Or were they perhaps to be determined by the planners who would administer future society? This was no academic issue, for *If I Were Dictator* had denied that democratic fora had any relevance in a planned society. Indeed, it is ironic that at almost the same time as Aldous Huxley was warning—in *Brave New World*—of the horrors of the technocratic vision, Julian should be advocating the reduction of parliament to "a sort of sounding-box" for the dictates of scientific rulers.[19] The problem still existed after 1945. Huxley did become more sensitive during the war to the need to make planning compatible with liberal democracy. But his notion of democracy was highly manipulative, still turning on an elitist notion of scientific experts. They were the guardians of "social values," as *UNESCO: Its Purpose and Philosophy* made clear in 1947. Experts were to be selected through a reformed educational system that would also serve to inculcate an acceptance of subordination to expertise among the majority judged incapable of ruling society.[20] Huxley's hierarchical view of society was still as strong as any Edwardian rationalist's.

Could this position be defended as scientific? Huxley's writings on humanism between the world wars attempted to do just this. From the 1920s, he struggled to defend an objective definition of "progress," couched in terms of moral value-frameworks, which would intellectually underpin his benevolent dictatorship. Indeed, his writings are a prime example of the intellectual torture caused after 1918 by the philosophical legacy of Edwardian rationalism. English intellectuals were finally experiencing the full force of earlier European doubt over the objectivity of moral values.[21] It thus became increasingly difficult for anyone who still hoped to avoid moral nihilism to sustain the Edwardian rationalists' intellectual disengagement from moral debate. Swinging from an instrumental image of science to one that tried to include moral judgments within its scope, Huxley desperately tried to bridge the intellectual chasm of late Victorian and Edwardian society.

What was his blueprint? The first two papers in *Essays of a Biologist* provided a sketch. One of the most striking aspects of Huxley's philosophy in the intellectual context of the 1920s was its synthetic sweep. The continued fragmentation of professional intellectual activity made synthetic thought seem to many impossible, even in principle. But Huxley's lifelong belief that the most valuable intellectual work was synthetic was underpinned by his strong belief that the universe itself was fundamentally a unity.[22] One way in which this unity was expressed was in the very notion of progress. Not surprisingly given Huxley's professional interests, this relied heavily on inspiration from evolutionary biology.

Huxley argued that modern science could identify a truly progressive trend within cosmological evolution. Progress was not inevitable, and it took different forms at the inorganic, organic, and human levels. The first of these is irrelevant for our immediate purpose. In the second, biological, phase only some of those evolutionary developments having survival value were truly progressive. These allowed for greater control over, and independence from, an organism's environment. On this basis, Huxley argued that humanity stood as the dominant biological species on earth. But the conceptual mental powers that formed the foundations of biological dominance also provided a new mechanism for progress at the specifically human level.[23]

In one sense, the focus of evolutionary change in the human sphere was social organization and traditions. Huxley claimed that the potential of the individual human mind had probably not risen greatly since the emergence of *Homo sapiens*. Instead, improved social organization and the rise of culture had permitted a greater realization of that potential. But the essential characteristic of human evolution was given, if ambiguously, at the level of the individual person.[24] Human beings experience a wide range of activities in diverse fields—practical, intellectual, and emotional—which, as Huxley put it, they simply find valuable. These "valuable experiences" were evolution's ultimate ends: defined culturally, they were to be attained by individuals.[25]

In *Religion Without Revelation*, published in 1927, Huxley made clearer the spiritual, moral dimension of these cultural goals. The book also drew out his intellectual connections with Edwardian ethicism. If, philosophically, scientific rationalists relied on a materialist metaphysics, ethicists stressed idealism, derived from the neo-Hegelianism of T. H. Green. This had been transmitted to Huxley's generation through, among others, the pioneering English academic sociologist L. T. Hobhouse, whose idealist interpretation of evolution was acknowledged in Huxley's early essays. There was also a more pervasive influence from Huxley's studying at Balliol, Green's old college at Oxford, in the years when the latter's pupils were in the ascendant. An even earlier influence was Huxley's mother, Julia, Matthew Arnold's niece. Her pantheistic faith not only inclined her son to a religiousness at odds with the stereotype of the hard-headed rationalist but also influenced the philosophical form that "evolutionary progress" eventually took. When one also remembers the ubiquitous presence in the Huxley household of Thomas Henry's eclectic rationalism, it is not too surprising that Julian tried to break through the hermetic seals of Edwardian intellectual debate on moral and political goals.[26]

Religion Without Revelation showed that even if Huxley rejected the intellectual basis of received Christianity, he was more than willing to accept many of its tenets as an attitude of mind, a "feeling of sacredness and reverence" in "the reaction of the personality as a whole to its experience of the universe as a whole."[27] Indeed, he accepted the contemporary findings of anthropology, which suggested that religion served a socially integrative function, and would therefore not merely fade away as earlier rationalists had hoped. This, of course, was at one with his organic conception of society bound together by shared moral values that later appeared in his politics. *Religion Without Revelation* also developed in more detail the specific value-frameworks, the kinds of "valuable experiences," highly regarded by scientific humanism. Classical music, for example, was one instance of a worthwhile aesthetic experience. As a contemporary theological critic noted, such value-frameworks involved "a real contact with the religious attitude and spiritual tradition of Christianity."[28]

Had Huxley succeeded in developing a truly scientific, objective, understanding of "evolutionary progress"? He clearly thought so, for the concept was taken as the basis of an objective ethical theory in his 1943 Romanes Lecture at Oxford University. It is easy to see why at this time the anchor of "evolutionary progress" should appeal to an audience afloat in a sea of contemporary moral uncertainties. Indeed, the attractions of such a position had already been publicized by C. H. Waddington in a 1941 supplement to *Nature*.[29] But these advantages were not universally appreciated. Both at the time and over the past forty years, Huxley's idea of "evolutionary progress" has been heavily criticized. There are two linked strands to these objections. First, there is the straightforward argument that it is unclear how the form evolution takes in its earliest phases can bear any relevance to its manifestation at the human level.[30] Second, it can then be argued that "progress" is not a strictly scientific notion. Instead, it is an interpretative evaluation of scientific "facts."[31] There is considerable force to both these sorts of criticism. But Huxley's critics since the 1940s have too readily ignored philosophical models of "science" and "evolution" that were prevalent in the 1920s and that go some way in explaining his stance.

Consider the second charge, of conflating descriptive and interpretive disciplines. It is important here to grasp the significance of the philosophical idealism associated particularly in the 1920s with A. N. Whitehead, Arthur Eddington, and James Jeans. Their philosophical accounts of science drew more upon relativistic physics than biology, but they stood as much in the tradition of British idealism as L. T. Hobhouse. As an eclectic thinker, Huxley must have been aware of this school's semipopular works.[32] These maintained an idealist epistemology that dissolved any clear boundaries between science and evaluative disciplines. This followed from the claim that the universe is essentially holistic; that is, there are no truly "independent" entities, no objects whose qualities can be understood without reference to the rest of existence. It followed that the natures of "objects" studied by scientists were in fact partly defined by the relationship of knowing-subject to object. Eddington, for instance, argued that scientists' supposition of stability in the physical world was "essentially a contribution of the mind" deriving from an "innate hunger for permanence."[33] From this idealist perspective, it was a short step to claiming that all types of human "values" shade scientists' perception of the world. Scientific knowledge could not provide neutral facts upon which evaluative interpretations were then made. It was too intimately bound up with the value-frameworks of its practitioners.

One cannot find this argument so baldly stated in any of Huxley's writings. Many of his comments on epistemology were superficial, inviting ambiguity, or simply confused.[34] In part, this was because an idealist image of science was so much at odds with the "no nonsense" account of rationalism. Committed to the Saint Simonian ideal of social control, Huxley could scarcely abandon its associated account of scientific knowledge as a reflection of an objective, manipulable world. In 1931 he proclaimed, for instance, that science should only attempt "to describe and to understand, not to appraise or assign values."[35] But there is a good deal of evidence that his thoughts on evolutionary progress did include an implicitly idealist perspective. Consider as an example this claim found in the 1923 essay, *Progress, Biological and Other*, discussing the scientific "fact of evolutionary direction."

> It is immaterial whether the human mind comes to have . . . values because they make for progress in evolution, or whether things which make for evolutionary progress become significant because they happen to be considered as valuable by human mind, for both are in their degree true. There is an interrelation which cannot be disentangled, for it is based on the fundamental uniformity and unity of the cosmos.[36]

Here were the essential elements of Eddington and Whitehead's idealist case: a commitment to the "unity" of existence, including human mind, in a way that supposedly explained the relation between value-frameworks and "reality" as revealed by science.

One of the many intellectual problems with this idealist argument was how to avoid a charge of arbitrariness.[37] Granted, there was no arbitrary element in the production of scientific knowledge, for this was necessarily tied to scientists' value-frameworks. But this merely pushed the problem back another stage, as Eddington recognized. The 1920s was a period of sharply competing moral, aesthetic, and political value-frameworks—in intellectual terms, there appeared to be a good deal of arbitrariness in their formation.[38] This was clearly unsatisfactory if one's search was for an objective ethic. Thus Eddington, for instance, moved in *The Nature of the Physical World* from personal to absolute idealism. He claimed that there is only one "objective" (or nonarbitrary) value-perspective. This is what actually forms the basis of the true world-picture. But to grasp this, one had to go beyond the myriad contradictory accounts given by science. Eddington was forced to abandon science as a reliable source of knowledge about reality. Mystical or intuitive apprehension of an absolute value-framework replaced scientific knowledge as the bedrock of moral wisdom.[39]

A similar retreat to mysticism can be found in Huxley's writings in the 1920s and beyond. It was not as clear as Eddington's, for the philosophy of neo-Darwinian evolution inclined Huxley against the philosophical notion of absolutes. He therefore partly accepted the modernist position that a plurality of value-frameworks existed. Indeed, occasionally he claimed that scientific humanism's particular values carried the mere appearance of absoluteness; they were "purely relative within the general scheme."[40] The corollary, however, was that humanists laid claim to the certitude of their own value judgments; "They cannot be disputed—they are simply experienced."[41] At the root of this account one therefore finds a common theme of modernism, the importance of commitment. Such commitment ontologically preceded discovery, for it was the basis upon which the social (and indeed natural) world was constructed. Now, if this idea of commitment is joined with the idealist epistemology that has already been outlined, one can see that Huxley's "evolutionary progress" stood in the same relation to "reality" as Eddington's scientific world views. It was indisputable in its own terms. But ultimately one was forced to recognize it as only one of any number of possible accounts, each based on a different normative framework to which one might be "committed."

This personal, pluralist idealism became transformed into an antiscientific absolutism in Huxley's humanism through the concept of "emergent evolution." This metaphysical scheme also served to unite what recent critics see as the separate phases of Huxley's account of evolutionary progress. Emergent evolution came to its fullest intellectual stature in England around 1920, with the publication of S. Alexander's *Space, Time and Deity* and the first of C. Lloyd Morgan's volumes on the topic, *Emergent Evolution*. Of the two, Lloyd Morgan, an animal psychologist, was the principal influence on Huxley. "Emergent evolution" shared many of the philosophical presuppositions underlying Eddington and Whitehead's work.[42] Lloyd Morgan's principal desire had been to maintain a philosophical conception of evolution that recognized the reality of novelty, while denying the idealist implications of pre-1914 vitalists such as Henri Bergson and Hans Driesch. There was a considerable irony in this, for emergent evolution's ultimate role in Huxley's writing was to deny novelty.

Emergent evolution was partly a reaction against mechanistic materialism, the philosophical account of evolutionary development that underlay, for instance, Herbert Spencer's thought. Spencerian evolution effectively rejected novelty, in the sense of the development of unpredictable forms. His universe was Laplacean, its state at any future time in principle predictable from a knowledge of its present standing and Spencer's so-called universal law of integration of matter. In contrast, Henri Bergson argued that life was distinguished from the mechanical workings of matter because it was informed by the élan vital. One therefore could not predict the development of life on the basis of even a complete knowledge of the workings of the material universe. Bergson's philosophical universe was dualist, the active essence of the élan vital being ideal or spiritual in contrast to passive matter.[43] Emergent evolution attempted to establish a middle ground between these two Victorian camps.

Lloyd Morgan argued that evolution *did* produce unpredictable, novel forms, although he rejected the idea of a separate philosophical substance to explain this. Development simply occurred and was inexplicable on the basis of scientific knowledge. Lloyd Morgan's philosophical image of science was in fact very close to the predictive, reductionist model central to the Saint Simonian, rationalist project. It could provide foresight of causal processes. It could understand the behavior of new forms once they had emerged. But the actual process of emergence stood outside its purview precisely because novelty demanded a breach with causality and total predictability. Lloyd Morgan therefore accepted a strong restriction on the scope of science. Its account of reality was an abstraction. The full truth about emergence in evolution could only be revealed through metaphysical insight.[44] Here of course one can see a strong convergence with the drift of Eddington's thought.

Part of this greater, metaphysical truth was that Lloyd Morgan was unable to accept that "mind" was an entirely novel product of evolutionary change. The concept was used in two senses. In the first, that of mental powers, Lloyd Morgan accepted that mind had appeared through evolutionary change. But in a second, metaphysical sense, he thought that "Mind" had always existed in the universe, as a "Potentiality" of substance. Substance was the building block of all evolutionary forms, a Spinozan entity possessing both physical and psychical attributes. The emergence of metaphysical, potential "Mind" as scientifically detectable "mind" was therefore not simply a chance result of evolution. In this instance, evolution was not about the production of genuine, qualitative novelty. Instead, it concerned the realization of qualities already defined by substance. Ultimately, evolution was nothing less than the realization of a moral archetype.[45]

Huxley did not develop his metaphysics in anything like Lloyd Morgan's detail. Nonetheless, there are sufficient hints in his writings to sustain the claim that emergent evolution formed a significant element of his philosophy. Huxley's first book, *The Individual in the Animal Kingdom*, a work of philosophical biology published in 1912, was explicit in recognizing Bergson's influence.[46] By the 1920s, such youthful enthusiasms had been replaced by advocacy of Lloyd Morgan's compromise position. The core of Lloyd Morgan's metaphysics was incorporated within Huxley's concept of the "world-stuff," which was similar to psychophysical substance. In the early *Biology and Sociology*, for instance, Huxley held that it was unlikely that mental processes had "sprung up during the course of evolution absolutely *de novo*." Rather, he argued, "if we wish to preserve our scientific [sic] sanity, our belief in the orderliness of the world," it was better to postulate that "something of the same general nature, the same category as mind must . . . be present in lower organisms and in the lifeless matter from which they originally sprang."[47]

Huxley did not consistently combine his advocacy of emergent evolution with an absolutist (or objectivist) account of value-frameworks. But again, there was enough in his writing to suggest that this combination informed his explicit statements. Take, for instance, this 1912 description of evolution, as a "line traced . . . towards . . . perfect individuality." The "crown of Life's progress" would be beyond the present stage of humanity, because a person was "individuality still tied to substance." The ultimate was the development of " 'disembodied spirits' . . . individuality without substance, free and untrammelled."[48] The sense of evolution as the realization of a moral archetype, indeed one defined wholly in spiritual terms, was overwhelming here. Furthermore, this general perspective survived Huxley's notional rejection of Bergson, appearing, for instance, in *Religion Without Revelation*. Human life "may perhaps be best thought of . . . as spirit realising itself in living matter."[49] It required only a minor extension of such comments to see all progressive evolution as the realization of the particular value-framework of scientific humanism:

> Truth, beauty, and goodness are the goals of . . . life . . . life, with human nature in its forefront, is the means of giving actuality to the ideal; . . . in this consists our true destiny. . . .[50]

Emergent evolution provided Huxley with a reasonably coherent philosophical framework. It linked the prehuman stages of "evolutionary progress" with human development and, ultimately, the possible transcendence of humanity. But it did so by implicitly abandoning scientific rationality and knowledge as the touchstone of ethics and progress. "Evolutionary progress" was a metaphysics at odds in almost every respect with the philosophical image of science at the core of scientific rationalism. Belief in the objective worth, the "intrinsic value," of certain experiences amounted to an unchallengeable claim to know a reality unknowable to science. It was an acceptance of a teleological, moral conception of the universe, on faith. Indeed, in this last respect, Huxley's attitude mirrored the later mysticism of Aldous.[51] The conclusion is inescapable. Huxley's attempt to translate the project of Victorian and Edwardian rationalism into the modern, hostile world failed to sustain the intellectual unity with which he was—rightly—so concerned.

The tragedy was that Huxley's failure to see the shortcomings of this intellectual project bolstered his social elitism, vitiating the potentially radical drift of his politics. One can only welcome his attempts to confront the instrumental rationality of "science" with questions about the wider goals of life. One may sympathize also with his professed aim to ensure that "science" created the material conditions for "spiritual" achievements. But freedom is equally a prerequisite of these goals. The tenacity with which Huxley maintained the spurious objectivity of his (and his class's) values implicitly threatened any conception of human beings as free political and "spiritual" agents. In this, at least, he was at one with the predominant drift of twentieth-century political thought.[52]

PETER STANSKY

Particulars in Huxley's Intellectual Climate

I must confess that I expected a little more of the particular in Colin Divall's excellent talk, for his title suggested, at least to me, more detailed consideration of the specifics of Huxley's background, although "climate" quite legitimately can stand for the interesting developments discussed in the paper. Some of those developments seem to me to be among the general intellectual currents of the twentieth century, rather than associated with England in particular. Indeed, it would appear to me that Divall has argued, and fairly convincingly, that Huxley did not make the transition suggested in the title if not in the paper itself, and that in many ways he remained a Victorian: as Divall put it, a "Victorian thinker fated to live in the modern age."

I am sure that it is not necessary to say much to this gathering about Huxley's family background, and the ideas and attitudes attached to it. To a considerable extent they helped propel him in his career, but they may also have been responsible for catching him in something of a time warp. His belief in evolutionary progress may have been more of a Victorian way of thought than of the twentieth century, and it is striking that a phrase by that earnest and agnostic Victorian John Morley, which Huxley read while visiting Colorado Springs, should have been so important to him: "the next great task of science will be to create a religion for humanity."[1] After all, it was his grandfather who coined the term "agnostic," with its implication

that the issue of God was not settled, but that questions of religion
were central. And if religion was important on the Huxley side, it was
even more so on his maternal side: the Arnolds. His grandfather,
Thomas Arnold the younger, converted to Catholicism and then back
to Anglicanism, then back to Catholicism again—the conversion to
Catholicism at the price of deeply distressing his wife and also costing
him both times his job. Arnold's daughter, Mrs. Humphrey Ward,
was very close to the Huxley children, and her immensely successful
novel, *Robert Elsmere*, ended with its hero embracing religion without dogma, or one might say *Religion Without Revelation*. And
Huxley's mother's pantheism was a great influence upon him. He
even had a proper Victorian "dark night of the soul" in his personal
crisis of 1912 that delayed his taking up his appointment at the Rice
Institute. This seemed to be the price that Victorians paid for appearing most of the time so strong. Huxley's crisis was parallel in a way to
the greatest fictional "dark night": that of Dorothea Brooke in *Mid-
dlemarch*. (I find it enlightening that the progressive Thomas Henry
Huxley, although he called on George Eliot, would not allow his wife
to do so because Eliot was "living in sin." He also forbade Oscar Wilde
the house after one visit. The Victorians could use respectability—
some would call it hypocrisy—as a cover to advance new ideas.
Certainly Julian Huxley used his secure class position to advance his
ideas, but even so at times he could not get away with it, as when he
was dismissed from his position as secretary of the London zoo
because of his attempts to make the zoo more accessible.)

In many ways, Huxley was a proper member of the Victorian
upper middle class, the "intellectual aristocracy." Colin Divall makes
a few tantalizing remarks about the degree to which Huxley was
caught by his class, but I wish he had told us a little more what he
meant. I also would have appreciated being told more about his
politics and how they related to his humanism. It seems to me that in
his politics he was able, to a degree that has perhaps not been given
sufficient credit in the paper, to come more into the twentieth century
than might have been apparent in what must be called his search for
religion. As an important figure in PEP (Political and Economic
Planning), he was, I feel, very much a part of the spirit of the 1930s,
an effort to counteract the apparent absence of planning in the government's efforts to cope with the Depression. I believe that in this

area he had moved on from merely being an Edwardian rationalist. And certainly in his concern with the environment and questions of population—though they had their roots in the late nineteenth century—he was an important founder, I should have thought, of the modern versions of those causes.

Presumably one does not want to make too much of a comparison with his brother Aldous. I think, however, that it is possible to regard Aldous as being more of a "modern." Aldous's concern with religion was both, I believe, more intense and more of this century. Those seven years between them—Aldous being the junior—may well have made a difference in how they viewed the world, with Julian the much more optimistic. He moved ever upward from the Eton Chapel, as he recorded in *Religion Without Revelation*:

> In spite of all my intellectual hostility to orthodox Christian dogma, the chapel services gave me something valuable, and something which I obtained nowhere else in precisely the same way. . . . Indubitably what I received from the services in that beautiful Chapel of Henry VI was not merely beauty, but something which must be called specifically religious. . . . But, once the magic doors were opened, and my adolescence became aware of literature and art and indeed the whole emotional richness of the world, pure lyric poetry could arouse in me much intenser and more mystical feelings than anything in the church service; a Beethoven concerto would make the highest flights of the organ seem pale and one-sided, and other buildings were found more beautiful than the chapel. It was none of the purely aesthetic emotions which were aroused, or not only they, but a special feeling. The mysteries which surround all the unknowns of existence were, however dimly contained in it, and the whole was predominantly flavoured with the sense of awe and reverence.[2]

Aldous writes about Eton Chapel with a very different tone in the opening lines of *Antic Hay* in 1923.

> Gumbril, Theodore Gumbril Junior, B.A. Oxon, sat in his oaken stall on the north side of the School Chapel and wondered, as he listened through the uneasy silence of half a thousand schoolboys to the First Lesson, pondered, as he looked up at the vast window opposite, all blue and jaundiced and bloody with nineteenth century glass, speculated in his rapid and rambling way about the existence and the nature of God.[3]

As this is the early Aldous Huxley, God does not exist. Using humor, satire, and cynicism, Aldous questioned, as Divall points out, the sort of planned centralized state that his brother envisioned in *If I Were*

Dictator, published two years after *Brave New World*. It might have been valuable to explore a little the contrast between the two brothers as a way of discussing Julian Huxley and how he did or did not make the transition from Victorian to modern. Aldous, shaped by the same family and class, felt the same needs as Julian. But his philosophy and form of religion were much more personal and mystical. Perhaps a question mark should have been placed after "modern" in the title of Divall's paper. Yet, whatever limitations one might see in Julian Huxley, there can be no question that he was a Victorian giant, one of those people so appallingly gifted in some ways, who made so many contributions during his lifetime and who can be studied now that his papers are available.

I might add in conclusion a way in which Julian Huxley would appear to be very modern, although not necessarily English: the study of the courtship ritual of the great crested grebe that he did with his other brother, Trev (whose youthful suicide indicates the price that may be paid for all those gifts at one's cradle). The sketches for the study were done just before Julian came to Houston. One of his sketches shows the birds doing their "penguin" dance as part of their courtship ritual. The male and female, with their identical adornments, indicate that they will build their nest equally. As Huxley has written: "The resultant paper, published in 1914, proved to be a turning point in the scientific study of bird courtship."[4] The point about these birds, as I understand it, is their extraordinary equality and mutuality. So in that respect, Julian Huxley might be seen, if it is legitimate to make such a conclusion from his observations, as much more forward-looking, with a more complicated idea of how society should conduct itself, than has been implied, I believe, so far.

MARTIN J. WIENER

The English Style of Huxley's Thought

Colin Divall's admirably lucid discussion of Julian Huxley's evolutionary humanism makes, I think, three particular contributions. First, although Divall himself does not bring this out as clearly as he might, his account highlights Huxley's ineradicable "Englishness." Second, it locates Huxley in the midst of the long and ultimately disappointing effort since the Enlightenment, and particularly since Darwin, to turn the cloth of "science" into a wardrobe of a philosophy of life and a program for social progress. Finally, Divall directs our attention to what every generation needs reminding of—some of the (to say the least) unattractive political implications of the attempt to govern human life by scientific "rationality."

Even before Huxley was born, Friedrich Nietzsche was complaining that the English resisted taking ideas to their logical conclusions, preferring to blur sharp edges of thought that might otherwise tear cherished values. This Anglo-Saxon inclination continued to be evident as Huxley grew up, in the way leading Edwardian scientific rationalists like Leonard Hobhouse and Graham Wallas tempered their explicit materialism with habits of thought drawn, often unacknowledged, from the Idealist tradition. While thrusting religion out the front door of their thinking, they took care to unbolt the rear door for the quiet reentry of "spirit" and ethics, now somehow immanent in the scientific enterprise or in the processes of nature itself.

Thus Hobhouse asked in 1904, "Is evolution a process making for the betterment, perfection and happiness of mankind, or a mere grinding out of the mechanical mill of existence of forms of life, one no better than another, the outcome of blind forces and destitute of any characteristics which can fill us with hope for the future of society?"[1] A few years later he answered his own question: In the midst of the ebb and flow of evolution as a whole and human history in particular, he insisted, "there *is* an onward movement discernible among the many changes that are valueless or worse, and this we may identify with the growth of mind."[2]

Such ill-defined "ethical" or "spiritual" materialism was, I suggest, even more common than Divall notes, constituting the norm rather than the exception among post-Darwinian British thinkers at least until the aftermath of World War I, and it constituted Huxley's formative mental environment. This blending of rationalism and idealism seems to have been nurtured for Huxley, as for many others, by the striking social pattern of intermarriage among the British "intellectual aristocracy"[3]—the merging of the moralizing and literary Arnolds and the scientific Huxleys was emblematic of this development. The frame of mind that often resulted permitted one to feel "objective" and "scientific," without losing hold of moral principles and confidence in their ultimate triumph. It enabled its holders to criticize the status quo without ceasing to feel connected to it. Partly in consequence, neither alienation nor revolution, so much part of the modern experience, could find as secure a purchase in Britain as elsewhere.

Huxley's thinking had, however, as Divall shows, a significance transcending national borders. Huxley carried on, in the decreasingly propitious circumstances of the twentieth century, the nineteenth-century effort to make "science" a program for social progress. This effort demanded, as Divall notes, that science be both objective and supportive of the "right" values. As professionalization of science and social science weakened moral discourse, and as "science" was employed in the service of increasingly divergent ends, it became ever more difficult to hold together "fact" and "value" in any convincing way. The tension that is noticeable among the late-Victorian rationalists became, in Divall's apt phrase, an "intellectual torture" for Huxley's generation. Thus Huxley continually modified and restated his evolutionary humanism, as he sought a formula that could stand. Not surprisingly, he never arrived at that resting point. However powerful and widespread the desire for a scientific "life-philosophy," it does not appear to be one that can be satisfied. Ironically, the more our lives are shaped by the products of scientific inquiry, the less science itself seems able to offer answers to the larger questions of life. Huxley's philosophical career, as distinct from his more strictly scientific work or his public work through UNESCO and elsewhere, seems a monument to this twentieth-century exhaustion of a Victorian quest.

That quest could have darker consequences, as Divall points out, than mere intellectual contortionism. Huxley's support for eugenics until 1935, and his impatience in this same period with the "irrationalities" and inefficiencies of democracy, suggest some of the illiberal and inhumane directions that scientific "progressivism" could take. It is indeed ironic, as Divall observes, that Julian Huxley was setting out his most elaborate and approving picture of a scientifically planned society at almost the same time that his brother Aldous was producing his powerful (and still compelling) technocratic dystopia. While they wrote, forms of technocratic dystopia were in fact beginning to be realized in the Soviet Union and in Nazi Germany. Scientism, it might be said, lost its innocence then. Thereafter, neither Julian nor, one would think, any other reasonably sensitive person, could summon the former naive enthusiasm for the reconstruction of human life by an elite claiming the authority of "science." Huxley's post-thirties modifications of his vision of a rationally planned society reflected a general disillusionment and backtracking by proponents of scientism, but not, as Divall makes clear, a repudiation of its basic orientation. One can question whether Huxley sufficiently absorbed the lessons of the totalitarianism of the Hitler-Stalin era. Even the more benign scientism of his UNESCO period may call for more critical scrutiny than it has yet received.

While Huxley's "English" style of thought is by no means to be scorned—the ruthless logic of Nietzsche is hardly to be preferred as a practical guide—the more general implications of Huxley's thinking about science and human life are rather disturbing. I have neglected the specific contributions to public enlightenment and social policy of his many-sided and productive involvement in the scientific and public worlds of his day. But considered as a natural and social philosopher, Huxley most exemplifies, it seems, a cautionary lesson—of the limitations of science, in theory and in practice. However much enlightenment and practical benefit the scientific enterprise has brought humankind, Huxley's career reminds us that it does not hold the answer to all human questions, and the attempt to place it beyond its sphere leads to disaster.

ROBERT OLBY

Huxley's Place in
Twentieth-Century Biology

Those of you who read poetry may have come across the following
lines:

> My crystal limbs and comet hair
> Glowed into birth I know not where
> Nor care I—enough for me
> My eternal self to be
>
> Thus for better or for worse
> I march across the universe.
> Who shall deny that I am I,
> The Self that knows not how to die?[1]

Or again, readers of the letters column in the *Times* might have found
a correspondent writing about the sky in Spitzbergen:

> Last comes the atmosphere. To him who knows Switzerland, and still
> more, I understand, to him who knows the Himalayas, the mountain sky
> here is a new thing. Gone is the dark of the sky of heights almost
> oppressive in its depth of blue, that hints at the blackness of empty space
> beyond; and in its place, the tender sapphire of English skies, English
> skies after rain, shading to delicate green upon the horizon. It is so pale, so
> soft, and the snow peaks stand out against it in a new and tenderer
> fashion.[2]

Both passages were written by a certain Julian Huxley—which Julian
Huxley? As his wife, Juliette (neé Baillot), has explained: "There
were, in fact, so many Julians within the tall carelessly dressed

wanderer bent on his various ploys, escaping from one activity by diving into yet another." "So many fingers in so many pies," she once said to him. "What a pity you haven't got a few more fingers!"[3] In truth, there was no single niche for this "Globe-trotter of Science," this whirlwind spirit, whose restlessness was driven by a "deep compulsion" to experience and triumph in fresh encounters.

Was he brilliant? Oh, yes. After all, he was a Huxley, the grandson of T. H. Huxley and Thomas Arnold. Did not the *Spectator* include him among the five best brains in Britain in 1930? Did he achieve greatness? Surely he did. Eighteen years ago the *Sunday Times* counted him among the "one thousand makers of the Twentieth Century" who have changed the form of our world and colored our thoughts. He was, declared the *Times*, "Public Scientist Number One," and in Holland he was described as the most important spokesman for neo-Darwinism.[4] But we should be wary of all this eulogy, and has not "Public Scientist" a connotation distinct from "Research Scientist," and has not "Spokesman" from "Architect"? There is obvious significance in the award to him of the Kalinga Prize "for distinguished service in popularizing science and scientific progress," rather than a Nobel Prize.

One might well question whether a self-respecting historian should even attempt to "place" a scientist, if by that is meant to evaluate the figure's contribution, giving him pluses for what he got right, and minuses for what he got wrong, working in a positive weighting factor for original research and theoretical developments, and negative factors for squandering time and energy on the popularization of science, parapsychology, and, some would say, worst of all, religion. Let us be clear, therefore, that this is not a merit-order exercise, but a contextual one. How does Huxley fit into the intellectual milieu of the biology of his day? What were the main features of that milieu? First, some preliminaries: Huxley was a naturalist, experimentalist, and theoretician. He was "at home" in the field and in the laboratory. Selous's observations of courtship in birds inspired him as much as the achievements of *Entwicklungsmechanik* in Naples, and despite his atheism he could appreciate Teilhard de Chardin's vision of evolution. In all his varied roles, however, his foremost concern was evolution. He upheld selection (but not so exclusively as R. A. Fisher) and Mendelian genetics (but not T. H. Morgan's narrow conception of this subject in terms merely of hereditary transmission), and like his grandfather T. H., he believed progress could be described in biological terms, but unlike him, he also held that a system of ethics could be constructed on the basis of our knowledge of biology. He belonged to a generation of British intellectuals in which was displayed a "flowering" of stimulating, industrious, and influential biologists, several of whom were very successful communicators who brought biology to a wide audience, who had political commitments, who took eugenics seriously, and who wanted some control over evolution. If we are going to judge Huxley, let us do so in terms of that generation and not our own. By Julian Huxley's time the biological sciences were well represented in such institutions as the Royal Society, the universities, and in research institutes for applied sciences. Nevertheless, the climate of opinion among both experts and the public was not supportive of wholesale selectionist evolution. Why was this so, and how was it altered? Let us begin by examining Huxley's own vision of this changing milieu.

In *Evolution, the Modern Synthesis*, Huxley offered his readers a sketch of the changing fortunes of Darwinian evolution. He pointed to the weakness in the theory as presented by Darwin: his assumption that "the bulk of variations were inheritable," his failure to realize that genetic recombination "may both produce and modify new inheritable variations," his partial reliance upon Lamarckian principles which, after Weismann's onslaught, were no longer credible. Indeed, so clamorous had Darwin's critics become, wrote Huxley, that one could speak of the "Eclipse" or "Death" of Darwinism. This reaction, Huxley tells us, set in during the 1890s. Younger zoologists were becoming discontented with the enthusiasm of their elders for phylogenetic constructions that led them ". . . (like some Forestry Commission of science) to plant wildernesses of family trees over the beauty spots of biology. . . . Evolutionary studies became more and more merely casebooks of real or supposed adaptations. Late nineteenth-century Darwinism came to resemble the early-nineteenth century school of Natural Theology."[5]

William Bateson's book *Materials for the Study of Variation* of 1894 was, Huxley believed, a symptom of the revolt and of what he dubbed "the period of mutation theory, which postulated that large mutations, and not small 'continuous variations' were the raw material of evolution, and actually determined most of its course." Bateson, Huxley explained, drew the most devastating conclusions from the results of Mendelian research, and in the period immediately preceding World War I "the legend of the death of Darwinism acquired currency."[6] In the two decades following World War I, claimed Huxley, Darwinism was reborn as the fruit of a process of synthesis that he described in the following words:

> Biology in the last twenty years, after the period in which new disciplines were taken up in turn and worked out in comparative isolation, has become a more unified science. It has embarked upon a period of synthesis, until to-day it no longer presents the spectacle of a number of semi-independent and largely contradictory sub-sciences, but is coming to rival the unity of older sciences like physics, in which advance in any one branch leads almost at once to advance in all other fields, and theory and experiment march hand-in-hand.[7]

The Darwinism thus reborn was modified Darwinism,[8] a "mutated phoenix risen from the ashes of the pyre kindled by men so unlike as Bateson and Bergson."[9]

What place did Huxley see himself filling in this scenario? He was, he explained, a "generalist" in an era of "overspecialization" when there were "too few polymaths about."[10] His ability to embrace and synthesize many fields he deployed beyond the limits of the natural sciences, for, as I shall argue, the aim informing and directing his many disparate activities was man's ambition to control and direct his evolution. Such a goal required that he should bring together biology, ethics, and even religion. His neo-Darwinian synthesis was thus *broader than the synthesis envisaged by other Darwinians*. What he called a "demon" driving him "into every sort of activity"[11] might be described in terms of his passion to convince his generation of the opportunities and responsibilities that lay in man's hands and his alone. The role of generalizer and publicist was, no doubt, in part forced upon him by his decision to leave the university sector, and, following his inevitable resignation from the post of secretary to the Zoological Society, to live by the pen.

However, his greatest strength was surely his pen, and, as John Baker has pointed out, the notice given by the national press to his work on the axolotl in 1920 was the stimulus that "launched him on a new part-time career,"[12] that of popular scientific writing. The experience of working with H. G. Wells on *The Science of Life* between

1926 and 1931 must surely have developed his skill in this role. Book 4, titled "The Essence of the Controversies about Evolution," offers perhaps the clearest, most readable, succinct, and informative popular account of the subject ever penned. It was here that he first expounded his own version of what later developed into the evolutionary synthesis.[13] In 1936 he outlined the same subject to the British Association for the Advancement of Science.[14] "The result," recalled Huxley, "exceeded my expectations. So many of my colleagues expressed interest and the wish that the address might be available in more extended and more permanent form, that I decided to essay expanding it into a book."[15] The resulting text won considerable acclaim. In *The American Naturalist* it was called "the outstanding evolutionary treatise of the decade, perhaps of the century. The approach is thoroughly scientific; the command of basic information amazing; the synthesis of disciplines masterly."[16] In the *Quarterly Review of Biology* the reviewer described how Huxley had "undertaken the imposing task of a general synthesis from the point of view of not just one field of biology but of many. In this task, he has succeeded remarkably."[17]

Huxley's account of the history of the theory of evolution raises a number of historiographical issues, which we should now explore. Chief among them is the concept of a synthesis of fields. Historians are hypersensitive these days about the slightest traces of "Whiggish" or "presentist" history. If they catch a whiff of it, they run a mile. The very concept of a synthesis suggests that the scientists in their specialties could see the way they were going and were working toward such a goal. In reality, a cynic might argue, they did not see the goal and the synthesis was a post hoc construction identified with the aid of hindsight. A more faithful model of the process involved, as William Provine has been urging recently, is that of the elimination of alternative theories. Thus, between 1915 and 1930, Lamarckian evolution, orthogenesis, evolution by mutation, and for some evolutionists (e.g., R. A. Fisher and E. B. Ford) evolution by drift, were disposed of, leaving only evolution by the natural selection of small mutations inherited in a Mendelian fashion. This would be a negative synthesis, i.e., synthesis by subtraction—the form taken by the synthesis had long been common knowledge. The alternatives, and with them the objections to the neo-Darwinian synthesis, were removed and the consummation of the long-proposed union could take place.[18] On this model the historiographical utility of the concept of a synthesis is lessened.

There is some truth in this model. Obviously, the alternatives mentioned (excepting, of course, drift) were severely undermined. Drift was preserved, and some abrupt production of distinctive adapted forms has been established. Nevertheless, the alternatives of Lamarckian and orthogenetic evolution are no longer serious contenders as modus operandi of evolution. However, the "successive elimination" model fails to account for the process of modification undergone by the surviving elements that became masters of the field. In the period 1915 to 1930 Mendelian heredity and De Vriesian mutation were profoundly modified, extended, and developed. The widespread existence of polyploidy, gene duplications, and multiple alleles was established. Equally the theory of speciation was developed, and the first decisive evidence for the action of natural selection was provided. Such results were the fruit of pure and applied research and controversy between the supporters of alternative evolutionary mechanisms. Finally, Mendelian heredity and Darwinian selection did not alone constitute the elements of the synthesis. Especially is this true of Huxley.

Two more methodological points need to be raised. One concerns the distinction between three aspects of a causal theory: the *existence* of the cause evoked by the theory, the *competence* of the cause to produce the phenomena of the sort and size required, and the *responsibility* of the cause for the particular phenomena to be explained. By *competence* we mean that we have evidence that the cause can account for the data we wish to explain. By *responsibility* we mean that the cause—such as the process of natural selection—has in fact produced evolutionary change in the past.[19] It was one task to formulate the lines along which an integrated theory of evolution might be constructed, and quite another to bring the conviction on both theoretical and empirical grounds that such a theory describes a process that has in fact been responsible for the production of all living forms. To illustrate this point, consider the following passage from August Weismann's contribution to the Darwin centenary of 1909:

> Even although we cannot bring forward formal proofs of it *in detail*, cannot calculate definitely the size of the variations which present themselves; and their selection-value cannot, in short, reduce the whole process to a mathematical formula, yet we must assume selection, because it

is the only possible explanation applicable to whole classes of phenomena, and because, on the other hand, it is made up of factors which we know can be proved actually to exist, and which, if they exist, must of logical necessity cooperate in the manner required by the theory. We *must accept it because the phenomena of evolution and adaptation must have a natural basis, and because it is the only possible explanation of them.*[20]

Here Weismann jumps from competence to responsibility simply because no other competent cause exists. Such a plea for acceptance, by default of any credible alternative, is no better than similar pleas for the existence of a Creator.

The second point concerns the fact that many of the claims made on behalf of a synthesis are vitiated by the associated assumption that the domains of the subjects brought together have remained constant. For example, what used to be called heredity and variation concern the same domain of subject matter as that now covered by genetics. In fact, heredity was considered by many biologists, though not by Galton and the biometricians, as concerning embryological development, as well as inheritance and variation. As Cowan has shown, Galton's great contribution was to separate heredity and variation from development and to "reduce" reversion and variation to the properties of the fundamental process—heredity. But the data of hybridization were marginal to Galton's program, and the domain in which Mendel's work was developed was that of species multiplication (the hybridization theory of species formation).[21] Several authors have suggested that the evolutionary synthesis would have been achieved long ago if only Darwin had known Mendel's work. In fact, Mendel's transformist views were not by any means like Darwin's, and Darwin's views on variation were fundamentally at variance with Mendel's.

Third, we may ask whether Huxley's picture of the development of evolutionary theory is reflected in the remarks of his contemporaries. Do we find the same self-conscious feeling that the 1930s marked a significant stage characterized by Huxley as the "rebirth of Darwinism"? In 1937 Dobzhansky did not say so. His style throughout *Genetics and the Origin of Species* was didactic, not historical, but when he wrote the introduction to Ernst Mayr's Jessup Lectures of 1941 he declared:

A new and significant trend has become discernible in biology during the last decade. The excessive specialization which had prevailed in the recent past seems slowly to be giving way to a greater unity; a science of general biology seems to be emerging. In a way this trend represents a partial reversal of a historic tendency of much greater duration. . . . During the last decade the conclusions reached by many of the specialists have begun to converge towards a set of general principles applicable to the entire realm of living matter.[22]

Other authors show an awareness of the opposition toward selectionism and of the difficulties that lay in the way of a successful application of genetics to evolutionary biology. Thus, Shull reckoned it required "twenty years to get even a moderately clear view of the mechanism whose more elementary fundamentals Mendel had postulated," and a further ten "to acquire certain of the details which mean most for evolution."[23] E. B. Ford, in his remarkable little classic *Mendel and Evolution*, opened his chapter on the bearing of genetics on evolution with the sentence: "Any attempt to apply the results of experimental genetics to the study of evolution has generally provoked a storm of criticism." The Mendelian theory was, he claimed, the only theory of inheritance "susceptible of proof and of exact study . . . . Yet it is to the application of this theory that so much exception is taken."[24] The thirties, it would appear, were a time of attempting to win influence on the basis of the achievements of genetics. By the forties, it was clear that this movement had largely succeeded. Although it was possible for Leonard Eyre to prepare a translation of Nordenskiöld's *Biologins Historia* (1920–24) in 1928,[25] and for E. J. Hatfield to bring out in 1930 an English translation of Radl's *Geschichte der biologischen Theorien* (1905–1909),[26] the success of such anti-selectionist texts would have been much less likely in the climate of the forties and fifties.

One concept of crucial importance in the development of the evolutionary synthesis was that of mutation. I would like to distinguish two versions of this concept and argue that the second version was crucial to the attainment of the synthesis.

The First Mutation Theory

As Dobzhansky and Mayr have pointed out, the term has been used in more than one sense. In 1869 Waagen used it to describe abrupt changes in the paleontological record,[27] but De Vries used it to refer to Bateson's "discontinuous" variations. It is well known that under De

Vries's enthusiastic leadership the mutation theory quickly emerged as a theory of species formation opposed to Darwinian evolution. This *first mutation theory* was announced in 1901 and quickly gained support, being identified with Hugo De Vries, T. H. Morgan (up to 1910), Wilhelm Johanssen, Louis Baringhem, Lucien Cuenot, William Bateson, and for a very brief period, W. E. Castle. Huxley referred to this phase of the theory as

> the period of mutation theory, which postulated that large mutations, and not small "continuous variations," were the raw material of evolution, and actually determined most of its course, selection being relegated to a wholly subordinate position.[28]

Garland Allen calls this the theory of macromutation. Its demise began in the second decade of this century, but the theory was not destroyed until the third decade under the impact of W. E. Castle and H. MacCurdy's study of hooded rats between 1906 and 1919, and the *Oenothera* studies of A. M. Lutz, B. M. Davis, and O. Renner, of which Huxley remarked in 1931:

> The net result of twenty-five years of research, summed up in Renner's monograph, is that no special storm of mutation is agitating *Oenothera's* germ-plasm. So far as we can see, the kind of sports it is throwing now it will continue to throw as abundantly as ever. . . .[29]

Renner's exposure of *Oenothera* as a "permanent heterozygote," the constancy of which is maintained by the nonviability of all its pure segregates, was followed by Muller's analysis of beaded inheritance in *Drosophila* in 1918. In a short but very important paper Muller attributed both cases to the presence of lethal, recessive genes.[30]

According to supporters of the first mutation theory, new forms showing adaptation and diverging from the parent species sufficiently to constitute a specific difference were produced by as little as one mutational step. They were happy that this denied to natural selection the *creative* role that Darwin had given it in the process of accumulating a series of small steps in a given direction. Selection had only to accept or reject the mutation when it appeared. It was active, not creative. Many of the supporters of mutational evolution considered the faith placed in natural selection as unpalatable as the faith placed in a providential creator. De Vries and Bateson realized that this mutational theory made it unnecessary to attempt to discover adaptive features in every specific difference. Underlying this view was the conviction of such authors as D'Arcy Wentworth Thompson and Bateson that many characteristics of organisms are determined by purely physical considerations. Variations in the patterns of symmetry, in the colors of flowers and butterflies' wings, and in the sizes and shapes of organisms were often the result of physical forces or chemical reactions.[31] Huxley inherited this tradition, and it encouraged him to conclude that the existence of many characters has not been dependent on their selective advantage.

The first mutation theory was thus an expression of the anti-speculative and anti-selectionist mood of the first two decades of this century. It also marked an increasing emphasis on experimentation, as William Coleman and Garland Allen have urged.[32] While Bateson stressed dependence on the experimental method, he also gave full play to his skepticism. After spending his life largely in the study of variation, Huxley declared of Bateson:

> he developed an increasing inability to satisfy himself how any progressive variation could ever occur. He crowned his scientific career by various lectures and addresses in which he reiterated his imaginative failure. This type of agnosticism was probably the negative aspect of a passionate and unquestioning faith in the implacable unteachableness and integrity of certain Mendelian units of heredity. . . .[33]

There is no mistaking the shadow cast by Bateson over British evolutionary biology. Huxley's account of the history of the subject, which we have described, used Bateson's publications as "markers" of the course of events. J. B. S. Haldane, in what is perhaps the most perceptive discussion of Bateson's ideas and approach, claimed that his

> fundamental notion of discontinuity in the evolutionary process which he enunciated seven years before the rediscovery of Mendel's work, will remain, though doubtless with some modifications, a component of any theory of evolution.[34]

Yet we do not find that Bateson's scepticism about natural selection put an end to belief in its all-powerful creative role, although he warned "those who could proclaim that whatever is is right" with all the scorn he could muster to base their faith "frankly on the impregnable rock of superstition and to abstain from direct appeals to natural fact."[35]

The Second Mutation Theory

According to the supporters of this theory, the majority of mutations are small. Mutations (normally understood to mean alterations of the gene but often taken to include rearrangements) are rare, but "recurrent" (see Table 1, p. 75).[36] This term appears in the literature from 1923 onward, a fact that I believe to be significant. These characteristics had the following important implications: (1) Natural selection would be required to create adaptive divergence by the accumulation of many small mutations. (2) Evolutionary divergence could not be directed by "mutation pressure." (3) It should be possible to estimate a statistical rate of mutation for a given gene under specified conditions. It is surely noteworthy that the concept of recurrent mutation as distinct from mutability is not found in the earlier literature, and is conspicuously absent from eugenic literature of that period. Clearly the recognition of "recurrence" would have posed a serious problem for those eager to purify our racial germ plasm! Indeed, Muller was aware of the problem presented by the accumulation of recessive mutations for those who wanted to produce a genetically sound stock. Like East and Jones, Muller advocated periodic inbreeding followed by selection to expose and remove recessive disadvantageous mutations.[37] In 1925 Morgan, Bridges, and Sturtevant explained:

> The actual frequency of mutation can scarcely be determined, even roughly, partly because it is so low as to present serious statistical difficulties, and even more because detection of mutations is so largely dependent on the personal equation of the observer.[38]

The first attempts at the estimation of mutability precede the 1920s, but as far as I can judge, were estimates of the rate at which a gene mutates and not the rate at which the same mutant form was produced. One might call the Second International Congress of Eugenics, held in 1923, a watershed, for there Muller introduced precision into the meaning of the term "mutation." He began his paper with the following words:

> Beneath the imposing building called "Heredity" there has been a dingy basement called "Mutation." Lately the searchlight of genetic analysis has thrown a flood of illumination into many of the dark recesses there. . . .[39]

At that congress also, Charles Danforth suggested the possibility of estimating mutation rates in man from the frequency of mutant forms in the population and their persistence. Under the conditions of equilibrium in gene frequency, he argued, the rate of loss of mutants in succeeding generations must be balanced by the rate of their production. This suggests that we can identify one further feature by which to distinguish these two phases of our subject. During the first phase, mutants were identified both in laboratory populations and in natural populations and attempts were made to measure "mutability"; but in the second phase, estimates of the rates of "recurrent" mutations were made, and it was these rates that offered decisive information for discussions of the mechanism of evolution.

Some Qualifications

This distinction between two phases and theories of mutation, I hope, serves to emphasize the important role that the concept of mutation has played in the elaboration of neo-Darwinism. Like all neat divisions, it surely deserves only qualified acceptance. Small mutations were discovered before the 1920s, and large mutations have not been written out of modern evolutionary mechanisms. Despite R. A. Fisher's determination to keep them out, one might almost say his perverse dogmatism on this point, his admirers, C. A. Clarke, E. B. Ford, and P. M. Sheppard, admitted them as the starting point in their theory of mimicry. John Turner, in his excellent account of this "Oxford School," has explained how they argued that natural selection of modifiers must have followed the abrupt emergence of a large mutation, modifying it from a rough resemblance to the form mimicked to a close resemblance. Not only does this represent a compromise between "mutationism" and "selectionism," but it was expressed by E. B. Poulton in 1912 and mentioned as a possible mechanism by R. C. Punnett in 1915. It is not Turner's point to unearth "precursors" of modern views, but to expose the extent to which Fisher's admirers departed from him. Turner writes:

> At the genetic level, Fisher's stand, which was fairly extreme even at the time, was followed by what could be called 'the softening of the modern synthesis' or 'synthesis by stealth'. The Oxford School, while remaining firmly opposed to a creative role for random drift, and therefore thor-

oughly adaptationist in this sense, compromised with the mutationism of Goldschmidt and Punnett.[40]

Not only were elements of the first mutation theory incorporated into the neo-Darwinian synthesis, but the hybridization theory of species multiplication found its way in a modified form into that synthesis. This theory, which originated with Linnaeus, was supported by J. G. Gmelin, Dean Herbert, and Kerner von Marilaun, but opposed by P. S. Pallas, J. G. Koelreuter, Carl von Gaertner, Carl von Naegeli, and Charles Darwin. Wichura demonstrated it in the willows (*Salix*), and Callander claims, I believe correctly, that Mendel believed it was a form of species multiplication, and he demonstrated it in the Hawkweeds (*Hieracium*).[41] In the twentieth century the discovery of apomixis and polyploidy made possible the formulation of mechanisms to account for the abrupt formation of a new species following hybridization. This was observed first in *Primula kewensis* (1914) and then in *Spartina townsendi* (1931).[42] The importance of polyploidy as a widespread phenomenon in the evolution of plants was thoroughly documented by C. D. Darlington in his *Chromosomes and Plant Breeding* of 1932.[43] After describing examples in his British Association address of 1936, Huxley concluded:

> From the standpoint of natural selection, species will then fall into two contrasted categories. On the one hand we have those in which natural selection can have nothing to do with the origin of the basic specific characters, but merely acts upon the species as given, in competition with its relatives. These include all species in which character-divergence is abrupt and initial. On the other hand we have those forms in which character-modification is gradual. Here natural selection may, on both deductive and inductive grounds often must, play a part in producing the character of the species. This helps to bring home the heterogeneity of the processes which we may lump together as "evolution."[44]

This brings us to inquire further into Huxley's view of the synthesis. What he wrote on the subject in 1942 differs markedly from what he wrote in the late twenties. Then, he was as keen to relate embryology to evolution as he was to relate genetics, and he did this through Goldschmidt's concept of *rate genes*. The subject figures prominently also in Huxley's 1936 address, but is definitely downplayed from 1942 onward. Let me attempt to explain why these rate genes were important to him. Although he shared de Beer's concern to give a developmental account of gene action that would cover the various phenomena attributed to recapitulation, his targets were also the objections to Darwinian evolution raised by such paleontologists as H. F. Osborn, who claimed the existence of internally driven trends, "orthogenetic" trends, in parallel groups such as the increase in size of body and the formation and increase of a horn in the Titanotheres. Huxley discovered an exponential relation between organ size and body size. Organs sometimes grew faster than the body—"positive heterogeny" (e.g., the deer antlers and beetle mandibles)—or slower—"negative heterogeny" (e.g., head size). His argument was that if large body size proved to be advantageous, the antlers would, by their positive heterogeny, become very large—an increase in size with (he thought) no adaptive advantage but attributable to this growth relationship. Thus many large mandibles in insects and horns in mammals may have no adaptive significance, and Huxley concluded: "This provides us with a large new list of non-adaptive specific and generic characters."[45] When he learned of the work being carried out at the Marine Biology Station at Plymouth on eye-color mutants of the shrimp, *Gammarus*, he eagerly set E. B. Ford to see what light it might throw upon the genetic control of the rate of development of eye color. At the Marine Biology Station, Mrs. Sexton had observed how even the color of the red-eye mutants darkened with age. Ford and Huxley were able to show that the different eye-color mutants were due to different rates of pigment formation. They concluded that the many eye-color mutants of *Drosophila* were probably produced in the same way, and they urged that wherever possible characters determined by genes should be "thought of as developmental processes with characteristic rates."[46]

Clearly Richard Goldschmidt's studies in physiological genetics had been an important inspiration to Huxley, but the system of genetic control revealed in *Gammarus* furnished evidence that Goldschmidt's view of the mechanism of the genetic control of development was too simplistic. His work was, nonetheless, classical. It was he, claimed Ford and Huxley,

> who has conclusively demonstrated rate Factors for sex determination in moths, and for melanin formation in moth larvae. Many other cases might be cited, but we are here only concerned to point out . . . how fruitfully the factorial theory of heredity can be extended by thinking of

the genes, at least in a large number of cases, as influencing not merely the character but also the rate of developmental processes.[47]

Seven years later Huxley and de Beer's textbook, *The Elements of Experimental Embryology*, appeared, which contained a section entitled "The Hereditary Factors and Differentiation." After describing the *Gammarus* story, they expressed themselves forcefully:

> It is necessary to think in terms of development before it is possible to discover what is the fundamental process with which a given gene is concerned. In such an analysis, the old concept of Mendelian *characters* will disappear. The visible character is not Mendelian in any real sense: it is the resultant of the interaction of a particular gene complex with a particular set of environmental conditions.[48]

Indeed, they perceived a shift in the interest of geneticists:

> Hitherto, neo-Mendelism has been concerned mainly with the manoeuvres of the hereditary units, and in large part with their manoeuvres during the two cell-generations in which the reduction of chromosomes is brought about. It is now beginning to concern itself with the mode of action of the hereditary units during the much larger number of cell-generations involved in building up the adult organism from the egg: and this task it can only accomplish satisfactorily in close contact with developmental physiology.[49]

The importance that Huxley attached to the work on *Gammarus* is evident from his correspondence with L. C. Dunn. He tried to establish a group to work on *Gammarus* at Columbia University on the occasion of his visit to the United States in 1932, when he brought stocks with him and left detailed instructions for their care. When he sent a copy of his *The Elements of Experimental Embryology* two years later, he explained that it should fill a need, the recently published *Enbryology and Genetics* by T. H. Morgan being "terribly one-sided, as he [Morgan] does not even mention the work of Child and so on. This latter we have tried to link up with what is ordinarily known as experimental embryology."[50]

We can surely conclude that Huxley, Ford, and their coworkers made a significant contribution to physiological genetics in their *Gammarus* research, which marked a departure from the approach of the Morgan school. Genetics was for Huxley not an independent specialty, but a resource that could be used to tackle problems raised by the critics of Darwinism and that offered a new foundation upon which to construct sound explanations of the phenomena previously attributed to the recapitulation of phylogeny in ontogeny. Morgan had introduced exact quantitative methods into genetics, but it was Goldschmidt who had opened the way to a study of the very aspect of the gene that Morgan had deliberately ignored—its mode of action. In 1932 Huxley was ready to launch a concerted attack on the nature of gene action. The physiological genetics he envisaged would bring together genetics, growth, and development.[51] His plan to settle in America and continue his work did not materialize, perhaps chiefly because his wish to marry Miss Waldmeier came to nought. He returned to England, no salaried employment, and no laboratory facilities.

Huxley the Outsider

Although Huxley had trained as a zoologist and taught the subject at Oxford and London, his academic employment came to an end in 1927. To an extent he became an "outsider" and therefore less closely identified with one particular school of thought or with zoology to the exclusion of botany. *Evolution, the Modern Synthesis* and his other evolutionary writings are noteworthy for the breadth of treatment and balance between the two kingdoms. This balance has important repercussions for, as Muller pointed out in 1925, the presence in plants of reproduction by vegetative propagation without the need for the intercalation of a sexual stage is what has given the opportunity for ploidy to occur. In higher animals, however, repeating vegetative cycles do not exist. The system of sex determination in animals, too, he held would not function in the polyploid condition. Thus the picture of evolutionary processes in the two kingdoms are not the same.[52] (Some examples of polyploid species formed following hybridization have since been observed in the animal kingdom.) Although Ernst Mayr's *Systematics and the Origin of Species* (1942) was intentionally confined to animals, he was still criticized for this limitation, and even more for drawing most of the examples from ornithology.[53]

We are accustomed to regarding zoology and botany as antiquated and artificial divisions of the subject that at the beginning of the last century was called "biology." Had not the discoveries concerning the protozoa and the algae (once called "infusoria") bridged the gap between the two kingdoms? Did not the cell theory go beyond the analogies of the eighteenth century and offer a structural unit common to both kingdoms? And now we have the universality of the genetic code and protein synthesis. Why do we need zoology and botany? Historically these subjects became more distinct as they were professionalized in the last century. Until genetics and biochemistry became established as academic disciplines, this separation was continued. There were obvious reasons for this. Higher plants and higher animals are different! The service roles and career opportunities of those working in these disciplines have also differed. For British botanists and zoologists of Huxley's time their chief careers were in the colonial services. Yet Huxley, though he became secretary to the Zoological Society, showed no academic insularity to botany.

Did Huxley's broad approach to evolution result in his presenting a pluralistic scheme of evolutionary mechanisms in contrast to the thoroughgoing selectionism of his Darwinian friends? This pluralism can also be found in J. B. S. Haldane's *The Causes of Evolution* and in the first edition of Dobzhansky's *Genetics and the Origin of Species*. Darlington was even more strongly committed to sudden changes due to mutation and to polyploidy. In 1932 he wrote:

> [M]any of the decisive steps by which great systems have arisen can be repeated under experimental conditions. It is not a question of waiting for a thousand generations to see how the change from sexuality to hermaphrodism, or from fertilization to parthenogenesis, from homozygosis to heterozygosis, or from diploidy to polyploidy and secondary polyploidy, can be produced. These are all irreversible steps, and they can all happen in one generation. Indeed they must so happen.[54]

Variations, he explained, arise in advance of the use to which they may be put, and they survive their use, as in the pollen of apomictics and the Y chromosome of many animals. Darlington recalled that in 1940 he tried to suggest to Fisher "that the principles of natural selection were not in fact going to operate with the absolute rigor he expected, but I could never get him to discuss it." Darlington added that he did not dare to suggest that his views were "more Wrightian than those of Wright himself, because our relationship would have broken off even earlier than it did."[55]

Darlington clearly had an important influence upon Huxley. Ford's

influence was more than that of the former student supplying him
with data, but Ford came increasingly under the influence of R. A.
Fisher, to whom Huxley had introduced him. In 1941 he reacted to
Huxley's inclusion of *drift* in the proofs of *Evolution, the Modern
Synthesis*, saying: "I distrust any theory of frequent nonadaptive
evolution. It is supported by Sewall Wright who, I am sure, does not
take into account the demonstration that mutation and recombination effectively neutral as regards survival value must be very rare."[56]
As the Oxford School progressed in its demonstration of selection in
situations where it had been thought absent, Huxley changed his
tune. Yet in 1963 when Huxley wrote the introduction to the second
edition of *Evolution, the Modern Synthesis*, he spoke at some length
on the important role of *genetic drift*, and went on to question Fisher's
thesis of balanced polymorphism in light of recent discoveries in he-
matology, enzymology, and immunology—multiple chemical forms,
yet with the same functions. These forms he attributed to small
mutations not subject to negative selection, which could thus ac-
cumulate to a frequency much higher than the mutation frequency.[57]

I conclude that Huxley did not share the Oxford School's determination to dispose of *genetic drift* just as they earlier disposed of the
first mutation theory. Nor was Huxley a willing and consistent exem-
plar of the trend toward the view that all evolution takes place under
the influence of selection, which Stephen J. Gould has so well described and called the "Hardening of the Synthesis."[58] In this respect
Huxley was distanced from the British school of Darwinists. There is
a further respect in which his conception of the evolutionary synthesis puts him outside the profession of evolutionary biologists, and
this is his view of progress.

Evolutionary Progress

We have argued that Huxley's outsider status was favorable to his
adoption of a balanced and broad approach to the subject of evolutionary mechanisms. Huxley's special family environment and traditions, as Colin Divall has explained, nourished in him an abiding
commitment to ethical considerations.[59] Here was another factor
besides his severance from academic employment that distanced him
from those evolutionists who would have nothing to do with talk of

progress. While denying that evolution has its own Bergsonian purpose driving it, Huxley was prepared to claim that the emergence of consciousness in man made possible the construction of a purpose for evolution. Consequently, one has the feeling that when Huxley described progress using empirical criteria, his choice of these criteria was dictated by his view of its purpose, and here ethical considerations come in.

Progress, claimed Huxley, can be defined in terms of "control over the environment, and independence of it. More in detail [these properties] consist in size and power, mechanical and chemical efficiency, increased capacity for self-regulation and a more stable internal environment, and more efficient avenues of knowledge and of methods for dealing with knowledge."[60] Huxley claimed that the natural selection of the modern synthesis could account for adaptation and for long-range trends of specialization; therefore, it could account for evolutionary progress, too. Evolution he pictured as a "series of blind alleys." All, save that of man, have "terminated blindly."[61] Man's future is therefore of central importance to the future of life as a whole. What we must not do is leave nature to take its course, for as he stressed, echoing his grandfather's views:

> Natural selection, in fact, though like the mills of God in grinding slowly and grinding small, has few other attributes that a civilised religion would call divine. It is efficient in its way—at the price of extreme slowness and extreme cruelty. But it is blind and mechanical; and accordingly its products are just as likely to be aesthetically, morally, or intellectually repulsive to us as they are to be attractive or worthy of imitation. Both specialised and progressive improvements are mere by-products of its action, and are the exceptions rather than the rule. For the statesman or the eugenist to copy its methods is both foolish and wicked. Not only is natural selection not the instrument of a God's sublime purpose: it is not even the best mechanism for achieving evolutionary progress.[62]

Having cleared the way for his great theme—man's control of his own evolution—Huxley used the final section of his 1936 address and of his *Evolution, the Modern Synthesis* to introduce a discussion of man's future. He mentioned the hope of raising the level of the performance of man's brain. Two approaches that he referred to were eugenics and extrasensory perception. Such faculties might become distributed as widely as "musical or mathematical gifts are today." By the work of Rhine, Salter, and others, he declared, these faculties were being forced "into scientific recognition." Even if man did not choose to improve his nature through the application of eugenic principles, there were other avenues of progress. He must explore his own nature fully in order to identify its potentials and its limitations. Natural selection had ceased to be of major importance as an agent of change. Instead it was the struggle between traditions and ideas and between nations and social groups embodying them that determined the evolution of man. Some aspects of these traditions, like the exoskeleton of insects, limited their evolutionary potential, such as the ideographic symbols of Chinese writing and the nonmetric systems of measurement of the British Commonwealth and the United States.[63]

For Huxley, man was the "organ" of evolutionary progress. In this "cosmic office" it was man's role to realize the highest possible spiritual experience. We need a Columbus to explore the geography of the mind so that we can be taught techniques "of achieving spiritual experience (after all, one can acquire the technique of dancing or tennis, so why not of mystical ecstasy or spiritual peace?)."[64] Apparently no one told him about yoga! Huxley admitted that other authorities regarded the idea of progress as a myth, but the scientific doctrine of progress, he declared,

> is destined to replace not only the myth of progress, but all other myths of human earthly destiny. It will inevitably become one of the cornerstones of man's theology, and the most important external support for human ethics.[65]

Huxley treated religions as "social organs whose function it is to adjust man to his destiny,"[66] and he classified as religious any system or teaching that concerned man's destiny. The evolutionary concept of progress was thus religious, but in contrast to old religions that helped man maintain his morale in the face of the unknown, religions today must "utilize all available knowledge in giving guidance and encouragement for the continuing adventure of human development."[67] In 1931 Huxley described the role of religious mysticism as submerging the ego in a greater being. Whereas earlier religions identified such a being with a god, the evolutionary "religion" of humanism identified it with the human species. Despite the struggle of World War I, Huxley glimpsed the process of coalescence of minds into superminds taking place. At the end of his vista of progress he pictured man "consciously controlling his own destinies and the destinies of all life upon this planet."[68] These views he expressed some sixteen years before he first met Teilhard de Chardin.

Biological Science in the Twentieth Century

Those who have surveyed the history of biology in the present century have stressed a variety of themes—the trend toward greater emphasis upon experimentation in contrast to the observational work of the naturalist; the deployment of reductionist methodologies (often but not always associated with a reductionist ontology); the "fertilization" of biology with the methods and concepts of physics and chemistry, which has been brought about by the intellectual migrations from those sciences, often in the "colonizing" spirit of a "hard" science taking control of a "soft" science. Ernst Mayr has emphasized the importance of the trend from "typological" to "populational" thinking in biology. He has shown how systematists and evolutionists as well as geneticists contributed to this transformation of the species concept.[69]

Some of the "actors" in modern biology claim that we have passed through a "molecular revolution" in biology, that they consider as significant as the quantum revolution that shook the physical sciences a generation before.[70] Whether or not we have had such a revolution in biology, how did Huxley respond to the discoveries of molecular biology? He appreciated the importance of physics and chemistry in biology, and would surely have applauded the success in genetic manipulation. He would have seen this technology as the most powerful of aids toward achieving human progress. However, we should remember that he was a biologist who worked with the whole organism, and as a biologist he had his priorities—they all concerned evolution, especially man's evolution.

In Huxley's introduction to the second edition of his *Evolution, the Modern Synthesis* (1963), he deliberately left until the end "the most important scientific event of our times—the discovery by Watson and

Crick that the deoxyribonucleic acids are the true physical basis of life, and provide the mechanism of heredity and evolution."[71] Some three pages later, having outlined the many avenues opened up by this discovery, he concluded:

> In general, however, the discovery of DNA and its properties has not led to important new developments or significant modifications in evolutionary theory or in our understanding of the course of biological evolution. What it has done is to reveal the physical basis underlying the evolutionary mechanisms which Darwin's genius deduced must be operative in nature, and to open up new possibilities of detailed genetic analysis and of experimental control of the genetic-evolutionary process. The edifice of evolutionary theory is still essentially Darwinian. . . .[72]

There is a sense in which Huxley was right. Clearly molecular biology has not undermined neo-Darwinism, but rather has supported it. Thus the machinery of protein synthesis cannot, as far as we know, work in reverse, thus making Lamarckian inheritance impossible to envisage in biochemical terms. At the same time there is a sense in which Huxley was wrong, for molecular biology has revealed the extent of those variations in protein sequence that have no known functional significance. The results have informed the debate on the extent of *neutralism* in evolution. Yet we should remember that Huxley had long accepted the existence of variations of no adaptive significance, a view he inherited from D'Arcy Thompson. For both men one determinant of morphological characteristics was molecular structure. Morphology may result simply from the molecular morphology of the chemical constituents and owe nothing to selection. That the new molecular biology should reveal fresh sources of nonadaptive variation may not have seemed to Huxley to call for any major reassessment of Darwinian evolution.

Conclusion

Huxley played an important part in the consolidation, dissemination, and popularization of the synthetic theory that we know as the neo-Darwinian synthesis. This he achieved through his own writings and through the work of his students, principally E. B. Ford. But Huxley's synthesis was broader than the synthesis of the neo-Darwinism architects, embracing not only embryology and growth, but also religion, progress, and ethics. His passionate pursuit of evolutionary humanism is in contrast to his unprejudiced and eclectic approach to biological research and debates within biology. These debates continue today, and we could do with more expositors and communicators like Huxley.

Table 1. Recurrent Mutations and Allelomorphic Series

Locus	Total Occurrences	Distinct mutant Types	Locus	Total Occurrences	Distinct mutant Types
apterous	3	1	lethal-a	2	1
ascute	4 ±	1	lethal-b	2	1
Bar	2 +	2 +	lethal-c	2	1
bent	2 +	2	lethal-e	4	1
bifid	3	1	Lobe	6	3
bithorax	3	2	lozenge	10	5
black	3 +	1	maroon	4	1
bobbed	6	1	miniature	7	1
brown	2	2	Notch	25 ±	3
broad	6	4	pink	11 +	5
cinnabar	4	3	purple	6	2
club	2	2	reduced	2	2
crossveinless	2	1	rough	2	2
curved	2	2	roughoid	2	2
cut	16 +	5 +	ruby	6	2
dachs	2	2	rudimentary	15 +	5 +
dachsoid	2	1	sable	3	2
Delta	2	2	scarlet	2	1
deltex	2	1	scute	4	1
Dichaete	3	3	sepia	5	1
dusky	6 +	3	singed	5	3
ebony	10	5	Star	2	1
eyeless	3	2	tan	3	2
fat	2	2	tetraploidy	3	1
forked	12	4	triploidy	15 ±	1
fringed	2	1	Truncate	8 ±	5
furrowed	2	2	vermilion	15 ±	2
fused	2	2	vestigial	7	5
garnet	5	3	white	25 ±	11
Haplo-IV	35 ±	1	yellow	15 ±	2
inflated	2	1			

Huxley the Biologist

J . A . W I T K O W S K I

Julian Huxley in the Laboratory: Embracing Inquisitiveness and Widespread Curiosity

My task is to discuss Julian Huxley's work as a laboratory scientist (Fig. 1). At first sight this seems to be an easy task—Huxley's scientific papers are readily available in journals, he discussed his scientific career in his autobiography, and there are biographies and obituary notices that deal with his contributions to science.[1] But Huxley is renowned for the diversity of his scientific activities; his research ranged from studies of cell aggregation in sponges to studies of bird behavior; his writings from erudite scientific papers to popular scientific journalism; his public speaking from technical symposia to radio discussion programs; his politics from the politics of the London Zoo to politics on a global scale as director-general of UNESCO. And even within the limits of my topic, Huxley's interests were many. He seems to have been inspired by the desire to learn all he could about biology, in the spirit of the question he claimed to have asked when aged four: "Why do all live things have natures?"[2] Nevertheless, one might expect to find a common theme that united Huxley's various laboratory researches and that would show that his contributions to different fields were in fact aspects of a unified approach to the study of biology. In particular, one might expect that this common theme would be evolution, for he resolved that "all my scientific studies would be undertaken in a Darwinian spirit and that my major work would be concerned with evolution, in nature and in man."[3] Now this is probably true, but, being so broad and sweeping a generalization, it

Figure 1. A picture of Julian Huxley at work in his Oxford laboratory in about 1920. The picture is from a newspaper article describing his work on thyroid-induced metamorphosis of the axolotl.

is not too helpful as a guide to his laboratory research. I think Baker's suggestion[4] that Huxley was concerned with ontogeny, that is with the growth and development of organisms, best describes Huxley's laboratory research. I will begin by describing Huxley's early background that led him to his interests in biology and experimental biology in particular. I shall then discuss some of Huxley's laboratory research but not in any systematic way—Huxley simply did too much.

Young Huxley and Science

It seems inevitable that Huxley would follow either the scientific example of his grandfather T. H. Huxley or the literary tradition of the Arnold side of his family and the example of his father, Leonard.[5] However, Julian's interest in natural history was aroused early and it seems that Julian and T. H. Huxley developed a special relationship. There is, for example, the delightful correspondence[6] between Julian and his grandfather concerning Charles Kingsley's *The Water Babies*. There was an illustration in the book showing T. H. Huxley and the great comparative anatomist, Richard Owen, examining a bottle containing the water baby, and Julian wrote the letter shown in Figure 2 to his grandfather. Clearly Julian was destined for science, and by the age of eight he began to study birds and flowers, and later butterflies and moths. At Eton he took biology in preference to German, his parents saying that he could always learn German by going abroad. Julian recorded that he owed his biology master, Mr. M. D. Hill, an "immense gratitude," for he was a genius as a teacher and so enthralled Julian that he decided to specialize in zoology. Even at Eton, T. H. Huxley reached out to influence his grandson, for it was T. H. Huxley who had persuaded the Eton board of trustees to build science laboratories.[7]

In November 1905, Huxley was awarded a zoology scholarship at Balliol College, Oxford. Huxley wrote later that "my first sense at Balliol was one of relief—relief at being free of the shackles of the smaller world of school, able to choose my own friends and venture into all kinds of adult interests."[8] He was in a privileged position as a consequence of his family background. Through introductions from his parents, he met such Oxford notables as Gilbert Murray, Sedgwick, and the Haldanes. He had already begun to move in that "intellectually adult world, the international intelligentsia,"[9] a world in which he was to spend the rest of his life. An early influence Huxley recalled with affection some sixty years later was that of Harold Hartley, his general tutor. Hartley encouraged Huxley to read

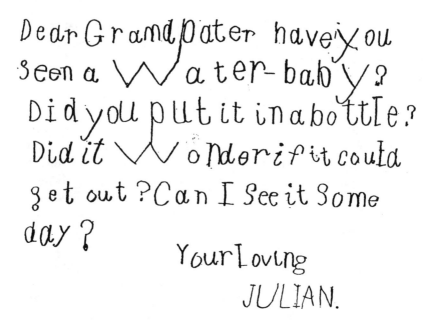

Dear Grandpater have you seen a Water-baby? Did you put it in a bottle? Did it Wonder if it could get out? Can I See it Some day?

Your Loving

JULIAN.

Figure 2. A letter written by Julian to his grandfather, T. H. Huxley, in 1892. Julian had seen a picture in *The Water Babies* of his grandfather and Richard Owen examining the water baby caught in a jar. (From J. S. Huxley, *Memories*, London, George Allen & Unwin, 1970, p. 23).

an autobiography of Pasteur that Huxley claimed opened his eyes to the way in which research was done. In his autobiography Huxley wrote, with evident approval, of scientific discovery resulting from "a mixture of intuition, pertinacity and occasional good-luck."[10]

Huxley was fortunate in the quality of his teachers. These included E. S. Goodrich for comparative anatomy, J. W. Jenkinson for embryology, and Geoffrey Smith for zoology. Baker described Jenkinson and Smith as "two of the most active and advanced minds in the biological world of their day,"[11] but, alas, both were killed in World War I. Jenkinson was perhaps the first English experimental embryologist, and his major achievement was establishing a program of experimental embryology at Oxford that produced a succession of remarkable biologists, including (in chronological order) Huxley, Gavin de Beer, J. Z. Young, Alister Hardy, E. B. Ford, and P. B. Medawar.[12] Jenkinson wrote three books dealing with embryology.[13]

Experimental Embryology was published in 1909 and was a comprehensive account of the contemporary state of experimental embryology. Jenkinson declared that "development is the production of specific form"[14] and that there were two main questions to be answered, one practical and amenable to experimental investigation ("What are the internal, what are the external conditions that determine the course of development?") and one theoretical. (Could a "mechanics of ontogeny . . . ever afford a theory which may be said to be complete either from a scientific or from a philosophical point of view"?[15]) Jenkinson's own research was concerned with external conditions, and he carried out experiments to determine the effects of factors such as osmotic pressure, centrifugation, and compression of embryos during cleavage on development. Nevertheless, despite his own interests in external factors, he recognized the importance of the work of Spemann and Lewis on lens development, describing it as "extremely interesting." He made the important point that the outcome of such dependent differentiation is a consequence of the interaction between the responding and the stimulating tissues. This foreshadows the distinction made much later by Needham between the processes of individuation and evocation in primary embryonic induction.

Geoffrey Smith, Huxley's zoology tutor, was a zoologist interested in a wide variety of subjects.[16] He had worked at the Naples Zoological Station and used embryological studies to examine, and disprove, at least to his own satisfaction, the biogenetic law. Smith's interests were in the internal factors affecting development, and as I will discuss later, some of his work may have had a decisive influence on Huxley's most important studies.

Except in that one respect there is little evidence that the work of either Jenkinson or Smith was a direct influence on Huxley's choice of research topic. Huxley mentions Smith and Jenkinson only briefly in his memoirs and rarely in his scientific publications. In *Elements of Experimental Embryology*,[17] he and de Beer acknowledge their debt to Jenkinson and refer to him as the "pioneer" of experimental embryology in Britain. But Huxley's first research topic in embryology was suggested by Paul Mayer at the Naples Marine Station rather than by Jenkinson. And while Huxley cited Geoffrey Smith's measurements of growth in his first allometry paper of 1924, Huxley

made clear in the preface to *Problems of Relative Growth*[18] that it was
D'Arcy Thompson's example that inspired him.

Nevertheless, leaving aside the question of specific research topics,
I am sure that Baker was right to say that "one can scarcely doubt that
it was the memory of J. W. Jenkinson's and Geoffrey Smith's teaching
that caused Huxley to devote the whole of his laboratory research to
experimental and analytical studies in ontogeny."[19] Experimental
embryology was the "hot" topic at the turn of the century as it was
being built on the foundations of the descriptive and comparative
embryology of the nineteenth century. The use of embryology to
support phylogenetic speculations and the biogenetic law had declined, and in 1909 Jenkinson could write that the aims of embryology were to understand "the increase in structure, the production of
form out of the relatively formless germ, and the gradual passing of
this into a new individual which is like the parents that gave it
birth."[20] Medawar wrote of a slightly later period that "students felt it
[experimental embryology] to be the most rapidly advancing front of
biological research,"[21] and E. B. Ford wrote that zoologists with
experimental leanings were "obsessed" with experimental embryology.[22] It is hardly surprising that this was Huxley's first area of
research.

Huxley and Experimental Embryology

Huxley's research career began at the Naples Zoological when he won
a scholarship funded by learned societies from a number of countries,
and he spent the summer of 1909 in Naples. Once again, T. H.
Huxley played a part in furthering his grandson's scientific education,
for T.H. had supported the establishment of the Naples Station. It
had been founded by Anton Dohrn, with whom T. H. Huxley had had
a long and close friendship, Dohrn staying with the Huxleys at
Swanage and the Huxleys visiting Dohrn in Naples. In 1873 T. H.
Huxley suggested that to raise money for the station, Dohrn should
make use of the Huxley name in any way he wished. Institutes,
societies, and individuals rented tables at which to work in the Naples
Zoological Station,[23] and T.H. much regretted that the condition of
English zoology was such that it was not possible to have an "English" table. In contrast, by 1898 American institutes and universities

rented three tables at Naples, and notables such as E. B. Wilson and T. H. Morgan spent time there. (Because of Morgan's experience at Naples he became an enthusiastic supporter of the Marine Biological Laboratory at Woods Hole and established the Corona del Mar Station of the California Institute of Technology.[24])

Like many scientists, Huxley fell under the spell of the Naples Station and established a lifelong friendship with Reinhardt Dohrn, who had succeeded his father as director. And like many young scientists, Huxley began his research career not knowing what subject to study.[25] Paul Mayer drew Huxley's attention to H. V. Wilson's remarkable experiments on the dissociation and reaggregation of sponges, published in 1907.[26] Wilson had produced single-cell suspensions by cutting sponges into small pieces and then forcing the pieces through fine cloth. Wilson found that these isolated cells formed clumps, the cells sorted themselves out, and tiny sponges reformed. Huxley[27] repeated these experiments using *Sycon raphanus*, a sponge that was able to live in the Station aquarium. *Sycon* behaved in similar fashion to the sponges that Wilson had used. Following dissociation, the cells reaggregated over a period of twenty-four to thirty hours to form clumps of cells, each clump being able to give rise to a new sponge. The later stages of the process included the development of spicules, first monaxons and then triradiates, and Huxley thought that the final stage reached by his regenerates was similar to the post-larval or olynthus stage of normal development— i.e., a gastral cavity formed, an osculum opened, pores developed, and the body began to elongate. Huxley did not deceive himself as to the degree of development reached by his sponges. They did not attain the "beautiful regularity" achieved in normal development, and the crowns of his sponges were "sadly lacking in order and beauty."[28] A new finding concerned the behavior of sheets of cells made up only of choanocytes, the flagellum-bearing cells that line the gastral cavity. Huxley prepared such sheets of cells by teasing the sponges apart and using a coarser gauze for straining the fragments. The choanocyte sheets rounded up, but to Huxley's surprise the flagella were directed outward rather than inward as in the intact sponge. "Of singular beauty and perfect transparency they looked at first sight like a choanflagellate *Volvox*."[29]

One justification for the study of embryology in the nineteenth

century had been that such study contributed to understanding the
evolutionary histories of animals, embryological data being used to
establish phylogenies according to Haeckel's biogenetic law or other
recapitulation theories. But *experimental* embryology rejected this
role and Jenkinson wrote that "a more intimate acquaintance with the
facts has made it abundantly clear that development is no mere
repetition of the ancestral series."[30] Furthermore, it has been remarked that the embryology Huxley learned from Jenkinson was "a
causal, experimental embryology utterly divorced from the reconstruction of phylogeny" and that Huxley's embryology in turn was
even "more rigidly experimental" and he never concerned himself
with embryological research on phylogeny.[31] This may be so, but
Huxley's discussion of his sponge experiments *was* concerned with
the phylogenetic argument. What makes his discussion interesting is
that, against a phylogenetic interpretation of his results, he advanced
a mechanistic explanation in line with Jenkinson's teachings.

Huxley pointed out the reaggregation of dissociated whole sponges
involved three processes: *reunion* of the dissociated cells (a phenomenon that does not occur in nature); *reorganization* (which does not
normally occur in *Sycon*); and *redevelopment* (that is similar to the
growth that occurs following metamorphosis in *Sycon*). There is no
larval stage in regeneration from single cells nor should one be
expected: "In regeneration," Huxley wrote, "the animal seeks to
reach this stage [redevelopment] by the shortest way it can find.
Since it is not starting from an ancestral state (as it is in the unicellu-
lar ovum) it is not necessary, or even convenient for it to pass through
ancestral stages."[32] Moreover, there was no reason to think that the
clear separation of the two layers of the sponge body at the end of the
reorganization stage was "reminiscent of an ancestral state." Instead,
Huxley offered an explanation in keeping with the viewpoint of
experimental embryology: "much more probably it is simply a stage
which is convenient or needful to reach as a point of departure for
further development," and he continued, "developing and regenerating sponges, like men, rising 'on stepping-stones of their dead selves
to higher things'. The simplest way to get order out of the chaos of
the reunited cells is for the various kinds of cells to sort themselves
out into their respective categories, not to crystallise out at once into
the adult organisation."[33]

But what of the *Volvox*-like choanocyte spheres? Given that precisely such a flagellated protozoan had been proposed as the ancestor of the sponges, was not Huxley's result striking evidence for such a hypothetical organism? Not necessarily, Huxley said, if it could be shown that the peculiarities of the choanocyte sphere "are in all probability the result of a reaction to external stimuli [shades here of Jenkinson], then their value as phylogenetic evidence falls nearly or quite to zero."[34] And he turned away from a phylogenetic interpretation to a mechanistic explanation. The choanocytes lose their flagella and coalesce into spherical masses, but such cells are "accustomed" to be in contact with the medium. In the embryo this can be done only by forming an internal cavity because the dermal cells prevent external contact. In contrast, the cells in the spheres make direct contact with the external medium and so direct their flagella outward. Hence the development of the *Volvox*-like structure is not phylogenetic but due to the peculiar circumstances in which the cells find themselves. Huxley goes on to describe the formation of the spheres from sheets of epidermal cells in terms of tensions in rubber sheets similar to the way in which His had modeled the development of embryonic structures some forty years before.[35]

On the other hand, Huxley reasoned that his results could not be used to argue *against* the choanoflagellate theory.[36] If this theory was correct, then the cells of the spheres should be able to give rise to the other cell types of the sponge, given that originally all these cell types would have been derived from the choanoflagellate ancestor. But the cells of Huxley's spheres remained stubbornly choanocytes, at least for the time he was able to keep them alive. However, what reason was there to suppose that the choanocyte of the modern sponge would have retained the characteristics of the primitive ancestral choanoflagellate? In other organisms such as *Clavellina*, the powers of regeneration belonged to the cells of the mesenchyme and not the phylogenetically more primitive ectoderm or endoderm. Huxley's conclusion was that "however curious and beautiful these spheres are in themselves, yet neither their structure nor their fate has any bearing upon the still-vexed question of the ancestry of sponges."[37] He had come to the same conclusion as Jenkinson, that development was a poor guide to phylogeny.

It was a number of years before Huxley published laboratory research again. In 1914 he published his classic paper on the courtship habits of the great crested grebe, and between 1911 and 1920 all his papers dealt with ornithology. He came to the Rice Institute in 1913 and, as he wrote in his autobiography, he became enmeshed in "the multifarious activities of starting a biological department from scratch."[38] Huxley did begin some research at Woods Hole in 1913 and he returned there to complete it before going home to England in 1916. He played his part in World War I and following it was appointed a fellow of New College and senior demonstrator in zoology at Oxford in 1919.

The laboratory research that he published in 1921 was concerned with the question of cellular differentiation and followed from the work he had done at Naples. Wilson in his studies of hydroids believed that specialized cells gave rise to "totipotent regenerative tissue," but Huxley believed that, while all tissues became dedifferentiated *morphologically* in the course of reduction, the cells did not dedifferentiate *physiologically*.[39] After the first "shock" of dedifferentiation induced by the unfavorable culture conditions, the cells were still capable of redifferentiating appropriately, their position within the cell mass being determined by their developmental commitment rather than their position determining their differentiation. But what distinguishes this paper is the way that Huxley firmly nailed his colors to the experimental mast when he discussed "normal" and "abnormal" phenomena. This was a controversial topic that had occupied American experimental biologists some years earlier, and Allen[40] has argued that the change to experimentally based research constituted a revolution in technique. Allen's views have provoked an interesting controversy,[41] and my study[42] of Ross Harrison, one of the greatest experimental embryologists, convinces me that Allen's views are largely correct. It seems clear that experimental biologists perceived the dichotomy between descriptive and experimental biology as a real issue between 1890 and 1920. In 1912 Harrison wrote of experimentation in biology that "there are many who would deny to this type of experiment validity in elucidating the phenomena of normal development, maintaining that the experimental conditions are too radically different from normal to be of value in interpreting the latter. . . . The old fear of getting hold of something abnormal must be cast aside as obstructive to progress."[43] The physical sciences had coped with the philosophical problems of experimentation as a way of research, so why could biology not do the same? Huxley makes these same points. Normal phenomena in development form an interlocking series where it may prove impossible to determine what is happening. However, "by varying the conditions, we may then throw light upon normal processes. . . . A word is also in order as to the use of the terms 'normal' and 'abnormal'. Abnormal is often used as if it were synonymous with 'pathological'." But the processes that Huxley was studying, while abnormal in the sense that they did not occur in normal development of the sponge, were "all perfectly healthy phenomena."[44]

The final paper that I want to discuss is interesting because in it Huxley makes use of the theories and experimental methods of C. M. Child. Child demonstrated that there existed "dominant" regions of high metabolic activity by studying the reactions of embryos and regenerating lower organisms to metabolic poisons. These dominant regions established physiological gradients that were revealed by differential susceptibility to such poisons along their length. These gradients were responsible for determining the polarity and symmetry of organisms or organs: "In short, these physiological gradients are characteristic features of axiate order and pattern; differentiation of regions and organs occurs in definite relation to them."[45] Huxley used these same methods to examine the relationship between zooid and stolon when the colonial ascidian *Perophora* underwent reduction in response to unfavorable culture conditions.[46] He concluded that his results supported Child's views. (The discussion in this paper is remarkable for a leap that Huxley made from dominance in morphogenesis to mental regression in man!) Child's work made a great impression on Huxley, and he used it as the theoretical framework for what I think is his greatest achievement in embryology, the writing, with Gavin de Beer, of *Elements of Experimental Embryology*.

Elements of Experimental Embryology

Huxley became the chief advocate of Child's ideas, or rather his own version of them, in Britain and Europe. In 1924 he published an article entitled "Early Embryonic Differentiation," in which he reviewed advances in knowledge since the publication of Jenkinson's book fifteen years earlier. In particular he discussed a possible scheme for their theoretical explanation, a scheme that united the latest results of Spemann on the primary organizer with Child's ideas on gradients. This, Huxley wrote, "enables us for the first time to give a coherent, formal account (however imperfect) of the early stages of development. . . . During gastrulation every portion of the germ has a definite relation to the germ's system of metabolic gradients. Of these there are (1) the primary apico-basal gradient; (2) the dorso-ventral, determined by . . . the active dorsal-lip region"[47] that is Spemann's primary organizer. The chromosomes "provide both the necessary complexity and the necessary specificity for development, while the two main gradients provide the differences between parts of the germ necessary for the start of localized qualitative differentiation and the activity of the dorsal lip the energy needed to set the processes in action."[48]

Elements of Experimental Embryology was essentially a further elaboration of these views, and while Huxley and de Beer limited the range of topics they covered—on the legitimate grounds that it was impossible to cover the whole process of development in a single book—they drew together a wide range of phenomena using what they called a "broadly biological" approach.[49] This approach tackles a biological problem by discussing "general rules and laws," in contrast to the physiological approach in which the research worker aims to analyze "the processes involved in terms of physics and chemistry." Huxley and de Beer were going to give the "results of the experimental attack on the problem of the biology of differentiation," but they were going to do this in a selective manner: "at present there exists in the subject a vast body of facts and a relative paucity of general principles. We have accordingly aimed at marshalling the facts under the banner of general principles wherever possible, even when the principle seemed to be only provisional."[50] And the banner under which Huxley and de Beer marched was developed from Child's metabolic gradients, extended in what they called the "gradient-field system."

Rather than analyze *Elements of Experimental Embryology* in detail, I want to use three contrasting reviews of the book to show how it was received by some of Huxley's contemporaries. The most straightforward review[51] was written by G. P. Wells, H. G. Wells's son. (Huxley had collaborated with G. P. and H. G. in the writing of *The Science of Life*, published in 1931.) It was a generally favorable review, but Wells seems to have failed to understand the distinction that Huxley and de Beer tried to make between a biological and a physiological approach. He clearly believed that the biological approach was inferior to the physiological approach, and he rather disparaged Huxley and de Beer's treatment. Embryology, Wells wrote, is "one of the backward branches of biologial science," and until the problems and results of embryology can be stated quantitatively, books such as *Elements of Experimental Embryology* will simply present "a vast amount of very entertaining facts" and will contain "more anecdote than law." (Needham[52] later castigated Wells for this review, saying that Wells revealed "a complete failure to understand the process of biological discovery.")

A very different opinion of *Elements of Experimental Embryology* was expressed in his usual forceful manner by E. W. MacBride. MacBride was professor of zoology at Imperial College, London. He was a comparative anatomist and embryologist, but by the 1930s he was an increasingly isolated figure. Ridley[53] has called MacBride "ossified" for his continued support of recapitulation, and in the 1920s MacBride had clashed with both Huxley and de Beer over MacBride's support for the inheritance of acquired characteristics and his vehement rejection of Morgan's genetics. (The clash between Huxley and MacBride continued in their private correspondence for some twenty years. As late as 1937 MacBride[54] was claiming that the case for Lamarckism was "cast iron.") This controversy seems to have been taken rather personally by MacBride, who wrote to Huxley that he resented the "patronizing attitude" adopted by Huxley and that he hoped that their argument would be one of "confronting evidence with evidence."[55] This is rather ironic in that MacBride's review[56] of *Elements of Experimental Embryology* is a classic example of a reviewer using innuendo and the setting-up and demolition of straw men to advance the reviewer's aims, in this case an attack on experimental biology. For example, having written that Huxley was too well known to need any introduction, MacBride went on: "Dr. De Beer is less well known . . . his acquaintance with experimental work is second-hand . . . his acquaintance even with general embryology is far from perfect."[57] MacBride's description of experimental embryology is not flattering. It aims, MacBride wrote, "at discovering how the egg grows to the adult condition, by pulling it to pieces, distorting it, grafting pieces of other developing eggs on to it and checking or stimulating its growth by chemical solutions, and drawing conclusions from the growth of the damaged egg."[58] And having denigrated the field in which Huxley and de Beer claimed to speak with authority, MacBride goes on to say that the authors are not even experts in the only field that does matter; his chief complaint is that the authors are not "good comparative embryologists." Finally, MacBride says, "the analysis of embryology embodied in *Elements of Experimental Embryology* is essentially unsound," especially the authors' attempt to bring Mendelian genes into "harmony with the facts of embryology" when it is quite clear that "the study of embryology provides the most scathing criticism of the whole idea that the Mendelian 'genes' are in any way significant for evolution."[59]

A more interesting but less entertaining review was written by C. H. Waddington, who criticized Huxley and de Beer for their loose usage of the term "field." Fields had been employed earlier,[60] notably by Gurvich and Weiss, but Huxley and de Beer achieved what Wolpert[61] has called a "brilliant synthesis," generalizing Child's gradients for a variety of developmental processes. Child's theories had been developed largely as a result of his work on regeneration in lower organisms such as planaria, and it was generally recognized that Child had made a "great contribution to biology" even by Needham,[62] who was unimpressed by Child's arguments that the gradients were of metabolic activity. Waddington thought that *Elements of Experimental Embryology* was a "considerable advance on anything which had existed before" and that "such synthetic work is of the greatest importance for the progress of experimental study of morphogenesis at the present day."[63] But the authors had not made clear what they meant by "field." Waddington pointed out that on the one hand the term had been used to designate the agent that brought about differentiation within its boundaries, and on the other it had been used simply to mean a place or location where something was happening. Nevertheless, Waddington thought that Huxley and de Beer had used the field concept with "such success that it is assured of a place in the present-day theoretical scheme. . . ."[64]

Child himself did not like the field concept: "reference to a field merely states experimental data in terms of an unknown, of a concept without definite content, and the field often becomes little more than a verbalistic refuge."[65] Nor does he seem to have been too pleased with the way in which his gradients became absorbed in Huxley and de Beer's gradient-field system. He seems to have felt that the term was redundant when he asked the rhetorical questions: "What is the field, as distinguished from the gradient or gradient system in it? Does the field determine the gradients or do gradients constitute the field?"[66] For Child, the field concept was of little consequence; what mattered were the gradients that composed it.

In 1975 Needham wrote that *Elements of Experimental Embryology* was a "stirring overture to a field of experimental morphology still of the highest importance and fascination today."[67] *Elements of Experimental Embryology* continues to be cited for its discussion of fields, and, while these citations are now made for historical reasons, they testify to the important role of *Elements of Experimental Embryology* in the development of embryology.

Relative Growth

I want to turn now to Huxley's studies of relative growth. Although Huxley was not the first to analyze the relative growth of different parts of the body, he devised a simple formula, capable of clear graphical expression, that was easily used and he applied it with considerable thoroughness. It is not clear why Huxley took up this subject at this time. As I mentioned earlier, Geoffrey Smith, Huxley's zoology tutor, made measurements of the claw and absolute body size in crabs, and of mandible length and body size in stag beetles,[68] and noted that the larger males formed relatively larger mandibles. As Huxley remarked in his first paper on relative growth, Smith "with characteristic insight" went on to generalize this observation: "on the whole, the large-sized species had secondary sexual characteristics which were not only absolutely but also relatively larger."[69] It is probable that familiarity with Smith's work played a part, but in the preface to *Problems of Relative Growth* Huxley acknowledged the inspiration provided by D'Arcy Thompson's classic book, *On Growth and Form*. And in the introduction to the 1971 reissue of *Relative Growth*, Huxley mentioned another source; he claimed that it was the first sight of fiddler crabs in 1913 that induced him to begin the study of relative growth, although his first publication in the field was not until 1924 and even then he did not use his own data.

T. H. Morgan[70] had measured the variation in abdomen width in female fiddler crabs and had concluded that those female crabs that had abnormally small abdomens showed changes toward maleness. However, Huxley's analysis[71] showed that while in males the abdomen grew at the same rate as the rest of the body, in females the abdominal growth rate was greater than the rest of the body and this was true even of the female crabs with the small abdomens. So these females did not have the growth pattern characteristic of males; it was just that their abdominal growth rate was slower than that of other females.

It was in his next paper[72] that Huxley made his most important and long-lasting contribution to laboratory science with the analysis of his own data on claw size and body size in the fiddler crab. He showed that if y is the weight of the large claw and x the total weight of the crab less y, then the plot of log(y) against log(x) was a straight line.[73] Furthermore, this relationship could be expressed in the form:

$$y = bx^k$$

This formula has become known as the simple allometry formula, the constant k being called the *constant differential growth-ratio* by Huxley. His analysis of the fiddler crab claw data is shown in Figure 3 and demonstrates the considerable simplification achieved using this analysis. Huxley summarized his work and presented analyses of an extraordinary variety of data culled from a large number of sources in his classic book, *Problems of Relative Growth*, published in 1932. Some of the more straightforward examples included growth of the mouse tail, skull growth in baboons, and stem growth compared with root growth in plants.

Huxley was well aware that the application of mathematics in biology was subject to the same shortsighted criticism that was leveled at experimentation in biology. A typical criticism, he wrote in the introduction[74] to *Problems of Relative Growth*, was that while mathematical formulae might be convenient to use in analysis, mathematics in biology could contribute nothing new and certainly nothing worth knowing about the biology of the phenomenon being analyzed. Huxley retorted that having a precise quantitative description of a biological event might be indispensable in thinking clearly about it. More important, the method of analysis might itself give new insights into the underlying processes, and Huxley believed that

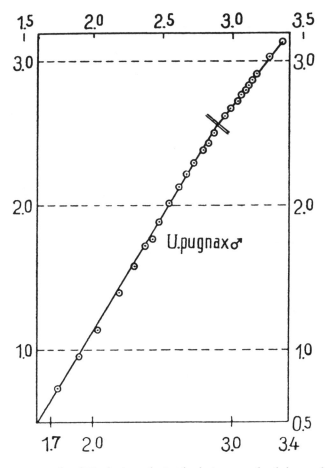

Figure 3. An example of Huxley's analysis of relative growth of claw and body in male fiddler-crabs. The logarithm of absolute chela weight (ordinate) is plotted against the logarithm of body weight (abscissa). The slope of the line is *k*, the constant differential growth ratio. The change in slope (marked by the double line) is probably related to the onset of sexual maturity. (From J. S. Huxley, *Problems of Relative Growth*, London, Methuen & Co., figure 3).

the allometry formula did reflect some fundamental principles of growth. However, J. B. S. Haldane[75] early made the criticism that if two parts of a limb, for example, obeyed the allometry equation, then the limb as a whole could not obey a function of exactly the same sort. This does not seem to have had a serious effect on the practical application of the allometry formula, but it did cause Huxley to adopt an empirical approach, saying that "the 'axioms' of growth which were put forward to justify the general use of the formula . . . should perhaps be considered as no more than the consequences of simple allometry where it has been found to occur."[76]

While the simple allometry formula is useful, it is not free of problems, and attempts have been made to replace the simple formula with more complicated analyses such as forms of multivariate analysis. A recent example is Gould's analysis[77] of the so-called "smokestack" individuals of the Bahamian land snail, Cerion, that I shall discuss later. Another source of difficulty has been the interpretation of the constant b in the formula, a constant that Huxley had dismissed as being of "no particular biological significance," being only the value of y when x is unity. Later White and Gould[78] concluded that b may be expressed in terms of a linear scale ratio, s, and that this can be used to compare similarity of form. Other difficulties mentioned by Reeve and Huxley[79] in 1945 (and also by Gould in his review[80] published twenty years later) included the interpretation of deviations from the simple straight line on the log-log plot, and the related problem of fitting lines to the data. In 1965, Alan Cock went so far as to say the "standard of reporting and analysing metrical growth data is still often deplorably low."[81]

Huxley expected that the simple allometric formula would make a valuable contribution to the analysis of growth, and he considered applications in systematics and taxonomy, comparative physiology, genetics, and embryology. It is impossible to do justice to the full scope of Huxley's treatment of this topic—after all, he took 250 closely argued pages to do it. Instead, I shall take examples from just three areas to illustrate the use of the analysis.

In taxonomy, absolute differences in size had long been recognized as unreliable criteria for distinguishing species, and Huxley showed that using percentage differences was little better. Instead, Huxley said that the *growth constants* of organs were taxonomically significant and that "these numerical values, if properly established, would be true taxonomic characters."[82] One example Huxley gave concerned the stag beetle, of which coleopterists had recognized at least five main types based upon morphological features of the mandibles. When the relative growth of the mandibles was plotted using the length of the beetle, a single continuous straight line was obtained, indicating that the different forms are the result of arbitrary classification.[83] The Bahamian land snail *Cerion* has a very wide range of forms that had been classified as separate species or subspecies until studies by Gould showed that the number of "true" species is much smaller. The so-called "smokestack" individuals have long, narrow shells with parallel sides. Curiously this form of shell occurs only in dwarf or giant forms, being absent in normal-sized populations. In a fascinating paper,[84] Gould shows that there is a negative relationship between whorl number and whorl size, and that this allometric relationship, together with the rules of construction of the shell, precludes the formation of the smokestack shape at normal shell sizes. But these rules can be broken when the final size of the shell is abnormally small or abnormally large, for example, as a result of injury to the shell or the persistence of juvenile growth. The important point is that allometric analysis shows that those individuals that might have been considered different species are forms of the same species.

Not surprisingly Huxley was alert to the evolutionary implications of allometric growth. He pointed out that for an organism possessing a part with high relative growth rate, the maximum size attainable by that organism would be limited because the part would become so large that it would prove harmful.[85] The classic example of this is the explanation given by Huxley for the extinction of the Irish elk, a deer that had antlers with a span of up to four meters. The antlers "showed positive allometry, but higher than that of other species. Thus they eventually became so large that they constituted a grave handicap during the ice age when there was less food available. In consequence the species became extinct."[86] Gould has shown from measurements of skeletons that this positive allometric relationship did hold for the Irish elk, but he has challenged the view that the huge antlers were only an undesirable side effect of selection for increased body size.[87] He pointed out that antlers are used primarily for display and that the antlers of extant species have to be swung from side to side to display the full palm. This would clearly be a severe problem for the Irish elk with antlers weighing 38 kg mounted on a skull of only 2.5 kg. However, the antlers of the Irish elk were rotated so that the full palm was shown to the front, and they were displayed without the head having to be moved. Gould suggested that the giant antlers of the elk were a consequence of selection for increased body size and for larger antlers that would be more effective in sexual displays.

Embryology seems an obvious area for the application of the analysis of relative growth, and Huxley believed that his demonstration of growth gradients was "the most important part of any contribution made by me to the study of relative growth."[88] "[O]f the growth gradients . . . I had no suspicion; their existence thrust itself upon me as new empirical fact . . ." as a consequence of his application of allometric analysis. However, the embryological application that I want to discuss is Needham's use of the allometry relationship to examine the changing proportions of substances during development. As Needham put it, "The metazoal organism, in its passage from fertilisation to maturity, passes through a long succession of stages which are characterised just as much by changing chemical composition as by changing morphological form."[89] Teissier,[90] using larvae of the waxmoth and mealworm, had shown that substances such as water, fat, and phosphorus followed the allometric formula, and Needham collected and analyzed a huge amount of data, as can be seen from Figure 4, showing changes in water content during development.[91] Similar relationships were found to hold for creatine, glutathione, glycogen, calcium, and ash.[92] Needham thought that the latter was particularly striking for "the close uniformity in the slopes of the straight lines, in spite of the enormous differences of morphological form and time taken to accomplish the changes depicted."[93] Needham's conclusion was that this type of analysis abstracted information from morphological form, from factors of nutrition, from the absolute values of the measured substances, and from the time factor. What then, he asks, is left? "Nothing but a series of ratios or relations, which may possibly be the same in all animals . . . in a word a chemical ground-plan of animal growth."[94] And in the same way that D'Arcy Thompson was able to transform morphological shapes, "so the chemical ground-plan is deformed within wide limits in time and space"[95] for different species during their development. Huxley, one feels, must have been pleased that the allometric formula could be used to analyze chemical as well as morphological form, although Needham later felt that this work, although interesting in itself, had not led to anything further.[96]

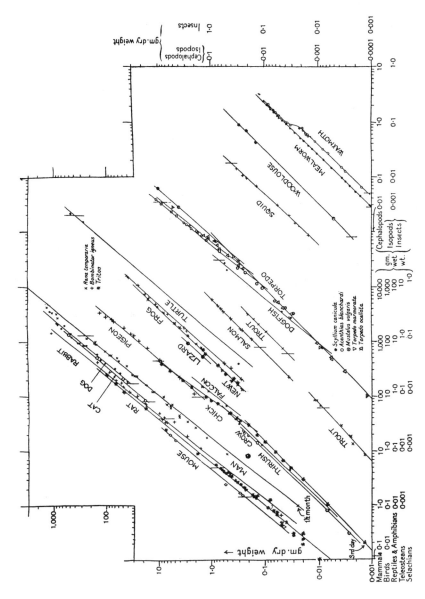

Figure 4. An example of allometry applied to chemical embryology. Joseph Needham plotted dry weight to wet weight and showed how similar the changes were for a wide range of organisms. (From J. Needham, "Chemical heterogony and the ground-plan of animal growth". *Biological Review* 9: 79–109, 1933, figure 4).

The Secrets of Life

I want to conclude by describing some of Huxley's scientific work that was not original and not in itself of any great significance.[97] It did, however, propel him into the limelight when the popular press caught hold of it, and I argue that the consequences of this work were perhaps the most far-reaching of any scientific work that Huxley performed.

In 1919 Huxley[98] was searching for research topics. He had just moved to Oxford and had begun teaching with the help of lecture notes that he had borrowed from James Gray. Huxley read of the work of Gudernatch, who had shown that premature metamorphosis could be induced in tadpoles by feeding them thyroid. This was a remarkable result, and thirty years later at the Cold Spring Harbor Quantitative Symposium on the Relation of Hormones to Development, Witschi recalled it as the "most spectacular discovery in the field."[99] Up to that time hormones had been thought of as internal secretions that regulated physiological functions, but here was an example of a hormone that stimulated a major developmental event. As Witschi remarked, this discovery had the secondary effect of stimulating the interests of embryologists in hormones. Further analyses of the role of the thyroid in amphibian metamorphosis were performed by Allen,[100] who showed that removal of the thyroid rudiment at the tailbud stage prevented metamorphosis. The tadpoles continued to grow, eventually developing into giant tadpoles that could then be induced to metamorphose by the administration of thyroid. Now the length of time that amphibians remain in the tadpole stage varies considerably from species to species. Bullfrogs, for example, remain in the larval stage for several years. Other species, notably *Urodeles*, remain in the larval stage more or less permanently and may become sexually mature and breed in the larval state. This phenomenon is known as neoteny and may have played an important role in evolution, including that of man. Huxley, by choosing the axolotl as his test subject, was able to examine the physiological control of an important and dramatic developmental event that had evolutionary significance.

Huxley reported his findings in a short note[101] to *Nature* in January 1920. He had begun to feed two young axolotls with ox thyroid on

November 30, and by December 15 the animals had already changed in color and resorption of the gills and tail fin had begun. By December 17 the degree of gill resorption had reached the critical stage, and on December 19 the axolotls left the water. In contrast, two specimens that had been left in shallow water in an attempt to induce metamorphosis by making them breathe air showed only minimal changes. The paper contains little discussion. Huxley pointed out that thyroid-induced metamorphosis was much quicker than that induced by air-breathing and commented that "many interesting problems present themselves, which it is hoped to work out as opportunity offers." One immediate problem was to obtain a satisfactory supply of axolotls, and Huxley appealed to his readers for specimens!

These experiments were not original nor did they make a significant contribution to biology. What they did do indirectly was to make a significant contribution to public awareness of scientific matters over the next thirty years or so. On February 16, Huxley gave a talk at the Linnaean Society in London at which he exhibited the metamorphosed axolotls. The London *Daily Mail* must have had a reporter at the meeting, for the next day a long article[102] appeared with the wonderful headline: "A Great Discovery . . . Thyroid Gland Marvels . . . Control of Sex and Growth . . . Renewal of Youth." The headline contains all those words guaranteed to catch the attention of the reader and the first sentence promised even more: "The secret of perpetual youth and renewed vigour, the determination of sex, and the curing of certain human diseases" were being revealed by Huxley in Oxford. (Other headlines were even more lurid—one read simply "Sex Secrets.") In fact the article in the *Daily Mail* was reasonably accurate, but the story gradually became more garbled as it was picked up by other papers. One can imagine how Huxley reacted to a story that claimed that "Professor" Huxley was hoping to cure mentally deficient children by administering thyroids taken from axolotls. Stories such as these must have confirmed the worst fears of friends like J. B. S. Haldane, who warned Huxley that he would lose his standing as a reputable scientist and be regarded as a quack.[103] Huxley responded with a full, illustrated article[104] in the *Illustrated London News* of February 28 and in so doing embarked on his career as a popularizer of science.

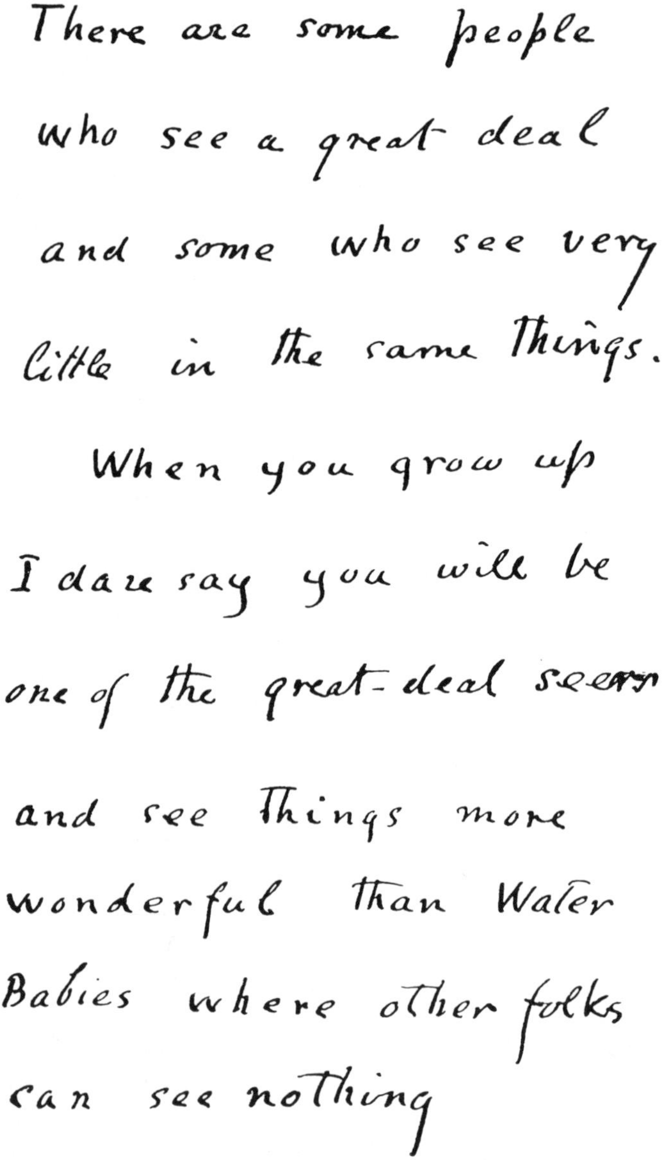

There are some people
who see a great deal
and some who see very
little in the same Things.
When you grow up
I dare say you will be
one of the great-deal seers
and see Things more
wonderful than Water
Babies where other folks
can see nothing

Figure 5. Part of T. H. Huxley's response to Julian's letter about the water baby (see figure 1). Julian commented that this carefully written letter was in marked contrast to his grandfather's usual illegible handwriting. (From L. Huxley, *Life and Letters of Thomas Henry Huxley,* London, Macmillan & Co., vol. II, 1900, p. 436–439; also reprinted in J. S. Huxley, *Memories,* London, George Allen & Unwin, 1970, p. 24–25).

Conclusion

In the preface to his autobiography, Huxley wrote that he had embarked on it to show that "embracing inquisitiveness and widespread curiosity can bring their own rewards,"[105] and certainly the range of topics that he studied testifies to his wide-ranging curiosity. The subjects I have discussed do not exhaust the research he carried out. I have not referred to his studies of amphibian metamorphosis and pigment responses to hormones carried out in collaboration with Lancelot Hogben,[106] nor to his and E. B. Ford's work on rate genes,[107] nor to his genetic studies on linkage in *Gammarus*,[108] nor to his work with Carr-Saunders on the inheritance of eye abnormalities produced by the injection of lens antibodies into pregnant animals.[109] Nor have I referred to Huxley's use of his research experiences in writing his fiction and poetry, most notably in "The Tissue-Culture King."[110] This story—with distinct undertones of H. G. Wells—involves experiments on tissue culture and experimental embryology being carried out in an African village and includes references to Alexis Carrel, Hans Driesch, Hans Spemann, and Ross Harrison!

One consequence of Huxley indulging in research on so many diverse topics is that he rarely worked anything out fully. *Problems of Relative Growth* is the exception that demonstrates the rule. Huxley fully recognized this when he wrote that "I seem to have been possessed by a demon, driving me into every sort of activity, and impatient to finish anything I had begun."[111] And finally, if he was to be remembered, he hoped that it would "not be primarily for my specialized scientific work, but as a generalist."[112] In discussing *The Science of Life* he referred to his gift of "synthesizing a multitude of facts into a manageable whole, aware of the trees yet seeing the pattern of the forest and drawing conclusions which gave the whole work vitality."[113] T. H. Huxley seems to have been prescient when he wrote to his grandson in response to Julian's inquiry about water babies: "There are some people who see a great deal and some who see very little in the same things. When you grow up I dare say that you will be one of the great-deal seers and see things more wonderful than water babies where other folks see nothing"[114] (Fig. 5).

ELOF A. CARLSON

Huxley's Interest in Developmental Biology

I agree with Witkowski's overall interpretation of Huxley as an experimentalist whose work was less innovative than insightful and whose collective experimental work does not fit Huxley's own mission to be guided in all his intellectual pursuits by the theory of evolution. I also agree with Witkowski's thorough analysis and assessment of Huxley's work on cell dissociations and aggregation, allometric growth rates, developmental gradients and fields, and the effect of hormones on development. My comments will stress instead some of the historical aspects that Witkowski's paper raised as I reflected on Huxley's life and work. My remarks deal less with the history of ideas than they do with the complex questions: why do we choose to follow the careers we do and how much are we influenced by our families, schools, mentors, and unpredictable events? While I cannot pretend to answer those questions, Huxley's life is rich in materials that touch on them.

Unlike Morgan, Muller, Spemann, Harrison, or Child, Huxley cannot be included among the great experimentalists whose body of work helped shape the direction of biological thought in the first half of the twentieth century. Yet Huxley's reputation may well outlast many of his contemporaries whose work, at the time, overshadowed his own. No doubt Huxley's reputation is enhanced by a remarkable family whose contributions continue to maintain a tradition of excellence and eminence. This alone would be insufficient for making

Huxley's work so enduring. I suspect his work as a generalist, as he liked to think of himself, has a lot to do with this—he was good at so many activities and he wrote several books that summed up in original and thoughtful ways many of the leading fields in which he worked. It is the specialist who tends to fade when new specialities replace the interest of scientists and only the most spectacular contributions of specialists survive with the names of their innovators attached to them.

It is particularly interesting that Huxley, whose central theme was evolution and who was an architect of the "new synthesis," selected developmental biology instead of genetics as his favored arena for laboratory research. It cannot be that he was unaware of all the attention addressed to genetics. Bateson had named the field in 1906. The rediscovery of Mendelism had caused both excitement and controversy in Great Britain. Why did he not choose to look into this opportunity? My guess would be that he would have walked into a damaging mess, with Bateson at odds with the biometric school. Bateson may have been an able experimentalist but his views on evolution would have seemed at odds with the Darwin-Huxley tradition that the biometricians tried to claim as their own. The biometricians might not have appealed to Huxley because they were strong on theory and weak on experimental studies. It was not clear to most biologists, other than DeVries and Bateson, that genetics had much to do with evolution. That may have been true until 1915, when Huxley clearly recognized the importance of genetics and deliberately went to Morgan's laboratory to recruit a student for his new department at the Rice Institute. I cannot imagine, either, that those walks in Houston with Muller to look at the crayfish and the carpenter ants would not have been accompanied by many discussions, with the zealous Muller trying to convert Huxley to his own genetic perception of evolution. Muller by this time had very clear ideas of what evolution should be like with modifier genes, complex associations of characters, recombination of factors, and the central role of the gene and its mutations as the driving force of evolution. If it is reasonable to see why Huxley's first research paper was on *Sycon*, it is not so easy to see why he continued the path of developmental biology after his return from military duty in 1919. Muller corresponded with Huxley throughout those years and Huxley was aware of many of the exciting ideas and experiments that were going on in genetics. Huxley evidently read avidly in genetics and made much use of these contributions to evolution in his summing up work in 1940.

I was struck by Witkowski's description of the Oxford school of zoologists that produced "Huxley, Gavin de Beer, J. Z. Young, Alister Hardy, E. B. Ford, and P. B. Medawar." All of them were popularizers and, to a degree not quite as extensive as Huxley himself, generalists. Neither Huxley not Medawar were set in a life direction as certainly as, say Muller by Morgan, Wright by Castle, or McClintock by Emerson. One clue to this Oxford phenomenon is Witkowski's remark that Huxley entered the world of adult intellectuals, such as "Gilbert Murray, Sedgwick, and the Haldanes."[1] One last impression is worth mentioning. Witkowski notes how Huxley might have felt obligated to pursue a developmentalist's approach to the study of differentiation and differential growth, problems that were the stuff of basic biology, rather than to the phylogenetic modelling that had led to specious evolutionary ideas. The obligation in this case was the death of both his early mentors, J. W. Jenkinson and Geoffrey Smith, during World War I. Europe lost many of its brightest scholars in that war. The United States was spared because of its later entry, and World War II, for all its carnage, had less hand-to-hand and trench warfare pitting scholar against scholar.

FREDERICK B. CHURCHILL

The Elements of Experimental Embryology: a Synthesis for Animal Development

"Generalizer" and "synthesizer" are two appellations often associated with Julian Huxley. The work that immediately comes to mind to justify such ascriptions is *Evolution, the Modern Synthesis*, which provided both name and compass for the resurgence of neo-Darwinism. Huxley endeavored, though less successfully, to fashion a summary twenty years later when he amassed disparate materials in a work titled *Biological Aspects of Cancer*. Such synthetic and summary endeavors were not, however, limited to the later periods of Huxley's career. Between 1927 and 1930 he assisted H. G. Wells and G. P. Wells in bringing together in one work the entire field of biology in their *Science of Life*. Some have argued that this work has yet to be surpassed as an introduction to biology. In 1934 Huxley joined forces with his former student and colleague Gavin de Beer to write *The Elements of Experimental Embryology*. This, too, was a synthesis of major daring and scope. It differed from *The Science of Life* in that it was intended to unify the many aspects of development and offer an explanatory model, rather than be a learner's guide, although it could be easily used as such. In this respect *The Elements* paralleled in expectation *Evolution, the Modern Synthesis*. Unlike this later work, however, it was unsuccessful in providing a comprehensive model that was intended to serve as a basis for further research in embryology.

It was indeed the case that in the third and fourth decades of the century embryology was an extraordinarily diverse and complex discipline. It combined experiment and description. It included many subfields, ranging from heteroplastic transplantations, tissue culturing, regeneration experiments on all common animal phyla and at all stages of development to implantation of organs for medical treatment, teratological and pathological observations, studies on the influence of single physical and biochemical factors on development, and attempts to associate the gene of classical genetics with developmental events. Embryology, in fact, was one of the most exciting areas in biology and called for unification. It says something about the ambition and daring of Huxley and de Beer that they saw the need and acted. Their objective was to embrace an entire domain of biology and to demonstrate how certain parts could provide the key to a unified explanation of the whole. That these parts in combination made logical and formal sense was an achievement in itself. That the synthesis was short-lived is beside the point. That Huxley had labored, albeit briefly, in many corners of the synthesized domain says a good deal about his varied laboratory activities. This last point is particularly relevant since Jan Witkowski rightly observes that "one consequence of Huxley indulging in research on so many diverse topics is that he rarely worked anything out fully."[1]

What I intend to do in this essay is to examine *The Elements of Experimental Embryology*[2] with an eye to elucidating its structure, extracting the major features of the synthesis, and evaluating the fate of such a bold presentation. At the outset it is important to comment on the relationship between Huxley and de Beer, who were, after all, co-authors in this venture.

Huxley and de Beer

The text of *The Elements* was a product of a long-term friendship and professional collaboration. The two men met at Oxford after World War I when Huxley had been appointed senior demonstrator in the Department of Zoology and Comparative Anatomy and de Beer was an undergraduate. Their association was a close one from the start, for de Beer lodged at the Huxleys'. Juliette Huxley leaves us with a delightful image of her competing with de Beer for Julian's attention at the breakfast table: "If he [Gavin] got in first," she somewhat peevishly writes, "we had a scientific discussion excluding me by its depth."[3] Over the next fifteen years Huxley and de Beer were to collaborate on a number of publications; *The Elements*, however, comprised the most substantial.[4]

In a pattern that was to be repeated in the writing of *Evolution, the Modern Synthesis*,[5] Huxley tipped his hand as to the nature of their embryological synthesis many years prior to publication of the full-scale book. In 1924 he published a short, three-page presentation in *Nature*.[6] By that time, Huxley and de Beer were in the midst of their combined studies on dedifferentiation and reabsorption of embryonic material,[7] but the principal elements of the envisioned synthesis were being developed in Freiburg and Chicago. During the spring of 1921 Hilde Proefscholdt had succeeded in her early attempts to transplant the dorsal lip of the salamander blastopore. Hans Spemann, her dissertation supervisor, immediately added a postscript to a paper of his describing the technique of heteroplastic transplantation, announcing her preliminary results and using for the first time the term "organizer."[8] In 1924, the same year that Huxley wrote to *Nature*, Spemann and Proescholdt (now Hilde Mangold) published their famous paper on the induction of a secondary embryo and on the organizer.[9] Across the Atlantic, Charles Manning Child had written his first books on axial gradients, and as early as 1920 Huxley revealed a keen interest in this line of work as well.[10] As he noted in 1924, both avenues of research along with recent developments in classical genetics were to become the cords binding together his and de Beer's embryological synthesis. More on this later.

At least twice again Huxley foreshadowed the future embryological synthesis. A chapter entitled "Growth of the Individual" in *The Science of Life* leaves the impression of a dress rehearsal before a nonprofessional audience.[11] In 1930 he published a note in *Die Naturwissenschaften*,[12] in which he spoke of "die Child-Spemannsche Ansicht der Entwicklung" and suggested that the designation was appropriate for a synthesis of the two research traditions. It is to the point that Huxley felt Child's work was not yet fully appreciated in Germany. The note was translated into German for publication by Spemann, but the favor did not imply his endorsement of the Huxley–de Beer synthesis. In later years Spemann wrote critically of Child's interpretation of metabolic gradients.[13]

The Huxley archives reveal how interested Huxley and his fellow countrymen were in the experimental work going on in Freiburg. By January 1924 Huxley had begun corresponding with Spemann. The seven letters between the two during 1924 and early 1925 are not informative from a theoretical point of view, but they show Huxley eager to make contact with his German colleague and publicize his accomplishments in England. Huxley solicited a paper from Spemann and issued him an invitation to speak at an English university. Spemann declined because of his poor English. The two met briefly in Amsterdam in March of the same year, and William Bateson issued Spemann a second invitation to speak before the Genetics Society in early 1925. Ultimately Spemann declined this engagement as well, for reasons that give some insight into Spemann's own feelings of national pride. With a pointed reference to French and Belgian occupation of the Rhineland, he explained to Huxley, whose confidence he had clearly won, that he had

> . . . zu viel Arbeit und dann—Sie wurden auch keinen Vortrag in Deutschland halten wenn wir einen Teil von England besetzt und zum versprochenen termin nicht hinaus ziehen [?] Besuchen Sie mich einmal hier.[14]

Although the Rhineland remained occupied until 1930, Spemann finally relented and traveled to England, where in November 1927 he delivered the Croonian lectures. Huxley was evidently very much involved in the arrangements for the visit.[15]

It is not clear that Huxley took up Spemann's invitation to visit Freiburg, but de Beer made a brief pilgrimage during the salamander breeding season of 1926. His reports back to Huxley reveal excitement and admiration for the work that was going on. The thought of combining Child's ideas with the organizer work was also clearly at the forefront of his mind. "The whole thing is: gradients as regards *direction, level* and quality of tissue induced. Vivat Child!" he effervesced to Huxley. Shortly thereafter it was, "If you have occasion to write to or see E. S. G. [Edwin Stephen Goodrich] you might suggest to him what an important thing it is to get the Organizer firmly implanted in England."[16] Returning home, de Beer also anticipated their embryological synthesis in a 1926 essay, "The Mechanics of Vertebrate Development,"[17] and he may have done so in his *Introduc-*

tion to Experimental Embryology of the same year. So the Huxley–de Beer synthesis was not a bolt out of the blue with *The Elements*. Both authors had wrestled for a decade with the thought of bringing together the process of induction, the notion of partial fields, the presence of axial gradients during initial egg organization, and the results of Mendelian genetics into a comprehensive explanation of prefunctional embryonic form.

Huxley's Research and The Elements

The synthesis that I will shortly describe was not solely developed from the separate work that Spemann and Child were doing on either side of the Atlantic. A tabulation of the references in *The Elements* helps us gain some sense of how Huxley's own laboratory research sustained the conclusions that he drew. If we look at a chronological listing and topical categories of Huxley's laboratory activities and publications,[18] we find that Huxley's diverse embryological interests all fed into his and de Beer's embryological synthesis (Fig. 1). Thus his work on dedifferentiation, done with de Beer; his experiments on morphogenesis, including grafting, temperature gradient studies and examination of early development; his interests in hormonal control, which propelled him to popular attention; his extensive examination of relative growth, including his book on that subject; and finally his sophisticated genetic experiments on rate genes in *Gammarus* all flow into the synthesis of 1934. (Huxley's student and biographer John Baker categorized *The Elements* as morphogenetic.) Huxley's laboratory research therefore does not appear as random or unconnected as a casual survey of his bibliography might suggest.

When we review Huxley's work chapter by chapter, another aspect of his research becomes clear (Fig. 2). From 1920 to 1932, Huxley had studied in the laboratory most of the separate segments of the synthesis that he and de Beer were to construct. Whether this was a purposeful strategy during the decade when Huxley was performing laboratory work is hard to say, but a reading of *The Elements* reveals that these citations were not forced attributions and that Huxley's research was timely and in keeping with similar work being performed elsewhere.

Finally, an examination of Huxley's full bibliography makes it clear

Year	Dedifferentiation	Morphogenesis	Hormonal Control	Relative Growth
		LABORATORY RESEARCH		
1912	1			
1913				
1914				
1915				
1916				
1917				
1918				
1919				
1920			9	
1921	10, 11			
1922			12, 13	
1923	15		14	
1924	18	19, 20		21, 22
1925			23	
1926	25	26		27
1927		34		30, 31, 33
1928				
1929				
1930		40		
1931				46, 43, 44, 45
1932				50
1933				
1934		55		
1935				
1936				
1937				
1938				
1939				
1940				
1941				
1942				

Fig. 1. Distribution of Huxley's major publications 1912–1942. Numbers refer to the bibliographical entries in John R. Baker's "Julian Sorell Huxley," *Biog. Mem. Roy. Soc.*, 1976, 22: 207–238. Major works during this time include #39 = *The Science of Life*, #50 = *Problems of Relative Growth*, #55 = *Elements of Experimental Embryology*, #72 = *The Uniqueness of Man*, and #73 = *Evolution, the Modern Synthesis*.

Rate Genes	Evolution	Behavior	Popular and Other	Year
		2, 3, 4		1912
				1913
		6		1914
				1915
		7		1916
				1917
				1918
		8		1919
				1920
				1921
				1922
		16	17	1923
				1924
24				1925
			29	1926
35		36	37	1927
38				1928
			39	1929
		41	39	1930
			48, 47	1931
49			51, 52, 53	1932
				1933
				1934
			57	1935
	58		59	1936
				1937
	60, 61, 62, 63		64	1938
	65, 66	67	68	1939
	69, 70			1940
	71		72	1941
	73			1942

The categories of research interests listed across the table are those used by Baker in his discussion of Huxley's career. Not all items 1–70 are recorded in this table. A more complete bibliography compiled by Jens-Peter Green may be found in J. R. Baker, *Julian Huxley, Scientist and World Citizen 1887–1975* (Paris: UNESCO, 1978), but this bibliography does not lend itself as well to this kind of analysis.

Citations to Huxley's Previous Work

Chapter	Title	References to Huxley
I	Historical Introduction	
II	Early Amphibian Development: A Descriptive Sketch	
III	Early Amphibian Developments: Preliminary Experimental Analysis	20, 34, 34
IV	The Origin of Polarity, Symmetry, and Asymmetry	1, 10, 25
V	Cleavage and Differentiation	
VI	Organizers: Inducers of Differentiation	
VII	Mosaic Stage of Differentiation	1, 10, 11, 13, 20, 25, 50, 50, 19
VIII	Fields and Gradients	1, 10, 25
IX	Fields and Gradients in Normal Ontogeny	34, 34, 40
X	Gradient Fields in Post-Embryonic Life	50, 50, 50
XI	Further Differentiation of the Amphibian Nervous System	
XII	Hereditary Factors and Differentiation	49, 49
XIII	Prefunctional Contrasted with the Functional Period	23, 50, 50, 50, 15
XIV	Summary	

Fig. 2. Citations in *The Elements* to Huxley's previous work. As with Fig. 1, reference numbers are found in Baker's bibliography. Clearly #50, *Relative Growth*, is an important reference point for Huxley and de Beer, but so too is #10, "Further studies on restitution bodies and free tissue-culture in *Sycon*" and #34, "The modification of development by means of temperature gradients" (1927). In all, there are thirty-one references to Huxley and coauthors. By way of comparison, *The Elements* cites Spemann forty-five, Child thirty-four, and Ross G. Harrison twenty-two times. T. H Morgan's embryological works are cited but not his genetic ones.

that, excepting a few recapitulatory statements published the subsequent year, Huxley's embryological research and writing came to an end in 1934 with *The Elements*. It would be wrong to characterize this successful text as a summary and synthesis wholly dependent upon the work of others. Huxley could draw upon many of his own experiences, and although his research on rate genes and relative growth may have been the only work with lasting import, his cumulative laboratory research allowed him to speak firsthand about many aspects of development. His and de Beer's synthesis was well informed by personal experience. At the same time, it should be emphasized that none of Huxley's research appears to have been designed to test the model that the two English embryologists produced.

Structure of The Elements

With these general comments as a background, let us look at the structure and contents of *The Elements* and see what they tell us of this embryological synthesis.

There can be no question that *The Elements* as well as being organized about a general message is a text rich in examples and references. In 442 pages the authors reveal a broad acquaintance with the world of experimental embryology, and, with their unique 36-page table combining bibliography and author index, they allow the reader to integrate specific contributions into the book's synthetic framework. The format, the clarity of the text, and the numerous and well-chosen illustrations make *The Elements* an attractive introductory textbook and a compendium of experimental work done during the previous thirty years. The first three chapters form, in fact, a prolegomenon of the issues surrounding morphogenesis with a focus on normal amphibian development and a review of some of the classical experiments of the field. *The Elements* still serves as a valuable guide. I will say more about its reception later; at the moment, it is worth emphasizing that Huxley and de Beer produced a readable and manageable survey of a complex field of experimental biology.

The Elements is obviously a good deal more than survey and introductory text. In keeping with so many of Huxley's other books, this work presents an elaborate theoretical synthesis in the sense that it gathers diverse experiments or observations around a general organizing model, which then serves as an explanation of disparate phenomena and as a program for further research. Huxley and de Beer were far more concerned with a general explanatory framework for early development than a survey of the vast array of previous results. This, of course, made the book more interesting to their peers, and rightfully it is more inclined to attract our historical attention. A question worth asking, as we examine this synthesis, is whether the construction of such a model is not typical of Huxley's whole mode of operation, whether applied to biology, to social thought, or to the development of worldwide strategies for the conservation of resources. It is a method that gives the initial impression of a hypothetico-deductive system: thus, a body of information leads to an organizing principle, the principle in turn leads to testing, more information, and calculated adjustments of the original model. The whole enterprise appears so rational, so in keeping with the positivistic movement championed by Huxley's grandfather, and so much a part of early twentieth-century biology. First appearances, however, are often deceptive, but before expanding upon that note of doubt, let us first look at the model that Huxley and de Beer developed in their book.

I find it convenient to divide *The Elements* into four distinct parts. The first comprises the three introductory chapters already described. The second is defined by chapters four through seven, which describe a sequence in development designated as "organizations of quite different type."[19] It is the authors' thesis that these types of organizations appear sequentially and are roughly defined by their chapters on egg polarity, on cleavage patterns, on the movement of cells from the stage of the blastula to neurula, and finally on morphological and histological differentiation. The third part, consisting of chapters eight through eleven, concerns demonstrations of embryonic fields and gradients in the regeneration of functional structures, in the pre-histologically differentiated germ, in post-embryonic life, and in the well-researched example of the amphibian nervous system. In other words, this part reviews the evidence for the morphogenetic gradients that were such a distinctive feature binding together the types of organization described in part two. The fourth part, chapters twelve and thirteen, examines the relationship between development on the one hand and Mendelian factors and functioning physiological systems on the other. It is the section in which the authors integrate their explanation of development with two other exciting fields in biology—genetics and physiology. They then conclude the main text with a brief summary chapter.

The Embryological Synthesis

The advantage of organizing *The Elements* in this four-part fashion is that it renders more clear the nature of Huxley and de Beer's synthesis. The story of development is told in part two. There they emphasized the establishment of axial gradients, which act upon the entire egg. The authors considered these gradients as primary in any developmental sequence and saw them initiated by external agents—the orientation of the oocyte, the influence of gravitation, or the point of penetration of the sperm. In the authors' words, the frog egg, for example, is a "machinery for realizing full normal bilateral symmetry" while "very slight differential action of various external agencies can act as a trigger permitting a particular plane of symmetry to realize itself."[20] In their analysis of cleavage Huxley and de Beer made it clear that degrees of viscosity, chemical differences, or other quantitative cytoplasmic differences organized around the axial gradients, not around the equivalent and equipotential nuclei, play the determinate morphogenetic role at this stage. From the blastula stage through neurulation and the tail bud stage, the authors found that Spemann's organizer and assorted inductors, acting within the established gradient systems of the cleaved egg, govern the next steps in development: "First, working as part of the gradient-field, the organizer may be figuratively said to sketch out the presumptive regions in pencil, and then, after invagination, the organizer goes over the same lines with indelible ink."[21]

After neurulation, Huxley and de Beer argued, the gradient system of the whole breaks up into a mosaic of partial fields or "areas of differentiation of organs." The location of each field is defined by the grid formed by the axial gradients and the organizer. Field formation is a gradual process guided first by a "chemo-differentiation" or invisible differentiation and later made visible by morphological and histological differentiation. What was initially a quantitative partition of the egg gradually became a qualitative process of differentiation. The embryo moves from external to general internal, then to increasingly localized determiners. Only at the stage of histological differentiation did the authors recognize factors of nuclear hereditary or Mendelian genes coming into play. The steps that the authors sketched out are difficult to identify in practice only because different organisms move through the sequence at different rates and the boundaries between the operation of one type of organization and another occur in different organisms at different embryonic stages.

If in part two Huxley and de Beer presented and documented "the sequence of types of organization," in part three they justified their overwhelming reliance on fields and gradients. Their inspiration came from the extensive experiments of Child on regeneration. Huxley and de Beer essentially argued that the rules deemed true in the regeneration of adult planaria, tubularians, and salamanders must also hold true in normal ontogeny. In the former, polarities, external factors, an apical or head region dominance, independence of isolated areas, and rate of reaction all pointed toward the determining influences of a gradient system; in the latter, analogous regulatory causes and effects called for a similar interpretation.

By means of analogy Huxley and de Beer moved to explain the organizing and regulating steps in early ontogeny. They saw a strong likeness between ontogeny and regeneration, but the two processes are not identical. Development and regulation in the egg are early embryonic reactions to changes and disturbances; they take place because the general gradient-fields are quantitative and inform the entire embryo. At early stages these fields generally dominate development; so the embryo is initially plastic. As the gradient fields lose their dominance to a mosaic of local and restrictive fields, the embryo becomes less plastic and loses its regulatory capacities. Regeneration occurs only later in development, after morphological and histological differentiation takes place. Then new, genetically determined responses develop, which allow various organisms to regenerate certain structures to varying degrees. That these regenerative responses must be guided by gradients of various generalities had been amply demonstrated by Child.

In what I have designated the fourth part, Huxley and de Beer turned their attention to showing how results of contemporary genetics and physiology might be integrated into their model of early development. Their effort to show the relevance of their work to genetics was particularly interesting, for in the same year Thomas Hunt Morgan published *Embryology and Genetics*. As is well known, despite its title, Morgan recognized that he failed to show how one discipline shed light on the other.[22] One of the basic problems, of course, was the nuclear equivalence of all cells. Approaching the interface of the two fields from the direction of their understanding of development, Huxley and de Beer delimited the sphere of operation of the genes. Essentially they saw early development as the effect of initial gradients, gradient fields, and inductions; they argued that only later developmental rates and the particulars of morphological and histological differentiation are the result of the activation of genes. "While the genes are by themselves incapable of initiating the processes of development and differentiation, it is obvious that they play an active part in the control of these processes, once development has been started, and their presence is essential."[23] In other words, the two authors circumvented the problem of nuclear equivalence by visualizing the partial gradient fields and chemical inductions as the external triggers for specific gene activation. The developmental system, not the cell-bound chromosome-restricted concept of the gene of classical genetics, was responsible for the phenotype.[24] Huxley's own study of rates of development in different stocks of *Gammarus* and of relative growth rates were important elements in the argument. An eagerness to see his own research within the larger framework defined by a number of fields was characteristic of Huxley's pursuit of biological models.

The Elements presents an exciting set of generalizations. Its authors visualized development as taking place in a sequence of different "organizations" and insisted that only through their study could biologists learn what the "term genetic characters really stand for." They understood experimental embryology as advancing in technique and method from the days when Wilhelm Roux had first insisted that experiments were the only guide to certainty. Hans Driesch, Theodor Boveri, Albert Brachet, and J. W. Jenkinson, among others, had then amassed experimentally ascertained facts and established epigenesis as the true description of developmental events. Hans Spemann, Ross G. Harrison, and Charles M. Child had provided the general principles for a theoretical framework, and now it was time to search for the physico-chemical bases of these principles. It was an optimistic vision of a rapidly maturing field of biology. Huxley and de Beer leave the impression that a "filling in" and "deepening" of the principles laid down in *The Elements* is all that would now be required.[25] Theirs was a synthesis, however, that was built on shifting sand.

Child and Axial Gradients

Child had developed his theory of axial gradients through a lifetime of regeneration experiments on the adult forms of hydroids and flatworms. The fundamentals of his position, however, were clearly staked out in his earliest work done within the American tradition of cell lineage studies. Unlike his fellow bench workers, such as E. B. Wilson, E. G. Conklin, and Frank Lillie, Child concluded that cleavage consisted of a quantitative rather than qualitative segregation of the egg and that the blastomeres did not represent an early step in the differentiation process. He endorsed Charles Otis Whitman's more general attack on the cell theory, and he subscribed to Hans Driesch's conclusion that differentiation was a "function of position." Adamantly rejecting Driesch's mystical vitalism, however, Child called for a mechanistic organicism: "The developing egg. . . is distinctly an organism at every stage."[26]

In 1900 Child began his regeneration studies, and by the next decade he associated regeneration patterns with axial gradients that organized the formative processes of the entire organism and that expressed the physiological unity of the individual. His experiments also showed that the dominance of the basic axes of organization could be manipulated under suitable experimental conditions so that the axes shifted or new ones formed in regions that were no longer under the dominance of the older axial gradients. By 1915 Child had also devised many physiological experiments that demonstrated that the rates of development, rejuvenation, senescence, and death could all be associated with metabolic gradients that correlated with the axial gradients. Thus he was convinced that the early formative processes of the organism must be derived from the physiology of the organism rather than from specific protoplasmic configurations or cellular contents. Throughout his subsequent career he repeatedly reminded his readers of the inadequacy of both a strictly biochemical and chromosomal rendering of organic form. By way of contrast, his axial gradient theory provided a "dynamic" or "physiological" explanation; early organic form was basically a product of living, integrated processes that began with primary axis formation.

There were shortcomings to this dynamic perspective. First, correlations in themselves do not demonstrate cause. Although his experiments repeatedly revealed metabolic (or physiological) gradients often determined by oxygen consumption or carbon dioxide production, Child failed to demonstrate definitely that they were causal factors of the axial gradients and other formative stages. Second, Child's chief documentation for the axial gradients came from regeneration experiments on adult organisms. It was problematic that his experimentally derived conclusions could be applied directly to early regulative stages of development. Third, when Huxley and de Beer adapted Child's axial gradient theory to their own theory of gradient fields and then became convinced that chemical differentiation preceded morphological or histological differentiation, there existed a question as to whether they were not calling upon gradient fields to be both the activators and the formative results at one and the same time.[27] Finally, there was a pronounced anti-cellular and anti-classical genetics bias in Child's message. It was symptomatic that in the same year as the appearance of Morgan, Bridges, Sturtevant, and Muller's *Mechanism of Mendelian Heredity* Child could write:

> Even some of our present-day speculations which attempt to assign actual topographic positions in the chromosomes to particular factors in heredity ignore completely the problem of the ordering and control of these factors which is involved in their assumptions. In fact, if we subject this group of theories to logical analysis we soon reach the point where it is necessary to assume the existence of something very like a superhuman intelligence as the underlying principle in all of them. They leave the essential problem unsolved, but their implications are anthropomorphic and teleological.[28]

By subscribing so wholeheartedly to Child's axial gradient standpoint, Huxley and de Beer implicitly adopted his organismic, anti-cellular, and anti-chromosomal interpretation of form.

Spemann and the Organizer

As we have seen, Spemann's organizer theory formed the other buttress of Huxley and de Beer's theory of gradient fields. From the turn of the century Hans Spemann dominated the development of experimental embryology in Germany. He began his experimental work with an examination of bilateral symmetry and rostrocaudal polarity in salamander embryos. Known as his constrictive experiments, this work was limited in scope but, when they produced partially duplicated salamander larvae, dramatic in results. Spemann was able to demonstrate the dominance of the dorsal region and the importance of gastrulation in the determination of the axial organs.[29]

In a longer series of experiments performed between 1901 and 1912 Spemann entered the realm of microsurgery with assorted ablation and transplantation experiments of the optic cups region of frog embryos. Through these he promoted the idea of lens induction in the epidermis by the subjacent cup of neural ectoderm. Further experiments by him and others, however, dispelled any "simplistic notion of a one cause–one effect mechanism."[30] The most important series of experiments of his career started in 1917 when Spemann focused on structure determination at gastrulation. Harrison had demonstrated the great value of using heteroplastic transplants (transplants between different species of the same genus) to reveal the fate of differently pigmented cells. Spemann saw this technique as a way of identifying the prospective neural and epidermal segments of the early gastrula, and the experiments helped him signify the dorsal lip of the blastopore as the "organization center" of later structures.

As mentioned earlier, it was in 1921 that Spemann's student Hilde Proefscholdt[31] implanted the dorsal lip under the epidermal flank of an older embryo of a different species. The induction of a secondary embryo, complete with neural tube, notochord, and somites—oriented in accord with the medial axis of the primary embryo—dramatically revealed the extent of the organizing capacities of the implanted tissue. Later variants of this experiment, including the preferred "Einsteck-method," gave Spemann, his students, and embryologists throughout the world the opportunity to examine the formative processes of gastrulation and neurulation. What became evident by the next decade was that the formation of the axial organs consisted of complex and multiple interactions. The blastopore lip when converted to caudamesoderm might create a unitary and whole embryo, but the axial organs, such as the mesodermal somites, might be chimeric in nature. Tissues with known presumptive fates could be manipulated to become organs normally of different germ-layer derivatives. The lateral portions of the blastopore could self-regulate so as to organize the medial axis of the embryo. Spemann intimated that there existed a hierarchy of organizers, which acted "as a chain of successive, casually related partial processes."[32] He spoke of "double assurance" when he envisioned an assimilative as well as a contact induction of the neural folds. Later he identified what he believed were "head" and "trunk" organizers. Some of Spemann's notions were soon shown to be inadequate. Nevertheless, by 1930, intensive inves-

tigation of the organizer principle revealed that axial formation consisted of a complex network of alternative possibilities that achieved a
unified outcome. Formation of the gastrula, neurula, and primitive
brain could not be understood in terms of a single pathway of actions
and responses. In Viktor Hamburger's retrospective words:

> What makes the organizer experiment unique is not the manifestation of
> a new principle but the unique constellation of important events at a
> critical period: the integration of self-differentiations, inductions, regula-
> tions, and self-organization, not just in the generation of "this or that
> organ" but in the creation of the axial organ system, and the unfolding of
> these activities during a critical relatively short time span, the gastrula-
> tion process.[33]

In addition to the increasing recognition that the processes involved were diverse and interactive, the understanding of the organizer as a unique and formative entity also radically changed. Originally Spemann had used the term "organizer" to designate the tissue
of the dorsal lip of the blastopore. At the 1931 Utrecht meeting of the
German Society of Zoologists he announced that he had destroyed
the cellular structure of the implant and still produced neural inductions and axial organs in the host embryo.[34] His students Hermann
Bautzmann, Otto Mangold, and above all, Johannes Holtfreter de-
vitilized the implant and again induced secondary embryos. Holtfreter tried an assortment of dried, coagulated, and frozen embryonic
tissues, which in the living state had no organizing capacities. Using
the Einsteck-method, he once more produced secondary embryos.
Various ether and water extracts, weak organic acids, methylene blue,
and cytolyzing agents were at one time or other implicated as the
primary inducing agent. It is relevant to our story that Huxley and
de Beer's own colleagues Joseph Needham and Conrad Waddington
became particularly active in investigating the biochemical nature of
the organizer. Both the normal and abnormal inductors rapidly lost
any morphological value and causal specificity. What originally had
appeared as a special structure associated with a precise event became
obscure in a welter of nonspecific, generic stimuli that in isolation
from the other events of early ontogeny had little developmental
meaning. Huxley and de Beer recognized these advances in embryology,[35] but the question naturally arises as to whether the changes
over a decade in the organizer concept materially altered their vision
of a gradient field system. My feeling is that it did not.

Although he had translated Huxley's brief notice in *Die Naturwis-senschaften* linking his name with Child's, Spemann was highly critical of Child's gradient theory. After a chapter's worth of argumentation in his Silliman Lectures he could conclude that "on the basis of the facts known at present, I cannot yet persuade myself that the gradient theory in the sense of Child, de Beer, and Huxley applies to the early development of the amphibian egg."[36] In 1939 the eminent embryologist Paul Weiss captured in a negative fashion the focus of Huxley and de Beer's synthesis when he critically reviewed Spemann's organizer theory:

> One would have been justified in retaining the term "organizer" in its full meaning if the progressive organization of a germ had been proved to be but a chain of successive inductions starting from this center and spreading over the rest of the germ like a blaze. Since we have seen, however, that the very development of the "organizer" itself is principally its own internal affair involving autonomous emancipation of partial districts, such as chorda, musculature, pronephros, inside the common cell mass, we realize that the problem of organization is actually contained in a nutshell in what happens inside of the "organizer" itself, regardless of what it does to other parts. Hence, to assume that the problem of organization could be fundamentally solved by exploring merely the inductive actions of the "organizer" upon surrounding parts of the germ, seems like looking for the fire outside of the house while the blaze is within.[37]

Weiss's words, "a chain of successive inductions . . . spreading over the rest of the germ like a blaze" describes a common misconception of the organizer theory; it unwittingly applies as well to the joining of the Spemann's ideas and Child's gradient system of which Weiss was also critical. If Weiss's words reflect a contemporary yet more sophisticated understanding of early development, we need to ask how it was that two such outstanding biologists constructed their synthesis on an infirm foundation. Was it simply that Huxley "rarely worked anything out fully," as Witkowski justly remarks? *The Elements*, I believe, reveals something more fundamental than distractions and haste.

Nature of the Huxley–de Beer Synthesis

By framing their synthesis around Child's gradient system and Spemann's organizer theory, Huxley and de Beer signaled the adoption of a particular perspective: First, they seemed convinced that their synthesis justified a rapid reduction of the events of axial organ formation to physiology and biochemistry. They did not adopt such a research program themselves, but they applauded its pursuit by others. Second, they deemphasized the structural components involved in early developmental activities. They concerned themselves neither with the subdivisions of the dorsal lip of the blastopore nor with the structure of the inductor and induced tissues of the neurula and tailbud stage. Third, by concentrating upon a sequence of "invisible" chemical differentiations preceding any visible histological differentiation, they minimized the importance of understanding the nature of the reacting systems. These were to loom more important as the organizer became transformed from a single entity to a "constellation" of events. Finally, Huxley and de Beer eschewed examining the formative activities within the cells themselves. Despite demonstrating their concerns for Mendelian genetics in chapter 12, their discussion of the classical gene was appropriately sanitized of the mundane world of chromosomes and other cellular structures. Huxley and de Beer dismissed the entire tradition of classical chromosomal genetics with the remark that "hitherto, neo-Mendelianism has been concerned mainly with the manoeuvres of the hereditary units, and in large part with their manoeuvres during the two cell-generations in which the reduction of chromosomes is brought about." They then added as an afterthought and self-justification, "It [neo-Mendelianism] is now beginning to concern itself with the mode of action of the hereditary units during the much larger number of cell-generations involved in building up the adult organism from the egg; and this task it can only accomplish satisfactorily in close contact with developmental physiology."[38] The equipotential parcel of genes in each cell simply reacted to the environment of chemical inductors and a gradient matrix.

I would like to suggest that the features in Huxley and de Beer's embryological synthesis, i.e., metabolic gradients, a biochemical organizer, a mosaic of chemical fields, and the unspecified reaction of genes in a chemical mosaic, reflected the explanatory aspirations of contemporary English biophysicists and biochemists. To employ one of Robert Olby's categories, Huxley and de Beer seemed to share methodological and philosophical commitments with the "dynamic approach" to the organism articulated by D'Arcy Thompson and

Joseph Needham.[39] In a different although related field Clifford Dobell, the foremost English protozoologist, revealed somewhat earlier a similar perspective in reaction to the traditional German interpretation of protozoa as single-celled organisms.[40] In both embryology and protozoology the English emphasis was placed on the organism as a dynamic, biochemically integrated whole rather than on the cell as the basic unit of life, that is, on multi-cellular organisms as an extensive society of these interacting units, and on the cleaving zygote as a transitional structural step from one structural state to the other. *The Elements* contrasts dramatically in these regards with the contemporary embryological texts by German embryologists, such as Waldemar Schleip[41] and Paul Weiss.[42] To rephrase Weiss's metaphor of the burning house, Huxley and de Beer were looking exclusively for the fire outside of the cell while the blaze burns also within.

R I C H A R D   W .   B U R K H A R D T ,   J R .

Huxley and the Rise of Ethology

In his rich and multifaceted career, Julian Huxley had many different claims to fame. In biology, he was particularly proud not only of his work on relative growth, his idea of clines, and his book, *Evolution, the Modern Synthesis*, but also of his early field studies on bird behavior. When Konrad Lorenz in the early 1960s referred to Huxley in a public lecture as one of the three founding fathers of ethology—together with Charles Otis Whitman and Oskar Heinroth—Huxley was pleased with the compliment, and he took it to heart.[1] In his autobiography Huxley described his paper of 1914 on the great crested grebe as "a turning point in the study of bird courtship, and indeed of vertebrate ethology in general." He also credited himself with "having made field natural history scientifically respectable."[2]

These were bold claims on Huxley's part. They are historically suspect, however, for two reasons. The first is that in 1914 vertebrate ethology was not sufficiently recognizable as an enterprise to even have a turning point. The second is that respectability is a relative quality that can only be judged in the light of local circumstances, and Huxley said little in his autobiography about what the local circumstances in this case were. To the extent that individual practitioners sought to make it a self-consciously scientific undertaking, the study of animal behavior in the early years of the twentieth century was a highly contingent and precarious venture. What we now call ethology—that biologically oriented, comparative, and naturalistic approach to the study of behavior that we associate with Konrad Lorenz and Niko Tinbergen—did not begin to become a coherent enterprise until the 1930s and 1940s, and even then its status was highly problematic. To gain a clearer picture of Huxley's relation to all this, we must situate his behavioral work in the context of the theories and the practices of his own day. Only in this way will we be able to appreciate what his work represented and what Huxley himself did— and did not—accomplish.

There are a variety of resources to aid us in this task. In the first place, from Huxley's own hand, we have several reminiscences, a series of scientific articles, an additional group of popular expositions, and a wealth of manuscript correspondence, notebooks, and diaries.[3] These materials, however, are not without drawbacks. Reminiscences have their characteristic deficiencies, and scientific papers, as Sir Peter Medawar has reminded us, have a tendency to "actively misrepresent the reasoning that goes into the work they describe."[4] Fortunately, we are able to supplement Huxley's published writings with the extensive and well-cataloged collection of his papers at Rice University, and also with other smaller collections, such as that of the Edward Grey Institute of Field Ornithology at Oxford. Lest one become overly sanguine, however, about how readily Huxley's field notebooks, for example, might illuminate the development of his thinking, a word of caution is in order. When Huxley donated his field notebooks to the Edward Grey Institute, David Lack was prompted to observe that great ornithologists seem to divide into two categories: those who keep "extremely neat" field notes and those who keep "appallingly untidy" ones.[5] Huxley proves to have been one of the latter.[6]

Beyond Huxley's own writings, there are additional sources to help us put his behavioral studies in their historical place. These include the excellent biographical memoir of Huxley by John R. Baker, several studies of the history of ethology, and, still insufficiently explored, a large quantity of published and manuscript writings bearing on the activities of other students of behavior in Huxley's day.[7] The latter materials are especially useful in reconstructing the context in which Huxley's ideas developed, given that his own efforts to identify his intellectual debts were at best uneven.

Huxley's activities as a birdwatcher and observer of animal behavior can be divided into three main periods. The earliest, from 1901 to 1911, was the period in which he first took up birdwatching as a hobby. The second, from 1911 to 1925, was the period in which he did his scientific studies in the field and published a series of five important papers on the courtship behavior of different bird species. The third, from 1925 until the end of his career, was the period in which he wrote a number of synthetic pieces on bird behavior and promoted behavioral studies through a variety of additional activities, though his main interests and efforts were admittedly focused elsewhere.

It is manifestly easier to recall a vivid personal experience than it is to identify the various intellectual influences that have shaped one's thinking.[8] This, at any rate, seems to have been true for Huxley with respect to his studies of bird behavior. When he described the development of his thinking in this area, he tended to stress first his personal experiences in the field, and then the *natural* development of his ideas, but not the extent to which his ideas or practices may have been derived from those of others.[9] Thus, while he recounted on several occasions how thrilled he was, at the age of thirteen or fourteen, to see a green woodpecker up close, and how this experience imprinted him on birdwatching for the rest of his life,[10] he never made explicit the ways in which his own development as a birdwatcher corresponded to a major shift that took place in the sensibilities and interests of amateur ornithologists in Britain at the beginning of the twentieth century.

The fact remains, nonetheless, that in 1901, the year Huxley began devoting his weekends, holidays, and spare moments to recording all the different birds he could spy, a new approach to birdwatching was emerging. While the young Huxley was still at the stage of hoping to add rarer and rarer species to his life list, other naturalists were turning their attention to more *common* birds, and to the detailed study of their habits. In an earlier period, the English amateur ornithologist had been represented by figures like William Brodrick, hawking over the moors and bringing back the trophies of the hunt to be stuffed and displayed in glass cases.[11] At the beginning of the twentieth century, in contrast, the English amateur was represented by figures like Eliot Howard, standing silently in a gorse bush, and "patiently enduring torments caused by insects," so that he could gain a detailed knowledge of the courtship behavior of the grasshopper warbler.[12]

This new approach was signaled initially not in the austere journals of the professional scientific societies but rather in *The Zoologist*, whose editor insisted that it was "not a forlorn quest" to attempt to study "the status of intelligence and the mental concepts in animal life,"[13] and also in such general periodicals as the *Saturday Review*. Two reviews that appeared in the *Saturday Review* in 1901 expressed quite clearly what the new approach to bird study involved.

The first of these reviews dealt with a new edition of J. E. Harting's *A Handbook of British Birds*. The anonymous reviewer allowed: "It is no reproach to [Harting's book] . . . to say that many new works of this kind are not likely to be called for by the bird-lovers of the near future." Thanks to the works of Harting and others, the reviewer explained, "the birds have all been admirably arranged and divided up into their proper groups and families: all are neatly ticketed." What remained for the future was to learn something about bird *life*. It was time to go out into the fields and woods, to discover "exactly how the birds build their nests, exactly why they sing: we want to see them at work and at play, and to know something of their loves and their hates, of their toilette, their travels, their relaxations, their etiquette, their marriage customs." It was time, in other words, to attend to that part of the bird world that was still unknown.[14]

That such attention was in fact already being given to the bird world, at least by one British ornithologist, was signaled a few months later in the *Saturday Review* in Warde Fowler's account of a new book entitled *Bird Watching*, written by Edmund Selous. To Fowler, who was himself a dedicated amateur ornithologist, what was especially impressive about Selous's work was his patience as a watcher. Patience, Fowler indicated, was the special virtue that, along with other qualifications, had to be brought to bear on "those problems of life and mind which will be the chief work of naturalists when that of collecting and classifying is gradually complete." Selous had devoted hours and days to watching the behavior of common birds. Wrote Fowler: "Reading Mr. Selous's book I feel that if I were beginning life again, I would give all my spare time to watching as he has watched. He has taken a new departure, and needs to be supplemented and tested. There is a wide field in front of the beginner who will follow in his footsteps."[15]

Warde Fowler was by profession a historian and classicist, an Oxford don who was noted for his scholarly studies of the political, social, and religious life of ancient Rome and who for twenty-five years served as sub-rector of Lincoln College. A few years after he wrote his review of Selous's book, Fowler was met in his rooms in Lincoln College and joined in his outings in the field by a rather timid young Oxford undergraduate who came to talk about birds with him.[16] The young man was Julian Huxley. Fowler and Selous appear to have been the two contemporary naturalists most responsible for shaping Huxley's early ideas about what could be learned from the study of bird behavior. A third amateur, H. Eliot Howard, figured prominently in the later development of Huxley's thought.[17]

Selous has received some attention in recent histories of ethology, but his work is yet to be adequately investigated.[18] Huxley acknowledged that it was one of Selous's books that first revealed to him "the enormous amount of interesting facts [about bird behavior] which still remained to be discovered."[19] It would be a mistake to assume, however, that Selous was content with simply reporting facts. He was always concerned to relate his observations to ideas and theories—particularly Darwin's theories of natural and sexual selection. His unparalleled field observations, first on the ruff and then on the blackcock, provided striking confirmation of Darwin's theory of sexual selection at a time when most biologists held that theory in low esteem.[20] He poured forth his facts and observations in vigorous prose, lacing his presentations with vitriolic comments about killers of wildlife, professional zoologists, and society in general, together with complaints about the difficulties and discomforts of field work. As he watched the courtship of the ruff very early one "bright, but bitterly cold morning," for example, he could not help but comment on those "learned ornithologists all over Europe" who lay "sleeping in their pleasant beds" but would "come down all the fresher for breakfast" and then from the comfort of their studies issue "bulls against sexual selection."[21]

Selous, as Huxley wrote of him, was "always . . . alert to see in small divergencies of behaviour one of the means whereby a species may become altered or may split into two."[22] One nice example of this is to be found in Selous's 1905 book, *The Bird Watcher in the Shetlands*, where he wrote:

> I remember once passing unusually close to a cock pheasant, which remained crouching all the while, though nineteen out of twenty birds would, I feel sure, have gone up. It struck me, then, that as all such

pheasants as acted in this way would have a greater chance of not being shot than the others that rose more easily, whilst these latter were constantly being killed off, therefore, in course of time, the habit of crouching close ought to become more and more developed, and pheasants, in consequence, more and more difficult to shoot. Some time afterwards I met with some independent evidence that this was the case, for a gentleman who shot much in Norfolk, remarked, without any previous conversation on the subject, that the pheasants there had taken to refusing to rise, and that this unsportsmanlike conduct on their part was giving great trouble and causing general dissatisfaction. That was his statement. He spoke of it as something that had lately become more noticeable, but only, as far as his knowledge went, in Norfolk, which I believe, is an extremely murderous country. [23]

More to the point with respect to Huxley's later thinking was the explanation Selous offered of the origin of courtship displays. Selous maintained that the key to understanding many of the striking features of bird behavior were to be sought "in the highly nervous and excitable organization which birds, as a class, possess, and, especially, in the extraordinary development of this during the breeding and rearing time. This nervous sexual or parental excitation produces all sorts of extravagant motions and antics which are at first quite useless, but on the raw material of which both natural and sexual selection have seized and are constantly seizing." The fact that *some* bird antics could not be seen as the result of sexual selection, Selous maintained, was no reason to discard the theory of sexual selection altogether. "By what agency the raw material has been shaped in any one case," he wrote, "is a question of the evidence in and relating to such case." [24]

It is impossible to determine precisely what Huxley derived from his earliest reading of Selous, which occurred at about the time he (Huxley) was leaving Eton. It is apparent, nonetheless, that without giving up recording the numbers of different *kinds* of birds he spotted, he began paying more attention to bird *behavior*, and that the thoughts he developed on bird *mind* corresponded closely to the views that Selous had expressed on the subject. [25]

The earliest record we have of Huxley's thinking on behavior is the manuscript of a paper entitled "Habits of Birds," which Huxley wrote as an Oxford undergraduate and presented in 1907 to the Decalogue Club at Balliol. The stated theme of his paper was how habits of birds bore on "various evolutionary theories." In fact, however, he did not address any of the major alternatives to Darwinian theory that were current at the time. Neither neo-Lamarckism, nor mutationism, nor orthogenesis received his attention. Instead, he used the details of bird behavior to illuminate and endorse Darwin's theories of natural and sexual selection.

Describing to his audience such distinctive performances on the part of birds as "drumming" (by the snipe and the lesser spotted woodpecker), Huxley explained that "the end or object of such performances" was twofold: it promoted "recognition between individuals of the same species," and it enabled the male bird to show off his special features to the female bird. Darwin, Huxley said, had "explained this by this theory of Sexual Selection—that the hen consciously or unconsciously selected as her mate the cock whose performance pleased her most."

In addition to endorsing Darwin's theories of natural and sexual selection, Huxley's paper advanced a number of other interesting points, including an explanation of bird song as an expression of emotions ("a kind of mental safety valve"); a recognition that individuals of the same species may display variations in their behavior; an appreciation that some forms of apparently purposive behavior in birds are in fact performed unconsciously; and a discussion of how in birds there is often a compromise between the need for protective coloration and the development of bright colors for sexual display.

Huxley concluded his paper with a paean to the good of the race or species. As he put it: "In Nature it is the evolution of the race that is the important thing—the individual is quite subservient to the race. . . . Man's reason has showed him that in union there is strength, & that he will only exhaust his inherent capabilities when every individual is playing his part merely as a unit of a glorious whole." Five years later, Huxley used a similar statement to conclude his first semipopular book, The Individual in the Animal Kingdom (1912), describing the state as "that unwieldiest individual—formless and blind to-day, but huge with possibility. . . ."[26] This focus on the benefit of the whole rather than the individual members of the species also informed Huxley's earliest scientific papers. When he discovered that female mallard ducks were frequently chased and trodden by many males and were often drowned in the process, he attributed this phenomenon, which he found highly repugnant, to "a disharmony in the constitution of the species."[27] The general notion of "the good of the species" was to remain a common theme throughout Huxley's writings on animal behavior and evolution.

Given Huxley's ornithological proclivities and his resolution, crystallized at the Cambridge celebration of the Darwin centenary of 1909, to follow in his grandfather's and Darwin's footsteps and to devote his scientific career to the study of evolution, one might suppose that the fit between field studies and Darwinian theory was a comfortable one that launched him happily on his career.[28] This was not, however, the case. In September 1909, Huxley went off on a year's fellowship to the Zoological Station at Naples. Though there is scarcely a hint of it in his autobiography, the time he spent in Naples was a period of considerable anxiety and unhappiness for him. Twelve years later, when Huxley's student Alister Hardy was in Naples and worrying about his scientific future, Huxley wrote Hardy a long and sympathetic letter revealing just how unsure of himself he had been when he started his career. He had gone to Naples, he told Hardy, wanting "to do something big, new, original." Two months of trying to make cultures of various protozoa, however, proved "completely in vain," and his work on sponge dissociations and then *Clavellina* did not satisfy him because it was work "based on other people's ideas." Furthermore, he was worried about his prospects as a researcher, because he doubted he was capable of doing good work in physiology. "Distrust of one's own powers combined with a good deal of ambition" left him distraught and miserable.[29]

As it was, when Huxley returned from Naples to Oxford in 1910 to become lecturer at Balliol College and demonstrator in the Oxford Department of Zoology and Comparative Anatomy, neither his research at Naples nor his responsibilities at Oxford gave him any particular reason to take up the study of animal behavior. What launched him into behavioral studies instead was something of an accident. In April 1911, during the spring vacation of his first year of teaching, he took a "reading-party" with him to the north coast of Wales. There he found that the reading he brought with him, Bütschli's work on the protozoa, was not nearly as interesting as watching the hundreds of birds that haunted the northern half of Cardigan Bay when the retreating tide left a "glistening sweep of sand and mud" behind it.[30] While he was pleased to see the grey plover and black-tailed godwit, two species he had never seen before, he was more pleased to have the chance "to study, under the most favourable conditions, the natural behaviour and home life of some of the commoner shore-birds."[31] This appears in fact to have been the first time he experienced the particular benefits of devoting "several spells of watching, day after day" to the study of *a single species*. The excitement of making his own observations on bird courtship left a greater impression upon him than any vicarious experience could have done.[32]

Huxley was particularly entranced by the courtship of the male and female redshank. He was struck not only by the beauty and excitement of the male's display, but also by the fact that the female in most cases rejected the male's advances by flying off. In a letter to his fianceé, Huxley said of the redshank's courtship:

> This is very interesting to me, as it shows how Darwin's idea of Sexual Selection was undoubtedly right in certain cases, such as this—the hen has the power of choice,—if she doesn't like the cock, away she just goes, & he always has to give up the chase eventually. Not only has she got it, but she exercises it a great deal—all the suitors so far have been rejected! It must be a queer kind of choice, I daresay, scarcely conscious at all, but very decided in its workings.[33]

According to Huxley, it was only after he had returned from Wales and had written up most of his observations that he discovered "a fairly complete account" of courtship in the redshank written by Selous.[34] This was an interesting declaration of independence on Huxley's part, for while he used it to claim autonomy for his own observations, it also serves to indicate that up until 1911 he was unfamiliar with what was arguably the most important paper on sexual selection in birds published in the previous half-decade. This, furthermore, was at a time when sexual selection—"the vexed question of sexual selection," as F. B. Kirkman called it—was a key concern of field ornithologists in Britain.[35] In any case, neither the observations Huxley reported, nor the conclusions he drew from them, departed significantly from those of Selous. The main thrust of Huxley's work was to underscore what Selous had already maintained, "namely, that the actions of the birds which lead up to each single act of pairing are explicable only on the Darwinian theory of

Sexual Selection, or on some modification of that theory." In the redshank, Huxley indicated, there was obvious *display* on the part of the male, and "an equally marked *power of choice*" on the part of the female. "Though the male in this particular species has the *initiative*," Huxley concluded, the *final decision* must rest with the female."[36]

The following year, Huxley again spent his spring vacation watching bird courtship. The bird on which he trained his binoculars in 1912 was the great crested grebe. Together with his brother, Trev, Huxley watched the grebes on the reservoirs at Tring, the Rothschild estate, for a period of ten days. Once more, upon returning home, Huxley found that Selous had in some measure anticipated him, having written a decade earlier what Huxley describes as "a welcome paper . . . which exactly dovetailed into my own observations."[37]

What struck Huxley most about the grebe was a phenomenon that, as far as he knew at the time, was unique to that particular bird. The bird had special structures—its erectile neck ruff and ear tufts—which, as Huxley put it, were "not only the common property of both sexes," but were "actually used in display, and used in exactly the same way by both sexes."[38] Huxley initially believed that since these structures were "only used in courtship," they had to be the result of sexual and not natural selection. He proposed using a word promoted by E. B. Poulton—"epigamic"—for characters that had arisen through sexual selection but that had come to exist equally in both sexes.[39]

Huxley's first thoughts on the grebe were published in 1912. Two years later he published his longer (and now famous) paper, "The Courtship Habits of the Great Crested Grebe." By this time, informed once more by the work of amateurs before him, he had come to interpret the grebe's behavior in a new light. Selous had anticipated Huxley in observing and recording numerous features of the grebe's behavior. Selous, however, had not used his own observations to call into question the theory of sexual selection.[40] In this respect, at least, it was not Selous who influenced Huxley most, but Eliot Howard, whose impressive two-volume work, *The British Warblers*, had in 1914 just been completed.[41] The problem that Howard stated quite clearly with respect to his warblers—and that he had indeed perceived even before he came to his famous idea of territory—was that much of the warblers' display behavior occurred only after the birds had paired up. Howard concluded that these displays could therefore not have developed through Darwinian sexual selection.[42] Huxley concluded that the same was also the case for the great crested grebe.[43]

In his paper, Huxley described at length the display behavior of the grebes, giving the postures or ceremonies he witnessed such names as the "cat-attitude," the "Dundreary attitude," "ghost-dive," the "shaking attitude," the "passive" and the "active pairing attitude," the "discovery ceremony," the "weed-trick," and the "penguin-dance." He interpreted all these complex postures and behavior patterns as expressions of emotion. Two things in particular impressed Huxley about these displays. The first was that they were mutual, i.e., both sexes displayed the same behavior patterns. The second was that they were by and large self-exhausting, i.e., they did not seem to serve as excitants to coition but instead were typically followed by periods of calm, where the birds simply engaged in "swimming, resting, preening, and feeding."[44]

Having concluded that the displays functioned neither in mate choice nor as a stimulus to coition, Huxley was left with the problem of determining what they were for, since it seemed clear to him that they had to fulfill some function. He decided they probably served "to keep the two birds of a pair together, and to keep them constant to each other." "From the point of view of the species," he explained, "it is obviously of importance that there should be a form of 'marriage'—constancy, at least for the season—between the members of a pair."[45]

Given the likelihood, then, "that many actions and structures solely used in courtship are of use to the species," Huxley concluded that these actions and structures were "maintained by Natural as opposed to Sexual Selection."[46] Instead of supposing that the actions and structures of the male and female grebe had arisen by sexual selection of the males followed by transference of the secondary sexual characteristics to the females (the explanation that Huxley suspected would have been Darwin's, and that Huxley himself seems to have been inclined to in 1912), Huxley attributed these actions and structures at least "in considerable part" to what he called "mutual selection." As he saw it, the various expressions of emotion that existed in the behavioral repertoires of the grebes' ancestors had over time been developed into the special rituals exhibited by the grebes of the present. In his words, "In the Grebe, [the surplus emotional energy of the bird in springtime] has been diverted into fresh channels through Mutual Selection, and thus pressed more quickly into the service of the species."[47]

In addition to noting how Huxley's observations on the great crested grebe led him to reflect on the relations between sexual and natural selection, one should note the extent to which Huxley stressed the subjective, *emotional* side of bird life in his interpretation of courtship behavior. In his words:

> Birds have obviously got to a pitch where their psychological states play an important part in their lives. Thus, if a method is to be devised for keeping two birds together, provisions will have to be made for an interplay of consciousness or emotion between them. . . . All birds express their feelings partly by voice, and very largely by motions of neck, wings, and tail; and not only this, but the expression can be, and is, employed as a form of language. This being so, we have here a basis on which can be reared various emotional methods of keeping birds of a pair together. As always, selection of accidental variations has led to very diverse results; so that we see this "emotional companionship" playing a part in many apparently very different actions of birds.[48]

Huxley's great crested grebe paper remains the single paper for which he is best known. With its promotion of the idea of *mutual selection*, it represented Huxley's first step away from the view that display characters used in courtship were necessarily the result of sexual selection. It also involved Huxley's first use of the idea of the *ritualization* of behavior, an idea to which he returned at the end of his career when he organized a symposium on the subject for the Royal Society of London.[49] Nearly four decades after Huxley published his grebe study, David Lack told him that it "must be one of the longest-lived of all papers. It's still constantly quoted (by me and others) in lectures." Likewise, Konrad Lorenz insisted to him how impressed he (Lorenz) had been with Huxley's observation that the grebes' activities could be at one and the same time "self stimulating and self exhausting."[50]

In view of this later praise of Huxley's paper and Huxley's autobiographical remarks on the paper's importance, it is worth noting that Huxley did not introduce his paper with any ringing manifesto, either for field biology in general or for animal behavior studies in particular. To the contrary, the tone of his introductory remarks was casual and lighthearted. He hoped the paper would "help to show what wealth of interesting things still lie hidden in and about the breeding places of familiar birds. A good glass, a note-book, some patience, and spare fortnight in the spring—with these I not only managed to discover many unknown facts about the Crested Grebe, but also had one of the pleasantest holidays. 'Go thou and do likewise.' "[51]

Huxley, in short, made no effort in introducing his paper to distinguish between what was an enjoyable hobby and what was serious biology. Furthermore, he did not stake his reputation on the work he was presenting. The paper was a safe one. No one could fault him for making such industrious and clever use of a holiday. The light tone of his remarks belied, however, the anxieties that continued to plague him concerning his scientific future. In the same year that his grebe paper was published, he experienced his first nervous breakdown.

Huxley's psychological difficulties over the next several years were a major trial for him. They did not prevent him from continuing his birdwatching, however, nor did they keep him from publishing in the *Auk* in 1916 a long and valuable paper on the ways in which birdwatchers, naturalists, and biologists could interact profitably with one another to resolve important biological questions.[52] He had another serious breakdown after the war, and for a time "was haunted day and night by the idea of suicide," but by the spring of 1921, when he wrote the letter to Alister Hardy mentioned above, he evidently felt he had satisfactorily resolved his intellectual and psychological crisis.[53] Interestingly enough, however, when he described to Hardy how excited he now felt about biological research, the research he described was limited to his experimental work in the laboratory. About the prospects of doing important field research, he said nothing. And this was despite the fact that he was about to set off to Spitsbergen to do field work with an Oxford University expedition.

One of the results of the Spitsbergen expedition of 1921 was Huxley's publication of a third major paper on bird behavior, that on the courtship of the red-throated diver.[54] This paper, though not as well known today as Huxley's great crested grebe paper, was comparable to the grebe paper in both quality and significance. In the paper, Huxley stressed once more that it was no longer possible to suppose "that Darwin's original theory of sexual selection [was] adequate to explain the origin of most of the sexual ceremonies and adornments . . . found in monogamous birds." The reason for this was straightforward: "These adornments are chiefly used in ceremonies which take place after mating-up has taken place for the season. There cannot therefore be a direct selection as between one male and another in respect to them."[55]

Citing the views of W. P. Pycraft, H. S. Sturtevant, and T. H. Morgan, Huxley allowed that the function of much display was evidently a stimulative one. Nonetheless, he was still inclined to believe that some mutual displays were truly self-exhausting. From a psychological standpoint, in other words, the displays were performed "for their own sake." From a biological standpoint, they had to be regarded either "as biologically functionless, as so many by-products of a mental organization of the type required to execute the stimulative forms of display," or else as instrumental in keeping the members of the pair together, when the duties of incubation and care of the young demanded this.[56]

Where Huxley's red-throated diver paper advanced most appreciably beyond his earlier work was in its development of an insight that Huxley may have gotten from the work of Eliot Howard. This was the recognition that the different modes of courtship in birds are the function of more than just the immediate sexual dimensions of courtship. What was important to realize, Huxley said, was "that the organism is a whole" and "that the form and extent of courtship, nay, in some periods its very existence, is due to causes which are not epigamic in origin, but connected with other fundamental biological needs in relation with the annual cycle of the animal."[57] This attention to the ways in which forms of courtship are related to other features of the life history of a species would become in the 1950s a hallmark of the behavioral-ecological studies produced by Niko Tinbergen and his students at Oxford.

In his red-throated diver paper Huxley also paid special attention to certain courtship actions that he felt could not be explained in terms of the "originally nonsignificant physical release of emotional tension." The behavior patterns he had in mind were actions such as preening, head-shaking, bill-dipping, and picking up nesting material—all of which appeared to have developed initially in connection with other functions and were only later "connected with sex in courtship displays." A certain type of preening in grebes and mute swans, he explained, had come to be employed in courtship in "a ritual way, and without any of [their] usual functional significance." He was confident that once attention was drawn "to this 'ritual' use of non-sexual actions during courtship activities," the phenomenon would be found to be "of very wide occurrence."[58]

While Huxley ended his paper with the view that much of Darwin's theory of sexual selection had to be rejected, since a large percentage of the characters and actions used in courtship could not be explained in terms of competition for females, he reinforced what he called "the second salient fact first clearly recognized by Darwin— the fact that the development of an epigamic character is dependent upon the emotional effect which it produces upon the mind of a bird of opposite sex."[59] In Huxley's words, "once epigamic characters come to be advantageous, the mind of the species (in the females in sexually dimorphic forms, in all individuals in those with mutual courtship) is exerting the indirect effect we have been describing upon the future development of colour, structure, and behavior in the race. This is the most important fact which Darwin perceived, and this stands firmer than ever in spite of the rejection of the bulk of the other part of his doctrine."[60] For Huxley, like many biologists and psychologists at the beginning of the century, but less like the behaviorists (and, to some extent, the ethologists), a concern with animal mind was always regarded as a necessary element of the study of animal behavior.

Huxley published two more papers on bird courtship, one on the avocet and the other on the oyster-catcher, as the result of a trip to the bird sanctuary of Texel in Holland in the spring of 1924.[61] He was especially pleased in the case of the oyster-catcher to be able to develop an explanation of the birds' piping ceremony. In his view, the piping was probably at first "a mere accidental by-product of the high emotional tone of the breeding-season, the bird having to 'let off steam' by expressing in action its general excitement." Later, the ceremony became stereotyped, serving a display and a threat function simultaneously.[62]

This explanation was identical in its general form, if not in its specifics, to the interpretation of display behavior Selous had advanced two decades earlier. It was also the same kind of explanation endorsed by Eliot Howard, with whom Huxley became good friends in the early 1920s. Huxley's interaction with Howard was one of the things that inspired Huxley at this time to think about writing a general book on bird courtship.[63] He made a start on the project in 1925. Unfortunately, the book "never got written."[64] Huxley's move to London in the same year and his ever-widening interests in other scientific and nonscientific problems led him to put his bird courtship book aside.

Huxley published several additional scientific papers on bird behavior, but no more original field studies. In 1930 he gave a lecture on the "Biology of Bird Courtship" at the Seventh International Ornithological Congress, held in Amsterdam. Four years later, at the Eighth International Ornithological Congress, held in Oxford, he lectured on "Threat and Warning Coloration in Birds." In 1938 he published two major papers on the standing of Darwin's theory of sexual selection and where it stood in light of contemporary research.[65] In addition to these writings he published (and in some cases transmitted by radio) a variety of more popular treatments of bird-watching and bird behavior.[66] He also contributed articles to the *Encyclopaedia Britannica* and wrote a long section on animal behavior for the book *The Science of Life*, coauthored with H. G. and G. P. Wells. In 1934, in collaboration with his naturalist friend Ronald Lockley, he made the film *The Private Life of the Gannet*, which later won an Oscar for the best short documentary of the year.[67] In 1936 Huxley was instrumental in founding the Institute for the Study of Animal Behavior, and he became the Institute's first president.[68]

Though Huxley's interests moved into other areas, his early work on bird behavior did have an influence on a number of other naturalists. This was true particularly in Holland. When Huxley visited Holland in 1924, his Dutch host, G. J. van Oordt, was surprised to find him interested in common birds rather than rare ones. Within a few years, however, Dutch ornithologists were distinguishing themselves for their detailed observations on the behavior of common birds. This was not entirely because of Huxley. A. F. J. Portielje, the charismatic zoo director who started using the word "ethology" in the titles of his papers on bird behavior as early as 1925, was not so much influenced by Huxley as he was by other writers, among them Oskar

Heinroth, the German zoologist who was to become the mentor of Konrad Lorenz.[69] On the other hand, the young Dutch ethologist and ecologist Jan Verwey cited Huxley at length, and the slightly younger Niko Tinbergen, for whom Verwey was an important role model, found the theoretical orientation of Huxley's bird papers particularly impressive.[70]

Huxley's role in the subsequent development of animal behavior studies was not limited to the example he set in his publications. It also involved his encouragement of other students of behavior—both amateurs and professionals—and his efforts to bring an evolutionary understanding of behavior to nonbiological scientists and to intelligent laypersons.[71]

In addition to gaining much from the amateurs Selous and Howard, Huxley in turn was helpful to them. Huxley helped Selous publish some of his bird diaries in the form of the book, *Realities of Bird Life*, which became the most visible source of Selous's thinking for ornithologists after 1927. Huxley also did his best to get Eliot Howard to put more of Howard's ornithological observations into print, not only prodding him directly but also exhorting Howard's closest scientific friend, C. Lloyd Morgan, to spur Howard on to publication.[72] At the same time, Huxley tried to wean Howard away from issues that seemed to Huxley to be unpromising. Feeling that Howard was spending too much time with the question of whether birds have "memory images" or not, Huxley wrote Howard, "I am hoping you will content yourself with stating the pros and cons and frankly saying it is not ripe for decision and then getting down to recording the quite invaluable facts which you and you alone possess. It is not fair on us outsiders to withhold them. Frankly I am looking forward to see you pouring out book after book of straightforward observation reserving a full interpretation . . . for the close." He also urged Howard to provide him with an account of some experiments Howard had done, in order "to show the lab-workers what field naturalists can do."[73]

With regard to Huxley's encouragement of younger men who became the leading scientific birdwatchers of the 1930s and afterward, mention should be made in particular of David Lack, Konrad Lorenz, and Niko Tinbergen. Huxley played a major role in finding Lack jobs. In Huxley's correspondence we find Lack thanking Huxley for having given him "encouragement in my bird watching at a time when official zoology at Cambridge merely frowned on it," and who told Huxley, "You were my only effective biology teacher. . . ."[74]

Huxley first met Konrad Lorenz in 1934, when the Eighth International Ornithological Congress was held in Oxford. What Lorenz remembered from this meeting was not so much the paper that Huxley delivered there but rather Huxley's great generosity in lending him a large trunkful of reprints and books on animal behavior. It was at this time that Lorenz first became acquainted with Huxley's paper on the great crested grebe. Later Huxley helped Lorenz in a variety of ways: sending him books after the war, providing him with critiques of his writings (beginning in 1948 with the introduction to Lorenz's big book on ethology that did not get finished for thirty more years), writing a preface to the English edition of *King Solomon's Ring*, encouraging Lorenz to contribute volumes on jackdaws and gray lag geese to the *New Naturalist* monograph series, and writing letters of recommendation—as, for example, to the Max Planck Society—as Lorenz sought to find a position where he could do his research.[75]

Niko Tinbergen, too, felt greatly indebted to Huxley for help and encouragement. As indicated above, Huxley's papers on display behavior in birds were read with great enthusiasm by the young Dutch ornithologists who were attracted to the study of bird behavior in nature. Tinbergen, as a young naturalist, was particularly impressed not only by the detailed observations Huxley offered in his papers but also by Huxley's attention to theory, especially in his papers on the great crested grebe and the red-throated diver. Interestingly enough, however, Huxley also provided Tinbergen with an additional sort of stimulus in 1930, when he lectured in Amsterdam at the Seventh International Ornithological Congress. From his own observations on terns the previous spring, Tinbergen recognized that what Huxley had to say about the male tern feeding the female at the nest was inaccurate. This helped inspire Tinbergen to publish his own observations in the first substantial scientific paper of his career, "On the Mating Biology of the Common Tern," which appeared in the Dutch journal *Ardea* in 1931.[76] The thrust of Tinbergen's paper, nonetheless, was quite supportive of Huxley's thinking. To Tinbergen it appeared that *Sterna hirundo hirundo* provided a good example of what Huxley called mutual courtship in the premating period. In this and later papers Tinbergen paid careful attention to Huxley's ideas.

In 1940, as Tinbergen began to elaborate the concept of "substitute activities" (later "displacement activities"), he wrote to Huxley about ritualization in bird behavior, and also about the possibility of establishing relations with the Institute for the Study of Animal Behavior.[77] Later, at the end of the war, it was to Huxley that Tinbergen wrote about the need to rebuild international ties among students of animal behavior.[78] In the 1950s, he thanked Huxley for his help in identifying sources of research funds. As indicated above, Tinbergen appreciated Huxley's ideas on the interrelations of functional systems in behavior. In addition he was aware of the value of Huxley's example as a popularizer of science, an example that seems to have encouraged Tinbergen in similar efforts of his own, both in writing and in filmmaking. In 1967, after devoting considerable effort to making a film on the "language" of the lesser black-backed gull, Tinbergen recounted to Huxley how he had been "completely bowled over" by Huxley's gannet film when he first saw it in 1934, and how the thought of this film had given him encouragement in his own filmmaking, "from the knowledge that even you had thought it worthwhile to give [full] energy, time, and thought to doing such a job as well as you could possibly do it."[79] Tinbergen was not prepared, however, to follow Huxley in all of his enthusiasms. When Huxley urged Tinbergen and other ethologists to consider more closely the *subjective* correlates of animal behavior, Tinbergen resisted. In his own efforts to make the study of animal behavior more scientific, Tinbergen had stressed the importance of taking a more objectivistic and physiological approach.[80]

Following Huxley's eightieth birthday, Tinbergen took the occasion to express to Huxley what men of Tinbergen's generation and interests owed to Huxley, thanking him not only for the moral courage and the penetration of his early work on behavior but also for his encouragement and stimulus of young people and his promotion of "intelligent understanding of the general problems of evolution and of Man as both a product and a carrier of evolution, with a mission with respect to his own destiny."[81]

On the basis of this necessarily abbreviated survey of Huxley's contributions to the study of animal behavior, what can we conclude about Huxley's role in the rise of ethology? The answer, I believe, is that although Huxley played a significant role in the study of animal behavior, it was not as great a one as he might have played, nor was it as decisive a role as he suggested in his autobiography. While his love of birdwatching and his concern with the problems of Darwinian evolution led him to produce some excellent field studies of bird behavior, he seems to have felt from the outset that this was not the most prestigious area in which to make a contribution to biology. He started his career hoping to make "big, new, original" discoveries in biology. He undoubtedly recognized soon enough that in addition to not having the prestige of laboratory work, field studies of behavior did not represent an area in which novel and major contributions could be made in a hurry. As Selous once complained, what generally befell the field naturalist interested in behavior was "many a weary wandering, many an hour's waiting . . . to see, and seeing nothing. . . ."[82] To Huxley, furthermore, it must have seemed that every time he made some fresh field observations, Selous had already anticipated him, not only in the observations themselves but also in their interpretation. Selous had chosen to devote his personal resources to watching birds almost fulltime. Huxley, with his professional aspirations and responsibilities, did not have this leisure. Huxley wrote to Eliot Howard in 1922, "Alas—I fear that in term-time I get less than no time to think over bird problems—I am full up with pupils, committees, lectures, & other research."[83]

Huxley's major papers on bird courtship, written over a fourteen-year period, were in fact the product of a total of at most forty to fifty days of field work—and partial days at that—mostly snatched from vacation time. As a teacher at Oxford, Huxley developed a course of lectures on animal behavior. In the early 1920s, some of his students, as members of the Oxford Ornithological Society, made field observations that Huxley incorporated in two short papers he published in British Birds.[84] It does not seem, however, that Huxley made field work a significant part of his students' formal instruction. Significantly, and in contrast, when Tinbergen went to Oxford a generation later, he concluded after several years there that "it is absolutely necessary for me to spend many weeks with my D. phil. people in the field during the first season, in order to train them while together kind-of-wrestling with the object and its problems. . . . Because in behaviour it proves so very difficult to acquire the 'biological' outlook (much more difficult than in most branches of physiology for instance), the training takes such a long time; only exceptional people grasp the type of approach within a year, and most of them need two years to develop it."[85]

Huxley, I would suggest, is best described not so much as a founder of ethology, but rather, as David Lack put it, as a bridger of the gap between Selous and Howard early in the century and the scientific birdwatchers of the 1930s. Huxley's primary accomplishment in this area was to make the contributions of writers like Selous and Howard more accessible to the scientific community. The cantankerous Selous had thrust his observations before the public in the form of scarcely digested field notes. Huxley, in contrast, was a master at organization, synthesis, and presentation. He worked to provide a coherent biological framework for the material and published papers in the more prestigious of the professional scientific journals, notably the journals of the Zoological Society and the Linnean Society of London—institutions with which Selous refused to have any association.[86] Warde Fowler identified Huxley's role and talents very well in 1916 when he congratulated him on the content and the literary quality of his recent scientific papers, saying: "Darwin never could be dull, & so it is with you. Selous & Howard are both rather heavily conscientious, but now someone has arisen to turn a search-light on their tracks."[87]

Huxley could in truth have played a much larger role in the history of ethology had he only written his book on bird courtship. If in terms of his field observations, methods, and individual ideas his behavioral work was largely continuous with the work of the English amateurs, he was still capable of much that Selous and Howard were not. He had both the training and breadth of perspective necessary to place behavioral studies squarely in a broad biological context. He displayed this in his paper in 1916 on birdwatching and biological science, where he identified the different points of view the evolutionist, the physiologist, and the psychologist brought to behavior studies, and where he observed how so many disputes in biology were due to a failure to distinguish between ultimate causes, immediate causes, and "mere necessary machinery."[88] He showed this again in the 1920s, both in 1924 in the special series of lectures he delivered at Rice Institute on "the outlook in biology" and 1925 in the introductory comments he sketched out for his proposed book on bird courtship.[89] The latter, had he completed it, would have been a singular contribution to the study of animal behavior. Neither Charles Otis Whitman nor Oskar Heinroth, the other professional zoologist interested in bird behavior in the early years of the century, was a field zoologist, however much each of them insisted on observing animal behavior unfold under natural rather than laboratory conditions. Furthermore, Whitman died without ever assembling his unrivaled knowledge of the behavior of pigeons, and Heinroth, though he described in detail the instincts of the different birds he and his wife reared by hand, was disinclined to be a generalist.[90] Huxley, in 1925, believed that the time had come to gather the data from "field observation, animal psychology & behavior, genetics, & comparative psychology . . . [and consider] the problem [of behavior] from a truly broad & unitary biological standpoint."[91] He failed, however, to carry through on the project. What is more, in the broad, synthetic book that he eventually did write, *Evolution, the Modern Synthesis*, he neither made behavior part of the synthesis nor offered guidelines to suggest how that might be accomplished.

Edmund Selous once wrote, "The habits of animals are really as scientific as their anatomies, and professors of them, when once made, would be as good as their brothers."[92] Huxley was one of the leading spokesmen for field studies of animal behavior in the early twentieth century, but he never aspired to be a professor of animal habits. He did not make the ultimate commitment of devoting his career to demonstrating the significance of such work. His interests and ambitions were too diverse. He acknowledged this in introducing his own autobiography, when he wrote: "I have been accused of dissipating my energies in too many directions, yet it was assuredly this diversity of interests which made me what I am."[93]

In recent years, historians of science have developed an increasingly rich sense of science as a social process. Through the influence of the sociology and anthropology of science, they have come to speak of scientists as actors and to appreciate that a scientist's success as an actor involves in no small measure his or her ability to induce others to take up the same role. Perhaps the most salient feature of Huxley's example as an actor in the study of animal behavior was that, with respect to the whole of his own career, he never treated his activities as a student of behavior as more than a bit part. It would be another generation before Selous's words were borne out—not in the career of Huxley but in the careers of Konrad Lorenz and Niko Tinbergen.[94]

JOHN R. DURANT

The Tension at the Heart of Huxley's Evolutionary Ethology

Professor Burkhardt's authoritative review of Huxley's role in the rise of ethology addresses at least three major issues: first, the wider context of Huxley's behavioral work; second, the principal theoretical concerns of Huxley's bird-courtship studies in the period 1912–1925; and third, the major contributions that Huxley made to the emergence of classical ethology in the 1930s and 1940s. Under the first heading, Burkhardt demonstrates Huxley's indebtedness to the amateur ornithologists Warde Fowler (whose influence was new to me), H. Eliot Howard, and Edmund Selous; under the second, he brings out the distinctively Darwinian preoccupations of Huxley's bird-courtship studies; and under the third, he emphasizes the importance of Huxley's personal role as a bridge between the amateur ornithologists of the turn of the century and the first generation of classical ethologists.

In this commentary I shall concentrate on the first two of these topics, namely the wider context of Huxley's behavioral work and the principal theoretical concerns of his studies on bird courtship. My aim is to render more comprehensible several striking and superficially unrelated features of Huxley's thought to which Burkhardt has drawn our attention: first, his anthropomorphic and psychologistic method of behavior study; second, his preoccupation with evolutionary arguments concerning the good of the species; and third, his concern with the relationship between natural selection and sexual selection. In relation to each of these features, I shall argue that Huxley's work may be understood in terms of a fundamental tension at the heart of his evolutionary philosophy of nature and that the existence of this tension suggests ways in which we might gain further insight into his biological thought.

There can be no doubt that Huxley was greatly influenced by a number of important but hitherto largely neglected late-Victorian and Edwardian amateur ornithologists. As Burkhardt clearly demonstrates, these ornithologists combined the most detailed and painstaking field observations of bird habits with the ambitious attempt to explain these habits in terms of the Darwinian theories of natural and sexual selection. In addition, however, it is worth reminding ourselves that the spirit in which these men generally set about their work was profoundly anthropomorphic. Edmund Selous, for example, is best described as an animal biographer. His observational diaries are a fascinating record of the attempt to gain an understanding of birds by empathizing with them to the point that the observer sees the world from his subjects' point of view.

Exactly the same qualities are to be found in Huxley's early behavioral writings. Huxley, too, was concerned to capture the most intimate details of his birds' lives (especially their sexual lives); and he, too, was willing to exploit his considerable feelings of empathy toward his subjects in order to gain greater insight into both the causes and the consequences of their behavior. This is clearly seen in Huxley's most famous behavioral paper, which continually draws upon his own and other people's experience in an attempt to understand the "courtship habits" of the great crested grebe. For example, he attempts a hypothetical account of the physiological processes underlying the mutual displays of grebe courtship; but having done so, he continues:

> This merely indicates the possible material mechanism; of the actual, we know next to nothing. However, by comparing the actions of the birds with our own in circumstances as similar as possible, we can deduce the birds' emotions with much more probability of accuracy that we can possibly have about their nervous processes: that is to say, we can interpret the facts psychologically better than we can physiologically. I shall

therefore (without begging any questions whatever) interpret processes of cause and effect in terms of mind whenever it suits my purpose so to do—which, as I just said, will be more often than not.

Let us take the parallel from human affairs. Far be it from me to go into the matter with a heavy hand; let us merely look at a few familiar facts in an unfamiliar biological light. . . .[1]

Selous's bird diaries and Huxley's bird-courtship papers appear to have been part of a larger and distinctively Edwardian genre of anthropomorphic animal biography. In popular culture, the genre was represented by short stories that combined the purveying of facts about animal habits with a considerable amount of moralizing about the human condition. The historian of popular science, Peter Broks, has shown that in the period 1900–1908 a single English popular magazine published no fewer than forty-seven such stories, with titles ranging from "The Autobiography of a Partridge" to "The Biography of a Bat." For one biographer, the habits of the beaver reflected a providential social harmony; for another, they represented a Darwinian struggle for survival. In each case, however, the beaver's biography was a ready source of illustrations of the author's preferred visions of the social order. Broks suggests that the narrative, story-telling form was particularly suited to this sort of moralizing, since it entailed a close identification of the narrator (and, hence, the reader) with animals.[2]

Selous and Huxley aspired to a great deal more than popular moralizing; however, their observational studies were an integral part of this literary genre. Selous wrote a number of romantic wild-life books for general audiences, and these (like his bird diaries) contained a mixture of factual and moralistic material. In *Bird Life Glimpses* (1905), for example, we read of nightjars, not only that their domestic habits are "very pretty and interesting," but also that their married life is "exemplary."[3] Similarly, Huxley was always ready to apply the facts of bird courtship to human affairs. In the grebe paper, we read that "flirtation" between one member of a pair and a third bird leads the deserted partner to drive the usurper away: "Thus all the anger of jealousy," Huxley observes, "is directed against the usurper, not against the mate—which again is very human!"[4] In a BBC Home Service radio broadcast about great crested grebes many years later, Huxley confessed that he "felt pleased at having laid bare some of the bases of human jealousy and human affection."[5]

Culturally, Edwardian animal biography was a favored source of familiar social stereotypes. Scientifically, it was a fertile source of new insights into the nature and causes of animal behavior. As I have argued elsewhere, the casting of birds as actors, even morally responsible actors, in their own social worlds did a great deal to foster the kind of observation and interpretation on which classical ethology was based.[6] For Selous, the brute beast was "a more intelligent, more emotional, more affectionate and generally fuller-feeling being that he has yet been acknowledged to be."[7] Similarly, for Huxley, while birds were not necessarily intelligent, they were emotionally complex and colorful characters; watching them was to enter into what he termed "a series of personal encounters."[8] Anthropomorphism set the tone for Huxley's observational accounts, generated many of his explanatory hypotheses, and underlay virtually all of his extrapolations from animal to human behavior.

Huxley's anthropomorphism was closely related to his explicitly psychologistic approach to behavior. Consistently, Huxley speculated about the mental lives of birds: courtship was an "interplay of mind" between two animals; its function was to express and to provoke "emotional excitement"; and so on. For Huxley, these were no mere metaphors; rather, they were literal statements of his belief that birds possessed minds of the same general character as our own. In retrospect, Huxley's papers appear full of fragmentary ideas that were to find mature expression in the work of Lorenz and Tinbergen. However, this wisdom of hindsight fails to do justice to the true distance between Huxley and classical ethology. To get from Huxley to Lorenz and Tinbergen it is necessary not merely that rigor and system should be added to occasional insight; rather, as Burkhardt notes, it is necessary that psychologism should give way to mechanism. Huxley readily adopted many of the ideas of the classical ethologists, but he never abandoned his interest in animal mind.

Partly, at least, Huxley's continuing interest in animal mind reflected the nature of his evolutionary world view. For Huxley, evolution was no mere biological fact; rather, it was a central philosophical principle applying equally in the domains of the physical, the biolog-

ical, the mental, and the social. Most of Huxley's intellectual endeavors were embraced within a monistic evolutionary philosophy
according to which matter, life, and mind were various expressions of
a single dynamic world-stuff. "Unity, uniformity, and development"
were the "three great principles" of Huxley's philosophy,[9] and they
demanded a definite stand on the question of animal mind: "[I]f we
are to believe in the principle of uniformity at all," Huxley wrote,
"we must ascribe emotion to animals as well as to men: the similarity
of behavior is so great that to assert the absence of a whole class of
phenomena in one case, its presence in the other, is to make scientific
reasoning a farce." Significantly, this argument appears in an essay
entitled, "Ils N'Ont Que De L'Ame: An Essay on Bird Mind."[10]

As the historian of science Colin Divall has noted, Huxley's vision
of the world was a curious combination of elements materialist and
idealist, rationalist and romantic, secular and religious.[11] Nowhere
was this mixture of elements more apparent than in Huxley's view of
the evolutionary process itself. For as Burkhardt notes, Huxley's
biological outlook was from the very outset thoroughly Darwinian;
but at the same time, and from an equally early stage, it was rooted in
a form of scientific humanism that looked to the evolutionary process
as a source of moral principles and even of spiritual inspiration. As a
Darwinist, Huxley was quite clear about the fact that the evolutionary process is utterly without purpose; but as a scientific humanist,
he was equally insistent that the evolutionary process is both materially and morally progressive. Huxley's problem was to discover
how, exactly, both of these things could be true. The situation may be
represented schematically as follows:

This diagram merely illustrates the ideological tension between purposelessness and progress at the heart of Huxley's vision of the
evolutionary process. I believe that this tension influenced several
different aspects of Huxley's evolutionary thought. In some cases the

influence was glaringly obvious; in others, however, it was rather more subtle.

Taking the most striking example first, it is quite clear that Huxley's most famous contribution to the evolutionary synthesis was in part an attempt to unite Darwinian science with humanist ideology. The inspiration for *Evolution, the Modern Synthesis* (1942) was an address that Huxley had delivered in 1936 on "Natural Selection and Human Progress." In the course of this address, Huxley noted the "common fallacy that natural selection must always be for the good of the species or of life in general. In actual fact we find that intraspecific selection frequently leads to results which are mainly or wholly useless to the species as a whole . . . [and] . . . may even lead to deleterious results."[12] On the face of it, nothing could possibly be clearer than this. Yet Huxley went straight on to reassert the fact of evolutionary progress; and astonishingly he concluded that, "if we cannot discover a purpose in evolution, we can at least discern a direction in the line of evolutionary progress. And this past direction can serve as a guide in formulating our purpose for the future."[13]

This precarious balancing act between purposelessness and progress was carried forward in *Evolution, the Modern Synthesis*. Despite its title, this book operated with two quite different and essentially unharmonized concepts of evolution: the one, a Darwinian conception rooted in theoretical population genetics; and the other, a humanistic conception rooted in the idea of progress. On the whole, the first of these conceptions prevailed in Huxley's discussion of small-scale evolutionary change, but the second asserted itself whenever he turned his attention to longer-term evolutionary trends. A key transition occurred halfway through the book, when quite without warning Huxley declared evolution to be "the process by which the utilization of the earth's resources by living matter is rendered progressively more efficient."[14] In his later evolutionary writings Huxley returned again and again to this fundamentally non-Darwinian definition as a way of incorporating the idea of progress into the heart of Darwinian evolutionary theory.[15]

A second example of the influence of ideological tension is provided by Burkhardt's astute observation that Huxley had a lifelong predilection for evolutionary arguments couched in terms of the good of the species. Many contemporary evolutionary biologists assert that the individual organism (as opposed to the social group or even the species) is the fundamental unit of selection. Moreover, they commonly suppose that in doing so they are merely reiterating the traditional position of both Darwin himself and the founders of the "modern synthesis." Leaving aside Darwin's views on the matter, we should note that many of the founders of the modern synthesis were more pluralistic in their views about this and other important theoretical issues than are their modern supporters. In this and other respects, as Stephen Jay Gould has noted, there appears to have been a noticeable "hardening of the modern synthesis" since the 1940s and 1950s.[16]

Certainly, Julian Huxley was a pluralist about evolutionary mechanisms. "If Darwin were writing today," he observed in *Evolution, the Modern Synthesis*, "he would call his great book *The Origins*, not the *Origin of Species*." Consistent with his generally pluralist outlook was Huxley's position on the question of the units of selection; for Huxley accepted a role for both inter-individual and inter-group selection. Turning once again to *Evolution, the Modern Synthesis*, we find that he specifically endorsed Sewall Wright's concept of "intergroup selection," and indeed extended the idea to include what he termed "social selection." Citing the work of the American student of animal cooperation W. C. Allee, Huxley noted that the bases for social life in animals were "deep and widespread, since what he termed "the aggregation itself" had become "a target for selection."[17]

The above remarks occur in a section of *Evolution, the Modern Synthesis* entitled "Adaptation and Selection Not Necessarily Beneficial to the Species." As a Darwinist, Huxley was well aware of the fact that natural selection did not necessarily produce biological improvement. In particular, he knew that selection acting at the level of the individual might favor characteristics that were positively harmful at the level of the group or the species. As a humanist, however, Huxley was anxious to show that as a matter of empirical fact selection does bring about the good of the species at least often enough to secure overall improvement in the long term. To establish this point, Huxley repeatedly invoked two arguments: the coincidence of individual and species interests; and group selection. Group selection, in particular, was crucial in the evolution of precisely those features of human societies that Huxley regarded as most progressive. "[W]e must distinguish clearly between the different ways in which progress may be operating in man," Huxley wrote: "In the first place it can appear . . . in the organization of the communities to which he belongs and on which natural selection seems mainly to act."[18]

For Huxley, the *locus classicus* of individual selection acting to the detriment of the species was Darwin's mechanism of sexual selection by female choice. Darwin had suggested that, by exercising preferences for particular kinds of mates, the females in some species had transformed the superficial appearance of males in precisely the same way that pigeon fanciers had transformed the superficial appearance of domesticated pigeons. Darwin had invoked female choice to explain the elaborate and often bizarre ornamentation of many males; in the case of a cumbersome contrivance such as the peacock's train, as Huxley himself observed, "a balance will eventually be struck at which the favourable effects slightly outweigh the unfavourable; but here again extinction may be the fate of such precariously-balanced organisms if the conditions change too rapidly."[19]

The theory of sexual selection by female choice was not generally well received in the late nineteenth century. Indeed, when Edmund Selous set out to demonstrate the fact of female choice in his studies of bird courtship, he perceived himself to be fighting a rearguard action against the consensus of expert opinion.[20] As Burkhardt shows, it was partly under Selous's influence that Huxley became interested in sexual selection. Yet Huxley's views on Darwin's theory were complex and not a little confusing. To begin with, as Burkhardt observes, Huxley appears to have accepted the Darwinian view more or less at face value; but in successive publications on bird courtship he rapidly downplayed the importance of female choice. Indeed, in two reviews of the whole question of sexual selection in 1938, Huxley gave Darwin fulsome credit for developing a principle the significance of which he then proceeded to minimize and marginalize at almost every turn.[21]

Why was Huxley both fascinated by and yet at the same time keen to downplay the evolutionary importance of sexual selection? Again, I believe that part of the answer has to do with his search for a coherent evolutionary philosophy of natural and social progress. I have already noted that, for Huxley, sexual selection was a graphic example of the way in which individual selection might generate characteristics positively detrimental to the well-being of the species. Significantly, Huxley's twin reviews of 1938 gave great prominence to the question of how far sexual selection may be expected to generate characteristics inimical to the welfare of the species. First, Huxley argued with great clarity that in certain cases (such as highly polygamous birds) intense sexual selection might indeed generate characteristics disadvantageous in the struggle for existence:

> Inter-specific [or species] selection obviously must promote the biological advantage of the species. Intra-specific [or individual] selection, on the other hand, though it may also act thus, may in certain circumstances favour the evolution of characters which are useless or even deleterious to the species as a whole . . . [For example] inter-male competition may promote specific as well as individual advantage, e.g. with most secondary sexual characters concerned with display and threat in monogamous species. Occasionally, however, characters useless to the species may be promoted even in such forms, e.g., the earlier appearance of males on the breeding territories in monogamous birds, notably those where territory is of biological importance (see Howard [14, 15]). With simultaneous arrival of the sexes, the available territory would doubtless all be taken up, so that total reproduction would be unaffected; but an early male will be more likely to secure territory than a late arrival. Male arrival is thus pushed back as far as possible, the limit being decided by food supply and climatic factors. . . .[22]

In this passage Huxley acknowledged that sexual selection could operate in direct opposition to natural selection, promoting the evolution of characters of direct disadvantage to individuals and species in the struggle for existence. The awkward thing about this process from Huxley's point of view, however, was that if this were to happen on a large scale, it would necessarily undermine any claims concerning the generally progressive quality of the evolutionary process. For this reason, I believe that Huxley may have been intellectually (or ideologically) predisposed to limit the evolutionary significance of female choice. This he did in two ways: first, he argued that many of the features that Darwin had attributed to sexual selection were in reality the result of natural selection for characteristics advantageous in the struggle for existence; and second, he claimed that even where Darwinian sexual selection was indeed operative, it frequently worked in the same direction as natural selection—that is, in favor of genuinely beneficial characteristics. Thus, he concluded a lengthy discussion of sexual and natural selection:

> We must, finally, discuss the selective implications of the facts. It is clear that Darwin's original contention will not hold. Many of the characters which he considered to owe their evolution to sexual selection do have value to the species in the general struggle for existence, and not merely in the struggle between males for reproduction. Broadly speaking, sexual selection is merely an aspect of natural selection, which owes its peculiarities to the fact that it is concerned with characters which subserve mating (epigamic characters), are usually sex-limited, and are often of allaesthetic type.[23]

Clearly, Huxley never rejected sexual selection. Instead, however, he relegated it to the ranks of minor and essentially insignificant evolutionary mechanisms. Given that this was the view taken by the one man among them who more than any other had made a special study of the subject, it is perhaps not surprising that the founders of the modern synthesis should have been conspicuous in their neglect of the Darwinian theory of sexual selection by female choice.

I have said enough to indicate, in broad outline at least, the way in which I believe the tension between Darwinian purposelessness and humanistic progression worked itself out in Huxley's evolutionary and behavioral biology. If I am right, then Huxley's views on theoretical issues as weighty as the definition of Darwinian fitness, the units of selection, and the status of sexual selection were all colored to some extent by his efforts to reconcile hardnosed Darwinism in science with the rosiest of progressivist ideologies. This being the case (and I claim only that the evidence is highly suggestive), then it will be worthwhile exploring other aspects of Huxley's evolutionary thought for similar signs of intellectual tension. A systematic analysis of Huxley's untiring search for a definitive biological definition of progress should yield interesting results; as should an approach from the opposite direction, via his voluminous writings on social evolution.

It may be appropriate to conclude with a thought that is intended as a backhanded compliment to the man we honor in this book. Arguably, the most remarkable and impressive thing about Huxley's intellectual career is that, over more than half a century of work around a single issue, he never succumbed to the temptation to give in to either of two obvious philosophical temptations, each of which (in principle, at least) might have resolved his intellectual dilemma: teleology, which by making evolution purposive would also have rendered it unproblematically progressive; and dualism, which by divorcing nature from human affairs would have made the purposelessness of evolution totally irrelevant to any alleged (or hoped for) progress in society. Unwilling to contemplate either of these possibilities, Huxley simply forged ahead with his quest for a "Darwinian credo" with which to replace what he took to be the discredited dogmas of conventional religion. This was always a sadly unpromising philosophical adventure, and Huxley fared no better or worse in it than anyone else. The theory of evolution by natural selection is a rich source of insights into organic origins, but it is a poor guide to the living of a human life, let alone to the conduct of a society. Huxley never accepted this, surely the most substantial point to emerge from a critical history of evolutionary thought; but at least he may be admired for having lived with the philosophical difficulties that flowed from this basic refusal without doing overt violence either to his science or to his most deeply held values.

SOLLY ZUCKERMAN

Comments and Recollections

I would like to begin by placing Julian Huxley within the framework of the social and intellectual world of London of which he was part, particularly during the 1920s and 1930s. I do so deliberately because ornithology, the field study of birds, was of little interest to most of the people with whom he then associated. In 1929 and 1930, when it ceased publication because of the Depression, there existed a journal of scientific humanism called the *Realist*. You can see where Julian Huxley stood by casting an eye down the list of contributors to the eight numbers of the journal that appeared. Among them were the two Huxleys, J. B. S. Haldane, Freud, and Bertrand Russell. There was barely a person who then mattered intellectually who was not among the contributors. But apart from Julian Huxley, I would say that not more than two or three were interested in field studies of birds. This group of intellectuals included some of austere intellect, such as Haldane and Lancelot Hogben.[1] Another member of Julian's world at the time was J. D. Bernal, a man with an extremely vigorous mind who had a far more embracing picture of the development of science than had Julian. The question now is, how important was the study of what we now call ethology in Julian's development?

Professor Burkhardt[2] has estimated that all of Julian's field work took up little more than fifty days. But fifty days presumably did not mean twenty-four hours a day; let us assume that it was three hours a day. But if all of us looked at birds for three hours a day over the course of our lives, we would still not have made observations of comparable significance to those that Julian did. It is the fact that he was able to make the kind of observations that he did that really distinguished him from other birdwatchers. We must remember, as I have said, that the scientific circle in which he moved was not really interested in what was, after all, merely a hobby, and a common one in England.

Professor Burkhardt has referred to Julian's claim that he made field studies of natural history scientifically respectable, and he has said that this is perhaps a plausible claim. But the question is, was it not always respectable? Who said that Edmund Selous and Eliot Howard were not respectable observers? What does the word "scientific" mean here? Scientifically respectable? As far as I am concerned, the best field observers did not have to have university degrees, and they certainly had not been through university departments of what we now style ethology. There are mole-catchers in my village who know more about the behavior of moles than any scientist. I would like to pose the question: does a general interest in animal behavior become more respectable scientifically when people call it "ethology"?

In those early days, Julian was not much interested in philosophy; he was interested in observation. Professor Burkhardt says that in his advice to Howard, Julian told him not to spend time worrying about birds having memory images. He said, rest content by telling us your observations. But Julian also urged Tinbergen, another ethologist, to consider more closely the subjective correlates of animal behavior. Tinbergen resisted, and I would say that it was wise of him to resist, because following that advice would have taken Tinbergen back to the kind of anthropomorphic animal behavior studies that were so common in the nineteenth century.

Insofar as Huxley became a philosopher, I would have called him a vitalist; if not a vitalist, a holist, or any similar term that was current in those days. The curious thing for me was that he was not interested (or if he was, he never showed it) in that sequence of great names that marked the development of empiricist philosophy in the United Kingdom, starting with David Hume and going on to Bertrand Russell and then to Karl Popper today. When I first met Julian, a philosophical debate was raging about the concept of holism, which was being promoted by amateur philosophers such as Lord Samuel, General Smuts and J. B. S. Haldane's father, J. S. Haldane, as well as by Kafka with his gestalt philosophy. The argument was between vitalism, or holism, or gestalt, and behaviorism and mechanism. Julian kept aloof, as we have been told, enjoying the pleasures of observation. He introduced his crested grebe paper with the statement, "This is all good fun, enjoy it while you can."[3] He was not interested in (at least he never referred to) the analytical studies of behavior—for instance, the kind of work that Rowan in Alberta was doing on the part played by sex hormones in bird migration.[4] And unlike J. B. Watson, he was very much against the idea of explaining bird behavior in terms of conditioned reflexes.

This brings me to the last point that I wish to take up: namely, what has been said about Julian Huxley having founded an Institute for the Study of Animal Behaviour. The fact is that no such institute was ever founded. There was never a hope that it would be, and Julian never set out a plan to show what such an institute might do. Here is what happened.

Apart from a few studies of animal behavior that I had started in Oxford, in those days the only other group working on the subject in the United Kingdom were members of a single department at Cambridge. In the whole country there was no proper facility where careful studies of animal behavior could be carried out. Since Julian was then secretary of the Zoological Society, the idea was that we would create a facility in the London zoo where we could carry out truly scientific studies of animal behavior. I arranged for two of my pupils to work there and to carry out experiments based on the idea that animal behavior could be dealt with objectively and not in terms of parallels with human behavior. I was concerned with the Cartesian separation of mind and body. What I wanted to do was see whether or not a means could be found whereby one could determine whether the capacity to make truly verbal and numerical abstractions was unique to the human mind. The abstraction I chose was "number," and I designed an experiment (which I discussed over a weekend with Bertrand Russell) in such a way as to ensure that if the experiment came out one way, the answer was yes, and if it came out the other, one could dismiss the idea that animals are able to make numerical abstractions (as opposed to reacting to the concrete differences in number). The students performed the experiments in a couple of small, hot, animal houses in the zoo. That was all there was to the so-called "Institute of Animal Behaviour." Julian was interested in my experiment, but he did not build an "Institute" in the London Zoological Gardens.

What Julian did build was a studio for painters and sculptors. Maybe he was wiser to do that than to build an institute for the study of animal behavior. He had a deep aesthetic sense, and through the zoo he came to know several painters and sculptors. One was Henry Moore. He gave Henry the skull of an elephant, not to show the difference between the skull of an elephant and that of a rhinoceros, but to show the beauty of different parts of the skull. The result is the brilliant series of drawings of the skull that Moore produced.

I believe that somehow or other Julian had a unique capacity to see what others had not seen before. Once, during a visit to my house, he came down to breakfast and said: "That's a very fine eighteenth-century picture you have looking north towards where the zoo is today." I said, "I have no such picture," to which he replied, "Yes, you have." I insisted: "I know every picture in the guest room. I should know them." So he brought the picture down and showed me that even though I had known the print for years, it was he who had noticed something that I had failed to see.[5]

WILLIAM B. PROVINE

Progress in Evolution and Meaning in Life

Progress might have been all right once, but it's gone on too long.
—Ogden Nash

If the title of this paper seems overly ambitious and pretentious, consider the more accurate (but unwieldy) one: "Progress in evolution, the modern synthesis in evolution, the foundation of ethics, and meaning in life." Julian Huxley addressed all of these ponderous issues in the final chapter of *Evolution, the Modern Synthesis*[1] (1942), which gave the name to the "evolutionary synthesis" of the 1930s and 1940s.[2] Huxley argued that a grand synthesis had occurred in modern evolutionary biology, proving that the process of organic evolution in nature was purposeless but progressive. Humans were at the pinnacle of evolutionary progress; and although no purpose or direction from God was detectable in evolution, ethics could be based upon an understanding of evolutionary progress. Moreover, man's place in nature, at the pinnacle of the evolutionary process, gave deeper meaning to human existence than did anything else.

I will argue that Huxley's idea of progress in evolution is merely the imposition of his cultural values upon evolution, that the modern synthesis in evolution is scarcely a synthesis at all and should be renamed the evolutionary constriction.

Julian Huxley and Progress in Evolution

Evolutionary biology raises fundamental issues about human culture. Almost simultaneously with his invention of the mechanism of natural selection, Charles Darwin opened his M and N notebooks on man, mind, and materialism. (See especially the careful new critical edition of Darwin's notebooks in Barrett et al.[3]) In these he jotted down his musings on the cultural, psychological, and metaphysical implications of evolution by natural selection. Most of the major evolutionists of the nineteenth century wrote and lectured widely on the larger implications of evolution. The literature from the period 1859–1900 on the cultural implications of evolution is enormous. Julian Huxley's grandfather, Thomas Henry Huxley, was keenly interested in the implications of evolution, but he rejected the idea that ethics could be based upon the evolutionary process, which he considered to be totally amoral and frequently disgusting. He presented this argument forcefully in the second Romanes Lecture in 1893,[4] when Julian was six.

From the beginning of his scientific career, Julian Huxley always and unwaveringly believed that evolution was progressive and offered hope and meaning to human existence. His first book, *The Individual in the Animal Kingdom*,[5] glorified from beginning to end the evolutionary progress of the individual, from the primordial protozoan to the human. Huxley even argued that "the State" was an evolutionarily advanced individual. He said that evolution tended to produce brains: "It is noteworthy that the course of internal differentiation has over and over again—in worms, in insects, in crustacea, in spiders, in molluscs, and in vertebrates—tended in the same direction—towards the formation of a Brain."[6] The book concluded with this note:

> All roads lead to Rome: and even animal individuality throws a ray on human problems. The ideals of active harmony and mutual aid as the best means to power and progress; the hope that springs from life's power of transforming the old or of casting it from her in favour of new; and the spur to effort in the knowledge that she does nothing lightly or without long struggle: these cannot but help to support and direct those men upon whom devolves the task of moulding and inspiring that unwieldiest individual—formless and blind to-day, but huge with possibility—the State.[7]

I think this passage helps to clarify Huxley's otherwise mysterious defense of the writings of Teilhard de Chardin when they finally began to be published in the late 1950s. Teilhard's notion of progressive evolution that led to the "noosphere" and later to the collective consciousness of the Omega Point[8] was congenial with Huxley's life-long beliefs about the progressive development of individuals in evolution.

Judging from his praise of Bergson[9] and of orthogenesis (directed evolution) two years later,[10] I suspect that in the early 1910s Huxley believed that evolution was not only progressive, but purposive. Later, he rejected any trace of purpose in evolution.

Evolution, the Modern Synthesis was an enormous compendium of research on evolutionary biology, mostly since 1900. The thrust of Huxley's argument in the book is easily seen in the first two paragraphs of chapter one:

> Evolution may lay claim to be considered the most central and the most important of the problems of biology. For an attack upon it we need facts and methods from every branch of the science—ecology, genetics, paleontology, geographical distribution, embryology, systematics, comparative anatomy—not to mention reinforcements from other disciplines such as geology, geography, and mathematics.
>
> Biology at the present time is embarking upon a phase of synthesis after a period in which new disciplines were taken up in turn and worked out in comparative isolation. Nowhere is this movement towards unification more likely to be valuable than in this many-sided topic of evolution; and already we are seeing the first-fruits in the re-animation of Darwinism.[11]

Not until the last chapter, "Evolutionary Progress," do the larger implications of the modern synthesis in evolution begin clearly to emerge. Huxley defined evolutionary progress "as consisting in a raising of the upper level of biological efficiency, this being defined as increased control over and independence of the environment. As an alternative, we might define it as a raising of the upper level of all-round functional efficiency and of harmony of internal adjustment."[12]

Nothing about this definition of evolutionary progress indicated that evolution was purposive. Indeed, Huxley stated that "the ordinary man, or at least the ordinary poet, philosopher, and theologian" always was anxious to find purpose in the evolutionary process.

> I believe this reasoning to be wholly false. The purpose manifested in evolution, whether in adaptation, specialization, or biological progress, is only apparent purpose. It is just as much a product of blind forces as is the

falling of a stone to earth or the ebb and flow of the tides. It is we who have read purpose into evolution, as earlier men projected will and emotion into inorganic phenomena like storm or earthquake. If we wish to work towards a purpose for the future of man, we must formulate that purpose ourselves. Purposes in life are made, not found.[13]

Without purpose of some kind, however, how could evolution serve as a guide to ethics or give a sense of meaning in human life? Huxley answered:

> But if we cannot discover a purpose in evolution, we can discern a direction—the line of evolutionary progress. And this past direction can serve as a guide in formulating our purpose for the future. Increase of control, increase of independence, increase of internal co-ordination; increase of knowledge, of means for coordinating knowledge, of elaborateness and intensity of feeling—those are trends of the most general order. If we do not continue them in the future, we cannot hope that we are in the main line of evolutionary progress any more than could a sea-urchin or a tapeworm.[14]

By his own criteria, most evolutionary change was not progressive. Indeed, Huxley concluded that of all animals, only humans retained potential for further progress. All existing single-celled organisms were automatically excluded from progressive evolution. "Only in the water have the molluscs achieved any great advance. The arthropods are not only hampered by their necessity for moulting; but their land representatives . . . are restricted by their tracheal respiration to very small size." Progressive evolution was impossible for cold-blooded animals; progress required both lungs and warm blood, "since only with a constant internal environment could the brain achieve stability and regularity for its finer functions." This left only the birds and mammals. "But birds were ruled out by their depriving themselves of potential hands in favour of actual wings, and perhaps also by the restriction of their size made necessary in the interests of flight." Huxley eliminated all mammals other than humans by arguments such as, "A horse or lion is armoured against progress by the very efficiency of its limbs and teeth and sense of smell: it is a limited piece of organic machinery."[15] Finally,

> The last step yet taken in evolutionary progress, and the only one to hold out the promise of unlimited (or indeed of any further) progress in the evolutionary future, is the degree of intelligence which involves true

speech and conceptual thought: and it is found exclusively in man . . . .
Conceptual thought is not merely found exclusively in man: it could not
have been evolved on earth except in man.[16]

With this view of evolutionary progress, Huxley came to the logical
conclusion: "Evolution is thus seen as a series of blind alleys,"[17]
except, of course, for the evolutionary avenue leading to humans.

Working from the foundation of the new synthesis in evolutionary
biology, Julian Huxley thus concluded, against the views of his grandfather, that organic evolution was progressive, and that evolutionary
progress provided the basis for ethics and meaning in life.

The Evolutionary Synthesis

What exactly was this "modern synthesis" that figured so prominently in Huxley's book? As a historian, I am immediately suspicious
when anyone describes his or her views as the "new" or "modern"
way of seeing things, to be sharply distinguished from the "old"
inferior ways. The "new" billing is often little more than scholarly
overstatement, an attempt to attract attention. Only two years earlier, Huxley had edited a volume that he titled *The New Systematics*.[18]
By all critical accounts then and now, this book had very little "new"
systematics in it. So it is fair to ask, what was this modern synthesis
in evolutionary biology.

There were about as many different versions of the evolutionary
synthesis as there were major evolutionary biologists associated with
it, augmented by a generous number of versions contributed by
younger evolutionists, historians, and philosophers.

There is a wonderful symmetry in the interpretations of the evolutionary synthesis by the biologists who participated in it. Each felt his
contribution to the synthesis was slighted and that he had to fight for
his proper place.

Ernst Mayr has said on many occasions that by the 1950s the role
of systematists and naturalists (his own work included) in the evolutionary synthesis had been terribly slighted by the geneticists, who
argued that the central figures of the evolutionary synthesis were
Fisher, Haldane, and Wright. Mayr began his campaign to focus
attention on the contributions of the systematists with his Darwin
centennial address, "Where Are We?" delivered at the Cold Spring

Harbor Symposium.[19] He organized his symposium on the evolutionary synthesis in 1974 precisely because he wanted to promote the role of systematists during the 1930s and 1940s. It may be difficult now to think that Mayr could have felt left out of the synthesis, but he did.

Sewall Wright wrote his strongly worded review of Mayr's "Where Are We?" address because he thought that Mayr was trying to read him and the other mathematical population geneticists out of the evolutionary synthesis.[20] Wright's feeling of exclusion had already been exacerbated by increasingly negative reactions to his belief in the importance of random genetic drift in evolution and by misunderstandings of his shifting balance theory of evolution.[21] Wright therefore believed he was left out of the synthesis by both geneticists and systematists. Much of the disagreement between Mayr and Wright concerns not so much differences in their views of evolutionary biology, but their differences in interpreting each other's role in the evolutionary synthesis.

Julian Huxley sought to gain his place in the evolutionary synthesis by writing *Evolution, the Modern Synthesis* and by defending this work on every possible occasion. He never revised the book, but reissued it as a new edition twice more with new introductions. In the introduction to the 1962 (second) edition, he clearly wanted the world to know that he was the real architect of the new synthesis (see also his address to the Golden Jubilee meeting of the Genetics Society of America in 1950). He never thought he was given proper credit for his role in the synthesis.

George Gaylord Simpson hesitated to come to Mayr's conference on the evolutionary synthesis, suggesting that the conference setting was not conducive to the elucidation of his role in the synthesis. After the conference, he was offended by the lack of appreciation of his contributions to the evolutionary synthesis, which he outlined more carefully is his autobiography, *Concession to the Improbable*,[22] and in his new introduction to the reprint of his *Tempo and Mode in Evolution*.[23] Simpson clearly believed that he was one of the truly major figures of the evolutionary synthesis, but that his role and that of paleontology had been slighted.

C. H. Waddington felt so slighted and left out of the evolutionary synthesis by Mayr and others that he published an entire book,

Evolution of an Evolutionist,[24] demonstrating the importance of his role in bringing embryology into the synthesis. G. Ledyard Stebbins has remarked on many occasions that the role of botany in the evolutionary synthesis has been overshadowed by zoological contributions, partly from the forcefulness of zoologists in taking their share of the credit, aided by the deep-seated animal chauvinism so ever-present in our culture. Among the other major figures of the evolutionary synthesis, Dobzhansky, Fisher, Ford, Goldschmidt, Darlington, Muller, Rensch, and Timofeef-Ressovsky all believed their work in the evolutionary synthesis was not, in hindsight, given the full deserved credit. The symmetry here is reminiscent of an academic department of top-notch scholars, each of whom believes that his or her work is not properly appreciated by the chairman and the rest of the department, no matter what rewards, accolades, and support are provided.

Among younger evolutionists, disagreement about the evolutionary synthesis is great. Niles Eldredge has written that the evolutionary synthesis did occur, but did not go nearly far enough. He argues that the hierarchical taxa at and above the level of species were not properly synthesized with other evolutionary processes in the 1930s and 1940s.[25] Stephen Jay Gould argues that the evolutionary synthesis hardened, in its later stages, into a pan-selectionist outlook that minimized the multiplicity of evolutionary processes, including biological constraints, nonadaptive mechanisms, extremely rapid speciation, and species level selection.[26] Motoo Kimura says that the evolutionary synthesis occurred and is the reason for the intensely negative reaction to his neutral theory of molecular evolution from many of the living architects of the synthesis.[27] Stebbins and Ayala[28] argued strongly that the evolutionary synthesis occurred and provides a robust basis for evolutionary biology today, a view also defended by Futuyma.[29]

Janis Antonovics, in his presidential address to the American Society of Naturalists in 1986, argued that the evolutionary synthesis did occur but hindered rather than promoted the advance of evolutionary biology.

> My thesis is that the Evolutionary Synthesis failed in many serious and insidious ways. I propose that the Synthesis had little direct impact on the progress of evolutionary biology as a discipline and that, at the conceptual

level, it may even have hindered rather than furthered our understanding of evolution. Many of the negative effects of the Synthesis have lasted to this day in terms of the institutional and conceptual structure of the field. I suggest that it is probably time that, rather than trying to finish the Synthesis as Eldredge (1985) has exhorted, we instead earnestly work to dismantle it. I suggest that only by achieving a Dys-Synthesis can we free ourselves of many of the methodological, conceptual, and even socio-religious difficulties that plague evolutionary biology.[30]

And finally, Mayr has just reevaluated the evolutionary synthesis in an essay titled, "On the Evolutionary Synthesis and After."[31] Here he argues that the recent critics of the evolutionary synthesis did not appreciate the firm foundation provided by the evolutionary synthesis, yet he also refers to the synthesis alternatively as a unification and as a consensus. Shortly, I will argue that this terminology is crucial.

One clear fact shines amid all this diversity of opinion about the evolutionary synthesis: all agree that *something important happened in evolutionary biology during the 1930s and 1940s.* Whatever it was had not happened by 1930, but had happened by the Darwin centennial in 1959. What exactly had happened? Can we characterize the evolutionary synthesis more precisely? I think we can.

Perhaps the easiest way to begin is by specifying what the evolutionary synthesis is not. First, it is scarcely a synthesis at all. According to Huxley and later to Mayr, many fields were part of the synthesis. Among these were ecology, genetics, paleontology, geographical distribution, embryology, systematics, comparative anatomy, and some mathematics. I can agree that there was a quantitative synthesis of Mendelian heredity and various factors that can change gene frequencies in populations. This was accomplished in the models of Fisher, Haldane, Wright, Hogben, Chetverikov, and others. Yet they disagreed, often intensely, about actual processes of evolution in nature, even when their models were mathematically equivalent. Beyond this genuine synthesis, the rest of the "evolutionary synthesis" mostly consisted of exercises in removing barriers, consistency arguments, and forging a consensus. A lot of hand waving was also involved, especially at the time Huxley coined the name "the modern synthesis."

I realize that Mayr and Shapere[32] and others have been willing to characterize consistency arguments and removal of barriers between fields as "synthesis," but I prefer to call these developments what they are: consistency arguments and removal of barriers, and to reserve for "synthesis" that which is actually synthesized. Of course consistency arguments and removal of barriers can be precursors, even necessary ones, before synthesis takes place.

Second, the synthesis is not characterized by startling or extraordinary new discoveries, concepts, or theories. Some candidates might be Fisher's "fundamental theorem of natural selection," Wright's shifting balance theory of evolution in nature and his surface of selective values, Mayr's founder effect and genetic revolution in relation to geographic speciation, Waddington's concept of an epigenetic landscape, and Muller's "ratchet." While each of these played a significant role in the period of the evolutionary synthesis, none fits the bill as a concept or theory around which the evolutionary synthesis was built or centered, in the way, for example, that Darwin built his theory of evolution in nature around the mechanism of natural selection.

Finally, the evolutionary synthesis of the 1930s and 1940s was not characterized by agreement on the mechanisms of evolution in nature. Although Fisher and Wright reached almost entire agreement on the mathematical consequences of their different quantitative models of the evolutionary process, they disagreed intensely about the relative weights of different variables in evolution in nature.[33] They could not agree on the relative roles of selection and random genetic drift, or on the population structure of natural populations. Fisher and Ford had little use for Mayr's concepts of founder effect and genetic revolution in speciation.[34] Late in the 1940s, as Gould has pointed out, there was a "hardening of the synthesis" toward a process dominated by deterministic natural selection; and this domination lasted until the mid- to late-1960s, when it began seriously to erode.[35] But this hardening came only at the end of the synthesis period, years after Huxley and Mayr thought evolutionary biology was "synthesized." The evidence is overwhelming that evolutionary biologists disagreed strongly about mechanisms of microevolution during the synthesis period, and even more strongly about mechanisms of speciation.[36]

If the evolutionary synthesis was not primarily a synthesis, was not characterized by important new discoveries or theories, did not generate agreement among evolutionary biologists about mechanisms of microevolution or speciation, and yet happened, then what was it? What did happen to evolutionary biology during the 1930s and 1940s that convinced its participants of palpable advance during this time?

A Different Interpretation: The Evolutionary Constriction

A revealing clue comes from the history of ideas about heredity in the late nineteenth and early twentieth centuries. In 1894, the French biologist Yves Delage, professor at the University of Paris, submitted for publication an enormous manuscript on ideas about heredity. It was completely up to date, with numerous references to literature published in 1894 (his introduction to the first edition was dated December 1894). The book, titled *L'Hérédité et les Grandes Problèmes de la Biologie Générale*,[37] was a huge compendium and analysis of all theories of heredity before 1894. Delage cited and referred to Focke's *Die Pflanzen-Mischlinge*,[38] in which Mendel's work was discussed, but Delage did not himself refer in any way to Mendel. This was no surprise for 1895.

The book was well received and Delage prepared a second, "revised, corrected, and augmented" edition published in 1903. The bibliography was unchanged, but the book was studded with new (specially starred) footnotes that referred to and discussed the literature from 1894 to 1901. Correns and Tschermak are nowhere mentioned. There is a substantial discussion of de Vries's theory of intracellular pangenesis and even some of his work up to 1900, but nothing about the rediscovery of Mendel. Bateson's *Materials for the Study of Variation*[39] was the most recent of his works discussed in the book. Mendel himself would have been appalled, since Naegeli's theory of the ideoplasm received fifty-one pages of analysis. Weismann's theory of heredity rated fifty-three pages of discussion. This book was a monumental achievement, poised on the edge before the rediscovery of Mendelian heredity.

Eight years after the second edition of Delage's book, three German textbooks on heredity appeared, one by Erwin Baur, one by Richard Goldschmidt, and one by Valentin Haecker.[40] Baur did not mention

Delage or his book, and the two others merely included a citation of the book in their bibliographies. By the third edition of his textbook in 1919, Goldschmidt had dropped all mention of Delage's book. None of the other textbooks on heredity published after 1905 (when the first edition of Punnett's *Mendelism*[41] appeared) even so much as mentioned Delage or his book on heredity. Included in this list are the textbooks of Lock, Bateson, Walter, and Castle.[42] If Mendelism had not arrived, my guess is that despite any existing national chauvinism, Delage's book would have been translated into English and German and would have been the single most cited book on heredity during the first two decades of the twentieth century. But Mendelism did arrive in 1900, and it overwhelmed the second edition of Delage's book even as it was published.

Mendelism shoved all the theories of heredity that came before 1900 into obscurity. Naegeli's theory, which deserved more than fifty pages of analysis in 1900, rated none after that. Instead, a theory that Naegeli rated as a combination of being overambitious, limited, and inconsequential became *the* theory of heredity. Ten years after its rediscovery, Mendelism so dominated the study of heredity that young persons scarcely ever heard about the earlier theories of heredity. De Vries's *Intracellulare Pangenesis*[43] received some attention, but mostly because of his role in the rediscovery of Mendelism and his mutation theory.

I think this example from the history of the study of heredity sheds some light on the situation in evolutionary biology during the same time. By the time of his death in 1882, Darwin had convinced biologists around the world that organic evolution had occurred. Louis Agassiz, who had retained his creationist beliefs, was already considered an anachronism at his death in 1873. But if Darwin had made evolutionists of biologists, he was much less successful in convincing them that natural selection was the primary mechanism of evolution in nature.

By 1900, the variety of evolutionary mechanisms advocated by biologists was bewildering. Most evolutionists emphasized mechanisms other than natural selection.[44] Nearly every theory of heredity that Delage had examined in his big book was associated with a theory of a mechanism of evolution. Primary among the reasons for the rejection of natural selection as the primary mechanism of evolution

in nature was the absence of convincing examples and outright abhor-
rence of the mechanistic implications of a mechanism that was purposeless and opportunistic. By my own rough estimate, the majority
of evolutionary biologists in the late nineteenth century believed in
one or another purposive mechanism of evolution.

What happened to this wide array of theories of evolution when
Mendelian heredity was rediscovered? Did theories of the mechanisms of evolution disappear along with the theories of heredity? Not
only did they not disappear, new theories of evolution emerged to
join those already existing, the most important being Hugo de Vries's
mutation theory (1901–1903). The array of theories of evolution
available in 1907 can easily be seen by consulting Vernon L. Kellogg's
masterful *Darwinism To-Day*.[45] And it can be compared with H. W.
Conn's *Evolution To-Day* or his *Method of Evolution*[46] to see that the
first years of the century had not seriously depleted the ranks of
theories of mechanisms of evolution. Conn's *Evolution To-Day* was
even reissued unchanged in 1907. A republication in 1907 of a compendium on heredity first published in 1886 was almost unthinkable.

In 1909, Delage and a collaborator, Marie Goldsmith, published in
French a review of theories of evolution. This book devoted one of
twenty-two chapters to Mendelian inheritance and its significance for
evolution. Since I was tempted to conclude that the obscurity of the
second edition of Delage's book on heredity occurred in large part
because of the chauvinism of English and American biologists, the
fate of his book on evolution is instructive—it appeared in English
translation in 1912 with publishers in both London and New York.
There was continuity of theories of the mechanisms of evolution
through the turn of the century.

This plethora of evolution theories suffered almost the same fate of
extinction as theories of heredity before 1900, but at a different time.
Mendelism did not seriously affect purposive theories of evolution
and did not vanquish evolutionary theories based upon the inheritance of acquired characters. Instead, just as Mendelism routed the
earlier theories of heredity, so the "evolutionary synthesis" routed
earlier theories of the mechanism of evolution.

The evolutionary synthesis was not so much a synthesis as it was a
vast cut-down of variables considered important in the evolutionary
process. Beginning in the late 1910s, the theoretical population ge-

neticists Fisher, Haldane, and Wright argued that evolution in nature could be modeled quantitatively.[47] They demonstrated clearly with their models that evolution within a population could be accounted for by quantitative relationships between a relatively few variables. The prestige of the mathematical models, the application of models by Dobzhansky and Wright to natural populations of *Drosophila pseudoobscura* and by Ford and Fisher to *Panaxia dominula*,[48] and application of the new genetics to systematics by Huxley and Mayr and to paleontology by Simpson, all combined with the dying out of an older generation of evolutionists to yield what seemed like a new way of seeing evolutionary biology.

What was new in this conception of evolution was not the individual variables, most of which had long been recognized, but the idea that evolution depended on relatively so few of them. So has gone much of science. Phenomena that appear so complex as to baffle the imagination, whether the motion of the heart, action of the tides, or periodicity of the comets, have been shown by science to depend essentially upon surprisingly few variables. Instead of a synthesis, evolutionary biologists during the 1930s and 1940s came to agree resoundingly upon a relatively small set of variables as crucial for understanding the evolution in nature. This I will now call the "evolutionary constriction," which seems to me to be a more accurate description of what actually happened to evolutionary biology.

The term "evolutionary constriction" helps us to understand that evolutionists after 1930 might disagree intensely about effective population size, population structure, random genetic drift, levels of heterozygosity, mutation rates, migration rates, and so on, but all could agree that these variables were or could be important in evolution in nature, and that purposive forces played no role at all. So the agreement was on the set of variables, and the disagreement concerned differences in evaluating relative influences of the agreed-upon variables. I agree with Gould that evolutionary biology "hardened" toward a selectionist interpretation especially during the late 1940s and 1950s. I see this as a further constriction of the evolutionary constriction (but I like the sound of "hardening of the constriction").

The evolutionary constriction drove from evolutionary biology all of the purposive theories of evolution that had been so common and popular before 1930. After the constriction, evolutionary biology was

utterly devoid of purposive mechanisms. Thus one effect of the constriction was to make the conflict between evolution and religion inescapable, or put another way, the previously respectable compatibility of religion and evolution became less tenable. At the 1925 Scopes
"Monkey Trial" in Dayton, Tennessee, the dean of American evolutionists was Henry Fairfield Osborn of the American Museum of
Natural History. For the trial, he wrote a little book titled, *The Earth
Speaks to Bryan*,[49] in which he argued that no conflict existed between
evolution in nature and his own deep-seated Christian beliefs.

> If Mr. Bryan, with an open heart and mind, would drop all his books
> and all the disputations among the doctors and study first-hand the simple
> archives of Nature, all his doubt would disappear, he would not lose his
> religion, *he would become an evolutionist.*[50]

Osborn's own theory of the primary mechanism of evolution was a
purposive force he called "aristogenesis." The evolutionary constriction eliminated this force along with Bergson's enormously popular
élan vital and a host of other purposive theories. The argument from
design, which had survived in evolutionary biology as long as Darwin's natural selection was supplemented by additional purposive
mechanisms, withered after the constriction.

Julian Huxley, the Evolutionary Constriction, and Progress

Julian Huxley's *Evolution: The Modern Synthesis* is the perfect exem-
plar of my thesis concerning the evolutionary constriction. His book
is anything but a synthesis. It consists almost entirely of a compila-
tion and discussion of the constricted set of variables, with little
digestion or synthesis of the variables. (To be sure, Huxley's set of
variables was larger than that employed by Dobzhansky, Wright,
Simpson, Rensch, or any of the major figures of the period.) Aside
possibly from Huxley's endorsement of the extrasensory perception
experiments of J. B. Rhine,[51] there is absolutely no hint of purpose in
any of the mechanisms of evolution presented in the book.

One of the great attractions of purposive theories of evolution
before the constriction was the possibility of having a robust notion
of evolutionary progress. The difficult trick was to have the progress

without the purpose. That is what Huxley desperately wanted. His solution was to specify anthropomorphic criteria to measure progress, or in other words to argue that whatever leads to humans in evolution constitutes progress. Huxley vociferously denied any hint of anthropomorphism, yet was blatantly anthropomorphic in his discussion of evolutionary progress.

What represents progress in evolution? Increasing brain size with respect to body weight, longevity of taxa, degree of adaptive fit with environment, reservoir of genetic variability, ability to adapt quickly to environmental change, degree of sociality, and any number of other criteria come to mind. The family of AIDS viruses fits some of these criteria better than hominids. Judging from particular criteria, many evolutionary changes may be rationally and coherently interpreted as progressive or regressive. The problem is that there is no ultimate basis in the evolutionary process from which to judge true progress. When there was purpose in evolution, there *was* such a basis, but the evolutionary constriction eliminated the purpose.

When Thomas Henry Huxley argued that evolution in nature provided no basis for ethics, he fell back upon the Judeo-Christian heritage. Julian Huxley gave up the Judeo-Christian heritage early. He seemed to think that the only hope for an ethical and meaningful foundation in life was tied somehow to progress in evolution. Thus, giving up progress in evolution was for him giving up a great deal.

With this perspective, Huxley's laudatory introduction to Teilhard de Chardin's posthumous work, *The Phenomenon of Man*,[52] is understandable. Teilhard's purposive view of evolution seemed ridiculous to most evolutionary biologists on the centennial of the publication of Darwin's *On the Origin of Species*.[53] Simpson,[54] Medawar,[55] and many others lambasted the book. Yet Julian Huxley was defending it, even after having argued for decades that no purposive forces exist in evolution. The key was progress. Teilhard believed there was progress in evolution and that it gave meaning to life. As Huxley declared in the first paragraph of his introduction to the book, Teilhard

is able to envisage the whole of knowable reality not as a static mechanism but as a process. In consequence, he is driven to search for human significance in relation to the trends of that enduring and comprehensive process; the measure of his stature is that he so largely succeeded in the search.[56]

Dobzhansky also defended Teilhard,[57] something that surprised those who were unaware that he had always been a religious person who wanted to find some ultimate meaning of life in evolution.

My assessment is that Huxley (and Dobzhansky) knew there was no purpose in evolution but nevertheless clung to the forlorn hope that even purposeless evolution might yield the same congenial meaning to life offered by purposeful evolution. The vast majority of evolutionary biologists now agree with the assessment of Simpson and Medawar, and look upon Huxley's endorsement of Teilhard as an aberration and wholly out of step with modern evolutionary biology.

When there was purpose in evolution, there could be real progress. The evolutionary constriction ended all rational hope of purpose in evolution. That is why I placed the quote from Ogden Nash at the beginning of this paper. It belongs here: "Progress might have been all right once, but it's gone on too long."

J O H N   B E A T T Y

Julian Huxley and the Evolutionary Synthesis

There is Huxley's view of the evolutionary synthesis, and there is ours. Ours extends beyond, and in some respects even diverges from, Huxley's. That may seem odd of us—even impertinent—since it was Huxley who christened the synthesis as such. However, our view must square not only with Huxley's but also with a number of other formulations and recollections of the synthesis, such as those of Chetverikov, Dobzhansky, Mayr, Simpson, Fisher, Wright, and (the historically neglected) Haldane.

There are two general theses about the synthesis at issue in Provine's contribution to this volume. The first concerns the so-called "hardening of the synthesis"—a trend well articulated and named by Gould,[1] and also strongly supported here and elsewhere by Provine.[2] The historical claim is that, at the time of the synthesis, natural selection was recognized as a significant, but by no means all-important, agent of evolutionary change. Other agents, especially random drift, were accorded considerable importance too. But in the aftermath of the synthesis, during the late forties, fifties, and sixties, a number of celebrated cases of evolution by random drift were successfully reinterpreted in terms of natural selection, with the consequence that selection came to be seen as *the* agent of evolutionary change. The synthesis thus "hardened" in favor of selection.

The second general thesis concerns the so-called "constriction of the synthesis." This thesis has gained considerable credence through the work of Bowler,[3] and Provine also strongly supports it here. The gist of this claim is as follows. The term "synthesis" suggests the coming together of many theories. But in fact the evolutionary synthesis effectively repudiated a large variety of Lamarckian, orthogenetic, and other theories of evolution (many of the twenty-odd alternative evolutionary theories described by Kellogg[4]). The synthesis thus reduced, rather than increased, the number of alternative modes of evolution that could be taken seriously.

These two theses may at first glance seem incongruous. One stresses the initial plurality of outlooks of the synthesis, the plurality then declining over time. The other stresses the plurality of theories *before* the synthesis, the synthesis itself being seen as the culling agent. The "constriction" of the synthesis must not have been too tight, or else there would have been no room for any further "hardening" of the synthesis.

We would like to be able to square Huxley's views on the synthesis with these two theses, since they describe the changing views of so many other people involved in the synthesis. (See also Beatty.[5]) In these brief comments, however, I propose to show that Huxley's views on the synthesis should at least make us careful about the ways in which we elaborate the two theses at issue. In particular, it is Huxley's unchanging commitment to the "multiformity of evolution" that, while not calling the two generalizations into serious question, helps us to articulate them more carefully.

Some background to Huxley's idea of the multiformity of evolution is in order first. *Evolution, the Modern Synthesis*[6] (hereafter *ETMS*) was written after the general principles of evolutionary genetics had been formulated by Fisher, Wright, and Haldane in the twenties and early thirties. Huxley's views on the synthesis are thus largely about what "remains to be done" in this enterprise. In this regard they are similar to (although they go way beyond) those of Dobzhansky and many others with regard to the (then) future course of the synthesis.

The general mathematical theory of Fisher, Wright, and Haldane delimited the possible modes of evolutionary change, but said very little about how evolution actually occurs. The general theory lacked specificity in two respects. First, while it was inconsistent with certain modes of evolutionary change, it was consistent with a wide variety of other modes. However, those other modes were not specifically

listed in the theory. Second, it did not address the relative "signifi-
cances" of the various possible modes of evolution (e.g., how frequently each mode occurs) in nature.

Thus, following the synthesis, a number of different forms of
natural selection were articulated (e.g., "frequency-dependent" selection, "balancing" selection), as were different combinations of
evolutionary agents (e.g., the particular combination of selection and
random drift that comprised the "shifting-balance" mode of evolution championed by Wright). Moreover, there was considerable discussion and disagreement about the relative significance of the various modes.

What "remained to be done" in evolutionary biology, then, was to
articulate all the various modes of evolution consistent with the
general theory of Fisher, Haldane, and Wright, and to document cases
of the various modes in nature with an eye to ascertaining their
overall relative importance. Dobzhansky expressed this view of the
state of the field in his classic textbook of the synthesis, *Genetics and
the Origin of Species*:

> Since evolution as a biogenic process obviously involves an interaction of
> all of the above [agents of evolutionary change], the problem of the
> relative importance of the different agents presents itself. For years this
> problem has been the subject of discussion. The results of this discussion
> so far are notoriously inconclusive. . . . One of the possible sources of the
> situation may be that a theory which would fit the entire living world is in
> general unattainable, since the evolution of the different groups may be
> guided by different agents.[7]

Dobzhansky's conception of what remained to be done was the
main message of Huxley's *ETMS*. From Huxley's point of view, the
multiplicity of theories of evolution pursued during the first couple of
decades of the century had given way to a different multiplicity of
possible modes of evolutionary change, all possible on the Fisher-
Wright-Haldane theory of evolutionary genetics and all having significant importance in nature. "Looking at the matter from another
angle, we are beginning to realize that different groups may be
expected to show different kinds of evolution."[8] Articulating, ra-
tionalizing, and documenting this variety of evolutionary processes
was the whole point of *ETMS*. He summarized his agenda early in the
book in a chapter titled "The Multiformity of Evolution."

His basic rationale in defending the multiformity of evolution was his reasoning that different groups of organisms—manifesting different developmental, physiological, genetic and group properties—would undergo different forms of evolution. In his own terms:

> Different evolutionary agencies differ in intensity and sometimes in kind in different sorts of organisms, partly owing to differences in the way of life, partly to the differences in genetic machinery. No single formula can be universally applicable; but the different aspects of evolution must be studied afresh in every group of animals and plants.[9]

Consider how this thesis is developed in the following passages:

> [W]riters of books on evolution ten or twenty years hence will . . . point out how the nature of this [genetic] mechanism governs or limits the evolutionary process, and how its variations affect the mode of evolution of their possessors. It is impossible for higher animals, whether arthropod or vertebrate, to evolve in the same way as do higher plants, owing to differences in their chromosomal machinery: noncellular and non-sexual organisms such as bacteria have their own evolutionary rules.[10]

> Nor will general organization and mode of development [ontogeny] be without its evolutionary consequences. The meristematic growth of flowering plants permits a fuller evolutionary utilization of many types of mutation than is possible to higher animals. In animals, allometric growth has evolutionary consequences which in their turn must be differently adjusted according to whether general growth is limited or unlimited. . . . The simple fact that most genes must act by affecting the rate of developmental processes is reflected in the evolution of vestigial organs, in recapitulation, in neoteny.
> The nature of an organism thus influences the mode of its evolution. This applies at every level. Within the individual, the microscopic machinery of genes and chromosomes, the mode of cellular aggregation and tissue growth; at the individual level, the type of reproduction, the way of life, the level of behavior, the method of development; beyond the individual, the size and structure of the group of which the individual is a unit, and its relations with other groups—all these, and many facts besides, have their evolutionary effects.[11]

Another indication of his pluralism is the variety of modes of speciation that he tried to provide instances of ("geographic," "ecological," "genetic," and many subcategories thereof) and of the variety of species concepts that he thought this necessitated.[12] Huxley acknowledged that selection was a pervasive mechanism of evolutionary change. But he was by no means, ever, *simply* a selectionist—for he was interested in a great *multiplicity* of forms of selection:

The generalized treatment of selection, as originally developed by Darwin and redrafted on a Mendelian basis by R. A. Fisher . . . must be particularized. Darwin made a significant beginning in his separation of sexual and natural selection, and Haldane . . . has carried the process a stage further by distinguishing various forms of intraspecific from interspecific selection. The analysis could, however, be extended on a fully comparative basis, with every effort to introduce quantitative treatment at the same time.[13]

I will return to the multiple forms of selection shortly.

But with all this emphasis on the heterogeneity of evolution, and the impossibility of generalizing about evolutionary processes or patterns, it is difficult to see what Huxley's "synthesis" amounted to. The multiformity or heterogeneity of evolution may have been the point of the book *Evolution, the Modern Synthesis*, but it hardly explains—indeed it renders quite problematic—the use of the term "synthesis." Thus Provine's evaluation: that Huxley was more of a "compiler" than a synthesizer. What more might Huxley have provided by way of "synthesis"?

Provine refers to a common interpretation of the synthesis, according to which it was mainly a matter of demonstrating the *consistency* of population genetics, systematics, paleontology, and more. *ETMS* certainly counts as a synthesis in this respect. But Provine wonders why the term "synthesis" is used here if only consistency is meant. I believe that Huxley intended something more, connected with his understanding of the term "unification":

> Biology in the last twenty years, after a period in which new disciplines were taken up in turn and worked out in comparative isolation, has become a more unified science. It has embarked upon a period of synthesis, until to-day it no longer presents the spectacle of a number of semi-independent and largely contradictory sub-sciences, but is coming to rival the unity of older sciences like physics, in which advance in any one branch leads almost at once to advance in all other fields, and theory and experiment march hand in hand.[14]

What Huxley is saying, and what he tries to show throughout *ETMS* (whether he succeeds is another matter!), is that viewpoints from fields like genetics and systematics are not only consistent but that they are mutually illuminating. Consistency would be an improvement on the previous contradictions that Huxley notes. But the only way to overcome the previous independence of disciplines that he regrets is to show *dependence*, which he attempts to demonstrate by showing how different fields can be mutually illuminating. It might be thought that unity and multiformity are incompatible, but that is not the case if, as Huxley intends, "unity" means mutual illumination.

Consider two examples from *ETMS*. In Chapter 4, "Genetic Systems and Evolution," Huxley first discusses the evolution of different genetic characteristics (e.g., the evolution of sexual vs. asexual systems of inheritance, the evolution of different mutation rates) and then proceeds to consider how taxa with different genetic characteristics may evolve by "their own evolutionary rules." Thus, evolution illuminates genetics, and in turn genetics illuminates evolution. Note also the theme of multiformity here.

Similarly, in Chapters 5–7 on speciation, Huxley discusses how taxa that have evolved different genetic systems (e.g., polyploidy) and different physiological and behavioral traits (e.g., different dispersal abilities) tend to speciate by different means. He then proceeds to discuss how different modes of speciation have different evolutionary consequences. And again the multiformity theme looms large— e.g., when Huxley stresses the plurality of modes of speciation, suggesting that "if Darwin were writing to-day he would call his great book *The Origins* not *The Origin of Species*."[15]

The hardening of the synthesis took place between 1942, when *ETMS* first appeared, and its reissue in 1963. Had *ETMS* actually been rewritten the second time around, the intervening hardening of the synthesis might have left a mark on it. But the republication featured only a new introduction, the sheer brevity of which makes it difficult for us to determine the extent to which Huxley altered his views about the importance of selection in relation to other agents of evolutionary change, like drift. But from the evidence that does exist, it does not seem that Huxley ever took as extreme a position regarding the importance of selection as many other influential evolutionary biologists in the sixties.

Consider, for instance, that the only derogatory remarks about drift in the new introduction are expressed mostly in the third person. For example, rather than fully acknowledging that certain cases of evolution previously ascribed to change had been overturned and were now (1963) seen as cases of selection, Huxley wrote that "Dr.

E. B. Ford informs me" of certain such changes in interpretation. And he adds, without committing himself, "In his [Ford's] *Ecological Genetics*[16] he gives numerous further examples of the inadequacy of drift and the efficacy of natural selection in accounting for local differentiation. . . ."[17] In fact, Ford pretty much inserted himself into Huxley's new introduction. An earlier draft of the introduction, which was sent to Ford for comments, had no such disparaging remarks about the importance of drift. Ford advised Huxley to "soft-pedal" drift in the republication, and Huxley acquiesced in correspondence, but not so explicitly in print. Ford certainly implored Huxley to be much more specific in this regard.[18]

Ford's book, published the year after Huxley's was republished, exemplifies the hardening of the synthesis, insofar as it explicitly draws conclusions about the all-importance of selection from those cases in which selection accounts had prevailed over drift explanations. Huxley's introduction contrasts sharply with Ford's book. To the extent that Huxley kept score of selection vs. drift accounts, he did not (at least not explicitly) draw any conclusions about the all-importance of selection. It is also worth contrasting Huxley's introduction with Mayr's *Animal Species and Evolution*,[19] which was also published in 1963 and which was every bit as typical of the hardening of the synthesis as Ford's book. Interestingly, in his otherwise enthusiastic review of Mayr's book, Huxley merely noted, without committing himself on the issue, that Mayr "dismisses drift as unimportant."[20]

On the other hand, natural selection does get the lion's share of attention in Huxley's new introduction. But it is important to note that Huxley did not present selection as a monolithic alternative to other agents of evolutionary change. Quite the opposite, he emphasized the wide diversity of forms of selection:

> Much theoretical and experimental work has been done on selection in general. In addition to survival (phenotypic) and reproductive (genotypic), sexual, and social (psychosocial) selection (see above), the following main types of natural selection are now usually distinguished. . . .
>
> *Normalizing*, centripetal, or stabilizing selection: tending to reduce variance, to promote the continuance of the "normal" type, and to prevent change in a well-adapted organisation.
>
> *Directional*, directed, or dynamic selection: tending to produce change in an adaptive direction.

Diversifying, disruptive, or centrifugal selection: tending to separate a single population into two genetically distinct populations.

Balancing selection: tending to produce balanced polymorphisms and heteroses in populations.

Selection for variability: leading to high variance in cryptic adaptation in certain conditions. . . . To which we may add

Post hoc selection, as when a viable new species originates suddenly by allopolyploidy.[21]

Let us return now to the two historical theses about the synthesis that are at issue in Provine's paper. Huxley's writings make it clear that although, in the course of the synthesis, a variety of modes of evolutionary change were repudiated or belittled, nonetheless, a great multiplicity of modes continued to be discussed.

In fact, so many modes of evolution have been articulated and investigated since the synthesis that one could well forsake some of them, like drift, and still be incredibly pluralistic about how evolution occurs. We cannot let the "constriction" and "hardening" of the synthesis blind us to the diversity of possible evolutionary agents being discussed in the fifties, sixties, and seventies and to the incredible room still left for controversy about the actual importance of those various modes of evolution. True, many of those alternatives are versions of natural selection, but many of them have been newly articulated since the synthesis, and there are, in fact, important and interesting differences among them.

I will close with some thoughts on Huxley's purported "progressivism" apropos his views on the multiformity of evolution. One cannot talk about *ETMS* without also talking about progress. In his introductory chapter on the multiformity of evolution, Huxley explained,

> In this book I shall endeavor to analyze some of the main types of evolutionary change. . . . and then to disentangle the various main roles (for they are numerous and diverse) of selection. This analysis will lead finally to a discussion of the problem of evolutionary progress—whether any such process exists, whether it is explicable on selectionist terms, and whether there is any prospect of its future continuance.[22]

There is a lot of talk about Huxley being a "progressivist," a "believer in progress," and the like. But in fact, as Provine noted, Huxley believed that progressive evolution—in his terms, evolution toward increased control over and independence from the environment—was an exceedingly rare phenomenon. It was by no means pervasive, by no means intrinsic to nature. Huxley believed that some forms of evolution by natural selection lead to "progress," especially in the case of humans. But he acknowledged so many other selectionist and nonselectionist forms of evolution (like random drift), most all of which have "nonprogressive" outcomes, that overall he was skeptical about the prevalence of progressive evolution. For Huxley, the multiformity of evolution rules out the prevalence of progress.

What rules out the prevalence of progress also reduces the probability of continued progress in humans. The key to continued progress, for Huxley, was eugenic intervention—or more generally, that humans take their future evolution into their own hands:

> Progress is a major fact of past evolution; but it is limited to a few selected stocks. It may continue in the future, but it is not inevitable; man, by now the trustee of evolution, must work and plan if he is to achieve further progress for himself and so for life.[23]

In this respect, Huxley's views on multiformity of evolution support (although they certainly did not inspire) his eugenic views.

Huxley the Statesman of Science

GARLAND E. ALLEN

Julian Huxley and the Eugenical View of Human Evolution

In a biographical appreciation of Julian Huxley on the occasion of his seventy-fifth birthday in 1962, the American geneticist H. J. Muller captured eloquently the ancestral impact of being "a Huxley":

> Perhaps more consciously than anyone else living, Julian Huxley, with his memories of sitting on the knee of his illustrious grandfather, Thomas Henry Huxley, could feel himself and all mankind to be the heirs of all the ages. For this benign yet fiery ancestor, pulling his long bushy eyebrows down into his mouth for his grandson's edification, was, even more than his retiring colleague Charles Darwin, the zealot and iconoclast who made evolution alive to the English-speaking peoples. Thus, Julian grew up feeling himself in a world that has moved forward for many millions of years.[1]

Muller touched on what appears to have been one of the great unresolved dichotomies in Julian Huxley's personal and scientific life: the conflict between heredity and environment, historical tradition and innovation, between ancestry and self. This conflict must have been acute for Huxley on occasion. Muller wrote that young Julian was often dismissed by envious contemporaries, who attributed his success more to the mantle of the Huxley name than to his real accomplishments. Heredity can be an oppressive burden, but being a Huxley, he could not deny its strength. Despite it all, however, he brought to his many concerns, especially those centered around eugenics and human evolution, a synthetic view that combined conflicting ideals and concerns in a new and more comprehensive whole.

Julian Huxley was, as the title of his important 1942 book (*Evolution, the Modern Synthesis*)[2] indicates, a synthesizer. But a true synthesis grows out of the necessity to reconcile opposing or conflicting tendencies. A number of such conflicts—scientific, social, and psychological—appear to have driven Huxley's synthetic capabilities. It particularly touched his views of human heredity and evolution, and perhaps surfaced most noticeably in his writings on eugenics and population control. The conflicts are numerous, but a few will suffice to suggest how they may have been influential in Huxley's approach to eugenics. For example, he saw clearly, and repeatedly discussed, the conflict between nature and nurture (heredity and environment), between preservation and innovation, between quantity and quality (of people on earth), between the individual organism and the population as a unit of evolutionary change, between technological progress and social degeneration. The central concern that many of Huxley's writings show for these topics—and especially his abiding interest in eugenics and human evolution—suggests that perhaps an important motivation came from the psychological dilemma of trying to understand his accomplishments as both part of, and yet separate from, those of his ancestry. Whatever the exact cause—and I do not want to push a psychological explanation too far—three interrelated topics form a major focus of Julian Huxley's biological writings throughout much of his career: heredity, environment, and evolution. The first two were linked in his abiding concern for eugenics, while the third provided the larger theoretical framework within which both could be pursued as meaningful scientific and social programs.

Julian Huxley's interest in eugenics, and its corollary population control, grew naturally out of both his intellectual and his social environment. No Englishman of the late nineteenth or early twentieth centuries, let alone a grandson of T. H. Huxley, could fail to see all biological phenomena in the context of evolutionary theory. At the same time, like many of his contemporary scientific colleagues, such as J. B. S. Haldane, Lancelot Hogben, J. D. Bernal, and Joseph Needham, Huxley was heir to new and radical concerns for understanding *and practicing* science in a social context. He brought to the study of eugenics a number of important traditions, ways of understanding problems of human heredity and evolution, that distinguished him significantly from eugenicists of the preceding generation. Like Haldane and Hogben in particular, Huxley represented that newer approach to eugenics fostered by the "progressive," or rational planning movement. Like his distinguished contemporaries, he also shared progressive awe, near-reverence in fact, for science as the method of salvation for humankind. He maintained a strong underlying commitment to genetic explanations of human social, personality, and mental traits, while simultaneously embracing a more sophisticated understanding of how such explanations might be applied to the improvement of human society. Huxley was not unique in such views, among his British or American counterparts from the 1930s onward. But he did weld together a particularly lucid combination of ideas that, while still essentially eugenical, marked a significant step away from the older, or, as Daniel Kevles has dubbed it, "mainline eugenics."

Before examining Huxley's eugenic beliefs, I would like to summarize briefly the various factors that influenced, or determined, the particular brand of eugenics that he espoused. This will serve two purposes. It will provide a general reference point through which each set of ideas can be related to one another in the concluding section of the paper; and it will provide an overview of the possible mix of causes that in others of Huxley's generation led to a modification of eugenic doctrines from their older, more rigid hereditarian, to their newer, more flexible, form. In the concluding section of the paper I will then be able to ask how interaction of these factors determined the course that Huxley's eugenical thinking followed.

Factors Influencing Huxley's Eugenical Views

A number of factors converged in Huxley to leave their stamp on the form of eugenical theories that he advocated in the 1930s and 1940s.

Perhaps first and foremost in importance is Huxley's intellectual background and concerns as a biologist. The focus of much of his work as a biologist was the evolutionary process, with attendant work in systematics, genetics, and animal behavior. As an evolutionist of the younger generation, Huxley was steeped in populational thinking. He thus saw eugenics less in terms of selecting individuals within family lines, and more in terms of the shift of statistical means within large populations. Such an approach gave his eugenical views a flexibility that was lacking in older eugenicists in both the United States and England.

A second factor was Huxley's perception of the direction in which the older eugenics movement was going. He could not accept its oftentimes blatant racial, class, or ethnic biases. He could see all too clearly that claims for the inheritance of I.Q., based on data from intelligence tests, could have little genetic or scientific import. To Huxley, many older eugenicists made pronouncements that exposed their ignorance of modern genetics. Moreover, the association of old-style eugenics with Nazi race theory became a major impetus for Huxley to seek a more scientific or what he considered balanced approach to eugenic ideals.

A third factor was the changing economic/social *milieu* in which Huxley emerged, as a young adult, in the period just before and especially after, World War I. This was the period of the decline of laissez-faire capitalism in both the United States and England, and its gradual replacement by concepts of economic and social planning. From his membership in the Political and Economic Planning (PEP) and the Next Five Year Group (NFYG) to his strong interest in the experiments in a planned economy being initiated in the Soviet Union, Huxley shared in the growing conception that economic and social crises such as those associated with the Great Depression, resulted in large part from lack of long-range organization and control. Similarly, in the biological realm, laissez-faire reproduction seemed to be leading to a calamitous deterioration of the natural qualities of the human species. Only scientific planning, guided by trained experts, could significantly reverse these trends. Reproduction could no longer be left to chance and individual whim.

A fourth factor was Huxley's espousal of a philosophy of scientific and later evolutionary humanism that profoundly influenced his approach to eugenical planning. Central to evolutionary humanism was his view of the uniqueness of man, which meant that eugenic practices must be shaped by that uniqueness. The crude application of Mendelian genetics to human personality and intelligence was misplaced science. What was unique about human beings was not their genetics—they were subject to Mendelian laws like other animals—but the existence of mind, which gave them two quite distinct, but equally powerful, forms of inheritance: biological and cultural. The problem for eugenics was to understand the interaction between these two forms of inheritance, not to single out one as *the* major determiner of the human condition. A further aspect of evolutionary humanism was its rejection of traditional religion as a source of ethics, values, or guidance in developing social policies. To Huxley and other evolutionary humanists, human beings held the key to biological progress, or deterioration, in their own hands. They could apply rational and scientific principles and make real advances, or they could remain ignorant of science and degenerate into class and nationalist conflict, carnage, and extinction.

A fifth factor was Huxley's exposure to, and influence by, radical and left-leaning colleagues who helped him see clearly the more conservative, reactionary biases that often permeated the writings of the older eugenicists. His long-standing friendship with Haldane, after the early 1930s a member of the British Communist Party, and with left-leaning socialists and radicals such as Hogben, Needham, and Muller, provided a sympathetic understanding of the necessity of radically remaking society before any eugenic research or social action programs could be effective. Such associations also alerted Huxley, after the mid-1930s, to the real dangers, as in Germany, of masking class and nationalist biases in scientific garb.

And finally there was the personal side—the psychological dichotomies between past and present, heredity and environment, ancestry and individual merit—that seem likely to have played a role in fashioning Huxley's brand of eugenical thought. Although I have no doubt that such personal conflicts played a part in the story, I will refrain from making much of a case for them at this time. I have a certain aversion to psychological speculation about persons who are safely beyond direct scrutiny. It is difficult enough to perceive what might be the psychological motivations operating in people we can talk to and probe in the flesh. It borders on the dangerously speculative when the subject is no longer among us.

Huxley and the Eugenical Approach to Human Evolution

While Huxley's interest in eugenics apparently goes back as far as his student days at Oxford, it is only in the 1920s, and especially the 1930s, that he emerged as a major writer on eugenical topics. It is beyond the scope of this paper to discuss the various changes in Huxley's formulation of his early eugenical ideas. Rather, I will focus on his views as they had crystallized in the mid-1930s, particularly in his essay titled "Eugenics and Society," originally given as the Galton Lecture before the Eugenics Society in London on January 17, 1936, and published afterward in the *Eugenics Review*. (It was later republished unchanged in Huxley's 1941 collection of essays, *The Uniqueness of Man*[3]; all references in the present paper are to the later version.) In many ways this essay epitomizes Huxley's eugenical view of human evolution. It brings together a large number of arguments against the older eugenics, formulating a new, more broadly based view reflecting contemporary genetic, evolutionary, sociological, and political theory. While he remained ultimately a eugenicist, his eugenics was decidedly more akin to the younger generation that included Haldane, Hogben, and Muller, than to the older generation that included Leonard Darwin, A. D. Derbishire, and Caleb W. Saleeby.

Old vs. New Eugenics

It may be well at this point to define the difference between what is meant by old, or mainline, and the newer, or reform, eugenics. That such a distinction exists has been noted by numerous students of both the American and the British eugenics movements.[4] The two can be distinguished by the age, educational background, and ideologies of their practitioners. In both the United States and Britain, the shift from old to new eugenics appears to have begun in the late twenties, continuing through the 1930s, gradually transforming itself after World War II into the movement for population control.

"Old style" eugenics was practiced by the generation contemporary with Karl Pearson (1857–1936) and Major Leonard Darwin (1850–1943) in Great Britain, and Charles B. Davenport (1866–1944) and Madison Grant (1865–1937) in the United States. It is characterized by a pronounced emphasis on heredity, almost to the exclusion of environment, by simplistic notions of human genetics (either Mendelian or biometrical), by overt race and class biases associated (usually) with conservative politics, and, especially in the United States, by a strong commitment, from the outset, to political

activism. The older generation of eugenicists tended to speak dog-
matically about the inheritance of personality or mental traits such as
intelligence, feeblemindedness, alcoholism, sexual deviancy, nomad-
ism, criminality, or rebelliousness. While most older eugenicists
were Darwinians of one sort or another, few were practicing field
evolutionists. By the 1920s and 1930s many of the older generation
had fallen out of touch with recent (post-World-War-I) developments
in genetics and evolutionary theory, especially mathematical population genetics. This factor, coupled with the rise of Nazi race theory
after 1933, may have been significant in bringing about the decline of
influence of the older style of eugenics in the 1930s.

At the same time, a newer, reform eugenics had begun to emerge in
both Great Britain and the United States. It was championed by a
younger group of biologists of the generation of Huxley (1887–
1975): including Haldane (1892–1964), Muller (1890–1967), and
Frederick Osborn (1889–1981). Reform eugenics was characterized
by a more sophisticated understanding of genetics and evolutionary
theory, by a greater appreciation of the role of environment—or at
least the interaction of heredity with environment in determining
human mental traits—by a greater awareness of the fallacies of race
and class bias, and finally by a more liberal—sometimes even radical—political outlook. It is into this reform tradition that Huxley
more clearly falls.

Huxley was persistently an outspoken eugenicist. The subject was
not one of merely passing interest. He was a member of the Eugenics
Society for many years, well into the 1960s (when he was president),
and he wrote prolifically on eugenical topics throughout the 1930s
and 1940s. Gradually, especially after World War II, however, his focus shifted to the issue of human population control, though viewed
always from the outlook of a eugenicist. Eugenics was thus to Huxley
a major concern throughout his long, varied, and productive career.

Huxley's Hereditarian Views

As a eugenicist, Huxley thought that many, if not most, mental and
personality traits were largely inherited, as well as more physical and
physiological traits such as brachydactyly, alkaptonuria, and stature.
Almost by definition, belief in an important hereditary component to

such social and personality traits was a central component of *all* eugenical thinking, older and newer alike. In this Huxley was no exception. In his Galton lecture of 1936, while questioning some of the tenets of mainline eugenics, Huxley assured his readers that he does not underrate the extent of genetic differences between human groups, either classes or races. After referring to some recent work on variation in human taste and smell capacities, which were found to be unevenly distributed among different populations, Huxley wrote:

> The existence of marked genetic differences in physical characters (as between yellow, black, white and brown) [human beings] make it *prima facie* likely that differences in intelligence and temperament exist also. For instance, I regard it as wholly probable that true negroes have a slightly lower average intelligence than the whites or yellows.[5]

Earlier, in 1930, he expressed similar views:

> Most mental defects appear to be due to a defect in the hereditary constitution, which is what we call "recessive," that is to say, it can be masked by its normal partner. So two people who are perfectly normal themselves may be carrying a factor or factors for mental defect, and when they marry some of their children will be defective.[6]

And, in the same paper, he continues:

> Some people are born talented, others are born morons; some inherit a healthy constitution, others inherit deformity or liability to disease. The United Kingdom would be a very different place from what it is today if its average level of intelligence were that of the stupidest 10 percent of its population—and very different also if it were that of the ablest 10 percent.[7]

Such statements, taken by themselves, are not a far cry from views put forth twenty years earlier by older-generation eugenicists. Like his predecessors and contemporary eugenic colleagues, Huxley displayed a noticeable germ of elitism as he viewed the question of genetic differences between groups. For example, in an oft-quoted letter to his coauthor H. G. Wells (1866–1946) in 1930, he wrote:

> As they stand the remarks about different social classes are to me untenable. You make sweeping assertions about the absence of differences between them which I really can't pass. I am quite willing to let you cut out my "sweeping" assertions about the positive differences between them, but let us point out the problem. . . . I really think we ought to say something on this point. It comes down to this; that the evils of slum life

are largely due to the slums, but to a definite extent *caused* by the type of people who, inevitably, gravitate down and will make a slum for themselves if not prevented.[8]

Further, in his 1936 lecture to the Galton Society, he argued in a clear display of class bias:

> I further anticipate that the professional classes will reveal themselves as a reservoir of superior-germ plasm, of high average level notably in regard to intelligence, and therefore will serve as a foundation-stone for experiments in positive eugenics.[9]

Without minimizing this brand of elitism, which, interestingly, Huxley was often quick to point out in others, I want to emphasize he was able to go beyond such views and incorporate into his analysis elements not present in older eugenic writings. These included his extensive knowledge of current work in genetics, evolution, and systematics.

Eugenics in an Evolutionary Context

One of the most important innovations in Huxley's reform eugenics is the fact that he placed the problem squarely in an evolutionary, as opposed to a purely genetic, context. At the beginning of "Eugenics and Society," Huxley stated that "once the full implications of evolutionary biology are grasped, eugenics will inevitably become part of the religion of the future, or of whatever complex of sentiments may in the future take the place of organized religion."[10] In other words, eugenicists must ask the same kinds of questions that evolutionists ask when they attempt to understand how selection operates to preserve or eliminate certain traits. The reference to "religion of the future" was what he later espoused as secular, or evolutionary humanism (see below). What are these questions, or evolutionary approaches, that Huxley felt must inform the eugenicist's work?

Importance of Equalizing the Environment

As it was for evolutionary theory in general, for Huxley too the chief question with which eugenicists had to grapple was determining which characteristics were genetically as opposed to environmentally controlled. Huxley took an explicitly interactionist approach to this issue. Admitting that many traits—for example, eye or skin color in humans, or flower color in plants—are largely independent of the environment, he emphasized that at some level every phenotypic trait results from the interaction of genes with the environment—both internal and external. Pointing out that the first step any geneticist must take to carry out proper breeding experiments is to control environment, Huxley stated:

> We shall only progress in our attempt to disentangle the effects of nature from those of nurture in so far as we follow the footsteps of the geneticist and equalize environment. . . . We must therefore concentrate on producing a single equalized environment. [11]

In a letter to his friend and former student at Oxford, C. P. Blacker, Huxley related this problem specifically to the pursuit of eugenical work:

> We can't do much practical eugenics, until we have more or less equalized the environmental opportunities of all classes and types—and this must be by levelling up. [12]

The reason for this goes back to evolutionary considerations. Selection always occurs in relation to *some* environmental component(s). In the absence of human direction, it is the *environment* that did the selecting. According to Huxley, if we try to compare two different populations that are subject to two different selection pressures, our results are meaningless:

> If the aim of eugenics be to control the evolution of the human species and guide it in a desirable direction, and if the genetic selection should always be practised in relation to an appropriate environment, then it is an unscientific and wasteful procedure not to attempt to control environment at the same time as genetic quality. [13]

What the exact effect of "levelling up" the human environment would be could not be predicted, Huxley pointed out, a priori. It depended completely on the type of gene-environment interaction that is involved for each trait. To make this point concrete, Huxley described several examples of phenocopy effects, one in the plant *Primula*, and the other in *Drosophila*. There is a gene for "abnormal abdomen" in *Drosophila* that shows its phenotypic effect only in a moist environment. In a dry environment the gene produces the

normal abdomen associated with the wild-type fly. A similar situation exists for a flower-color mutation, and temperature, in *Primula*.
The mutation produces red flowers when grown at normal temperatures, but white flowers when grown at high temperatures. Phenotypic variation, and its underlying genotypic foundation, therefore, depends for its expression on environmental conditions, though
not in an easily predictable way. If this were true for relatively simple
traits such as flower color, Huxley maintained, how much more true
must it be for human personality and mental traits?

epigenetics

Environmental Uniformity and Genetic Diversity

The converse of the phenocopy argument was not, however, that
uniform environments necessarily produced more uniform phenotypes. Taking *Drosophila* again as an example, he pointed out that in a
genetically heterogeneous population for normal-abnormal abdomen, phenotypic uniformity (i.e., all normal abdomen flies) would
result from equalizing the environment toward dry conditions, while
phenotypic diversity (some normal, some abnormal abdomen) would
result from equalizing the environment toward moist conditions. A
nonhomogenized (heterogeneous) environment, containing both
moist and dry conditions, would in this case encourage the greatest
amount of expression of genetic variability. It is the particular type of
gene-environment interaction that determines the effect a diverse or
uniform environment has on phenotype. There is no blanket generalization by which such effects can be judged a priori.

   To a staunch eugenicist, especially with Huxley's evolutionary
leanings, genetic uniformity would ultimately lead to a dysgenic, not
eugenic, state. He could agree later with Muller about the deleterious
effects of unwanted variation—i.e., genetic load—but he could also
see that a fallacy lay hidden in the claims that by equalizing the
environment a more uniform expression of genetic potential would
result. In Huxley's eugenical vision, the future evolution of the
human species depended on a two-pronged eugenic approach: (1)
elimination of unwanted genetic variability (genetic diseases, mental
defects, and so on) through negative eugenical measures, coupled
with selection for desirable variations through positive eugenics and
(2) encouraging the maintenance of nondeleterious genetic diversity

throughout the population as the basis for future evolutionary advance. Equalizing the environment had its own social and moral rationale. Production of phenotypic or genotypic uniformity was not one of them.

Thus, prior to the field observation by Dobzhansky that natural populations contained a large range of genetic variability, Huxley claimed that such diversity must be present and must play a crucial role in evolution. Diversity is the raw material on which selection acts. Populations that tend toward genetic uniformity lose their flexibility, Huxley claimed, and were on the road toward evolutionary extinction. I would argue that this view, while not unique to Huxley, was a new argument for eugenicists in 1936. It may well have grown out of Huxley's field and natural history background, especially his work on bird behavior. In general, an appreciation of variation among organisms seems to be more clearly tied to field, as opposed to only laboratory, experience, and is a part of the larger naturalist-experimentalist dichotomy that figured prominently in early twentieth-century life sciences.

Applying to eugenics the idea of environmental and genetic diversity, Huxley pointed out, for example, that to make educational opportunity more favorable to all classes might actually increase, rather than decrease, the currently observed diversity in achievement between such groups. Quoting his friend Lancelot Hogben, Huxley argued:

> The effect of extending to all classes of society the educational opportunities available to a small section of it would presumably be that of increasing variability with respect to educational attainment [i.e., performance]. The effect of depriving the more favoured of their special advantage would be to diminish variability in educational attainments. Either policy would result in an equalization of environment; but equalizing it by making it more favourable would bring out genetic differences more fully, while the reverse process would mask them.[14]

On the other side of the coin, Huxley chastised mainline eugenicists such as A. M. Carr-Saunders, who a decade earlier had claimed that children from poor neighborhoods did less well in school because they were genetically inferior in intelligence.[15] Without equalized environments, no such conclusions were possible about the causes leading to difference among social classes, or between racial and ethnic groups in basic intelligence.

As pointed out earlier, Huxley truly believed that differences in educational achievement or personality type were likely to be caused by genetic differences. But unlike mainline eugenicists, he was keenly aware that this was merely supposition—possibly bias—on his own part, and not a demonstrated biological fact. When he claimed that "I regard it as wholly probable that true negroes have a slightly lower average intelligence than the whites or yellows," he also added: "But neither this nor any other eugenically significant point of racial difference has yet been scientifically established."[16] The reason for this is that in humans culture and learning have an overriding effect on genetics:

> But—and this cannot be too strongly emphasized—we at present have on this point no evidence whatever which can claim to be called scientific. Different ethnic groups have different languages and cultures; and the effects of the cultural environment are so powerful as to override and mask any genetic effects.[17]

Huxley recognized how eugenical claims for hereditary differences between groups could be thinly disguised appeals to racial, ethnic, or national prejudices. He strongly attacked the Nazi race hygiene movement as blatantly racist and nationalist, having no grounds whatsoever in biology or genetics:

> The Nazi racial theory is a mere rationalization of Germanic nationalism on the one hand and anti-Semitism on the other. The German nation consists of Mendelian recombinations of every sort between Alpine, Nordic and Mediterranean types. The theory of Nordic supremacy and initiative is not true even for their own population: it is a myth like any other myth, on which the Nazis are basing a pseudo-religion of nationalism.[18]

In a number of other ways Huxley demonstrated his reform eugenic position. He argued strongly against the use of the existing "intelligence" tests as tools for measuring innate mental differences between groups. Such tests, he claimed, measured learning and achievement, not inborn ability. In unequal social and economic environments, intelligence tests measure primarily what opportunities individuals have had. One could not distinguish intellectually inferior or superior human groups using such tests:

> The results of intelligence tests applied to different ethnic stocks are for the same reason devoid of much value. Intelligence tests are now very efficient when applied to groups with similar social environment; they

become progressively less significant as the difference in social environment increases. Again, we must equalize environment upwards—here
mainly by providing better educational opportunity—before we can evaluate genetic difference.[19]

Quite different from mainline eugenicists, especially in the United
States, Huxley rejected the I.Q. score as adequately reflecting any
genetic components. Besides, as he pointed out on a number of
occasions, even if such tests did reflect genetic differences, the distribution curves for various ethnic or social groups show such overlap
that they could hardly be used to make any significant differentiation
between the various populations.

The Concept of Race

It is in Huxley's rejection of the concept of "race" as applied to human
beings that his evolutionary, and particularly populational, perspective on human genetic differences becomes most evident. Many of us
are aware of Huxley's views on race from the 1939 geneticists' manifesto, and again from the UNESCO statement on race of 1950, in
both of which his authorship was prominent. Yet much earlier, in the
mid-1930s and early 1940s, Huxley championed the idea that as a
biological concept "race" is meaningless. At the same time that sociologist Otto Klineberg published his pathbreaking work on black-
white intelligence in 1935, Huxley and his coauthor, anthropologist
A. C. Haddon, had already made similar points from an anthropological and evolutionary standpoint. In *We Europeans*,[20] Huxley and
Haddon argued that the concept of "race" as used in reference to
human beings was highly inconsistent. Sometimes, they pointed out,
"race" was used synonymously with nation, or with ethnic groups (as
"the German race" or "the Jewish race"). Sometimes it was used
synonymously with family line (as the Wälsung "race" in Nordic-
Teutonic mythology). Sometimes it is used in a semi-biological way
as synonymous with subspecies. Even in the latter case, they felt, the
application was superficial and incorrect. Unlike other species in
which true, largely isolated subspecies have diverged from a common
ancestor, the human species has been panmictic too long for any such
level of divergence to have occurred. As Huxley and Haddon wrote in
1935:

In man, migration and crossing have produced such a fluid state of affairs that no such clear-cut term, as applied to existing conditions is permissible. What we observe is the relative isolation of groups, their migration and their crossing.[21]

They went on to urge that the term "race" be dropped with respect to human geographic groups. (Huxley proposed substituting the terms "ethnic group," or "peoples" in its place.) For even more obvious reasons, there can be no distinction in any general sense between "superior" and "inferior" races. These sorts of judgmental views, Huxley and Haddon agreed, are the basis of racial problems, and always turn out to be the result of one geographic group trying to dominate another. To Huxley (and Haddon) "racial" problems are social, not biological in origin, and have to be solved by social, not biological methods.

In taking the position he did on the concept of "race," Huxley again demonstrated his evolutionary and populational perspective. By viewing organisms as members of populations having geographic ranges, containing a considerable amount of variation (some of which is hidden, or cryptic), distributed (usually) along a normal curve, and evolving by a gradual shift in gene frequencies, Huxley was clearly thinking in terms of populations of organisms, not merely collections of individuals. Species to Huxley were not "types." The older generation of eugenicists, especially in the United States, thought much more in typological, or essentialist terms (to borrow terminology from Ernst Mayr). They tended to see subgroups, such as Negroes, "Mediterraneans," or "Slavs" as homogeneous, having a fixed and innate set of characteristics. Older eugenicists saw intercrossing between groups as destroying genetic "purity" and hence dysgenic, whereas Huxley saw it as a source of increased variability, and hence eugenic. For Huxley the human species represented a gene pool of immense diversity that defied classification into distinct, biologically separate "races" or types.

What Are Eugenicists to Do? Huxley's Eugenical Program

If environment was so important, and if mental and personality traits were only influenced in some undetermined way by genes, and if human subgroups such as social classes or races were not real biolog-

ical groupings but only human creations, what was the eugenicist to
do? How, in a word, was a eugenicist to differ from a social reformer?
To mesh his reform position with a continued belief in the importance
of eugenics, Huxley produced a grand synthetic argument that incorporated several articles of faith and several concepts from standard
evolutionary theory.

The main article of faith was his continued belief, *even though he
acknowledged that there was no scientific basis for it*, that many
human personality, mental, and behavioral traits were controlled to a
significant degree by genes. A second, subsidiary article of faith was
the belief that human groups *differed* genetically in personality, mental, and behavioral traits. Although present and future environmental factors might well mask these differences, to Huxley they were
still there, and if allowed to perpetuate themselves uncontrolled could
lead to the degeneration of human intellectual ability. In terms of
evolutionary theory, he had a ready explanation for why true genetic
differences might exist across class or ethnic lines. Huxley made
a distinction between what he called "pre-selective" and "post-
selective" influences. Pre-selective influences were those that predisposed an organism, or group of organisms to select one environment over another. The classic example Huxley used was cave fauna.
Animals with poor eyesight might more frequently wander into
caves by accident, and, once there, less easily find their way out. It
was even possible, so Huxley thought, that the dark environment of a
cave might be more comfortable for animals with "sensitive" eyes.[22]
Pre-selection simply meant that there was some nonrandom process
involved in determining which organisms entered caves and which
did not.

Once established in the cave, however, post-selective influences
would come into play. Post-selective influences were those acting on a
population subject to a given environment, thus favoring certain
evolutionary trends over others within that environment. For example, once a population of organisms with poor eyesight was established within a cave, post-selective forces could decrease eyesight in
favor of the senses of touch or smell. Thus, once a trend is established, the direction it follows subsequently becomes a function of
post-selective influences.

From a eugenical standpoint, Huxley argued that pre-selective influences might well have been at work in the initial sorting of people into social classes. Those who were inherently lazy, or, as he put it "stupid," might "pre-select" into the lower classes, while those who were inherently more intelligent or energetic might "pre-select" into the higher classes. Once in these different social and economic environments, post-selection would begin to work in different directions. Thus, over many generations, some genetic differences might well come to exist between social classes. As Huxley put it:

> With the passage of time, more failures will accumulate in the lower strata, while the upper strata will collect a higher percentage of successful types.[23]

A similar process might work by creating or maintaining ethnic or other geographically distinct subpopulations.

In terms of eugenic programs, the above argument would seem to suggest the same plan of action as that advocated by the mainline eugenicists: reduce the rate of reproduction of the lower classes, or poorer ethnic groups, and increase the rate of reproduction of the upper classes, or wealthier groups. Huxley was too sophisticated socially and politically, however, to admit that such a simplistic approach would necessarily achieve the desired eugenical end. The net result of pre- and post-selective influences would be good eugenically, he felt, if what we judged as social or economic success at the present time were really correlated with biological value. As he wrote, true selection into social classes

> would be good eugenically speaking if success were synonymous with ultimate biological and human values, or even partially correlated with them; and if the upper strata were reproducing faster than the lower. However, we know that reproduction shows the reverse trend, and it is by no means certain that the equation of success with desirable qualities is anything more than a naive rationalization.[24]

Here, then, was the crux of the matter. Pre- and post-selection undoubtedly operated, but the results they yielded would always be a function of the given socioeconomic environment in which they were set. Unlike mainline eugenicists who took the structure of the human social environment of the day as a given, Huxley, along with Haldane, Hogben, Needham, and others, recognized that such environments are totally manmade, and are not a "fact of nature." There was nothing fixed, or writ in stone, about human social systems. While eugenic solutions could indeed work, they could only be progressive to the extent that we recognized what sort of environment we were selecting for, and then selected those characters that would be most adapted for it. In several lengthy passages toward the end of his 1936 paper, Huxley spelled out this relationship clearly. Starting with the general eugenic assumption that human beings can be subject to genetic selection like any other species, Huxley stated:

> There is no doubt that genetic differences of temperament, including tendencies to social or antisocial action, to co-operation or individualism, do exist, nor that they could be bred for in man as man has bred for tameness and other temperamental traits in many domestic animals; and it is extremely important to do so. If we do not, society will be continuously in danger from the antisocial tendencies of its members. [25]

Now, knowing that selection is always practiced in a particular environment, Huxley argues:

> Our eugenic ideals will be different according as we relate them to a slave order or a feudal order of things, a primitive industrial or a leisure order, a this-worldly or an other-worldly order, a capitalist or a socialist order, a militarist or a peaceful internationalist order. [26]

This meant to Huxley that eugenicists must necessarily consider the social environment, and in this sense become as concerned with social reform as the most radical nurturist. In effect, he stated, eugenicists had three choices with respect to the social environment in which eugenics is to be practiced: (1) they could assume that the present social order, while evolving, will always remain largely the same as it is now, and adjust genetic selection accordingly; (2) they could imagine an ideal social environment and select for values commensurate with that ideal, hoping the environment actually changes in the desired direction; or (3) they could mount a joint attack on both the environment and the human germ plasm through eugenic practices (both positive and negative), recognizing that it is possible to work for a more harmonious environment in which selection can lead to a truly positive, progressive end. It was clear, Huxley thought, that the third alternative was the only rational course to follow.

What sort of environment would Huxley advocate in which the new eugenic selection should be practiced? It was clearly not the competitive, individualistic, capitalist society of late-nineteenth- and twentieth-century Britain:

> It seems clear that the individualist scramble for social and financial promotion should be dethroned from its present position as a main incentive in life, and that we must try to raise the power of group-incentives. Group-incentives are powerful in tribal existence, and have been powerful in many historical situations. . . . [Here he cites the ancient Japanese, Nazi Germany, and the USSR.][27]

After denouncing the group incentives currently developed by the Nazis as militarist and nationalist, he pointed to the Soviet Union as an example where a more humane, productive use of group incentive was then being instituted. Group incentive includes a number of more specific traits that should be the basis for eugenic selection: altruism, readiness to cooperate, sensitiveness, sympathy, enthusiasm.[28] These should replace the traits that are currently selected for in capitalist society: "egoism, low cunning, insensitiveness, and ruthless concentration." At the moment it was not possible, he stated, to select for these traits, because "the expression of such genes is so often inhibited or masked by the effects of the environment."[29] Indeed, the present social structure of capitalism, with its attendant nationalism, was, in Huxley's view, dysgenic. Both lead to artificial social stratification (as opposed to natural, biological stratification!), and inevitably to war. War itself is highly dysgenic, Huxley thought, since it leads to the death of the most eugenically fit members of society. The only environment that had the potential of being truly eugenic was one that was equalized, economically and socially, for all members of society. Until such a uniform environment is created, the long-range effects of eugenics will never be realized. So, to Huxley, as to his more radical associates, the eugenicist can no longer avoid being both a eugenicist and a social reformer. In Huxley's view, the one without the other does not make sense.

It may be worthwhile to inquire just how Huxley envisioned eugenics, even in the best of environments, might actually be carried out. What were the methods to be used in the practice of eugenic selection? In terms of positive eugenics, Huxley, like many other eugenicists, had little to suggest. The correlation between increase in educational levels and socioeconomic status and decline in birthrate had been widely recognized since the first decade of the century. In

Huxley's view, equalizing the environment would mean raising the socioeconomic levels of the lower classes, and thus would be expected to eventually reduce their overall fertility. But short of this indirect approach, positive eugenics (incentives to professional families to have more children) had proven a considerable failure.

Birth Control and Sterilization

In terms of negative eugenics, however, Huxley had more distinct ideas. He was a staunch advocate of birth control—especially the dissemination of birth-control literature and encouraging the use of birth-control devices. He wanted to see the dissociation of sex and love from procreation, looking toward the day when decisions about procreation would be based on genetics and health of the offspring:

> It is now open to man and woman to consummate the sexual function with those they love, but to fulfill the reproductive function with those whom, on perhaps quite other grounds, they admire.[30]

Huxley thought that eventually such an ideology would develop, but it would have to overcome bitter opposition in the meantime:

> This consequence [of eugenics] cannot yet be grasped. It is first necessary to overcome the bitter opposition to it on dogmatic theological and moral grounds, and the widespread popular shrinking from it, based on vague but powerful feelings, on the ground that it is unnatural.[31]

There was another problem, ultimately more difficult to deal with than public opinion about birth control. This was getting those who most needed birth control to understand its importance and to care enough to put it into practice. This group, Huxley (and others) referred to as the "unteachable class," or the "social problem group." Huxley felt that the birth-control advocates' most difficult task would be to reach these groups. He estimated that the "unteachables" comprised 5 to 10 percent of the current (1930) British population. The problem of reaching them, he claimed, was a result of their innate worthlessness:

> They are either too lazy to come to the clinic at all, or when they get there they are so stupid they can't learn the methods, or they are so shiftless that they won't trouble to use them properly when they get home. They [the unteachable class] are unteachable just because they are in one way

or another undesirable, because they are stupid, or lazy, or shiftless, and yet they are just the people we most want to get at.[32]

Huxley believed that the "unteachables" contributed the greatest percentage of hereditary defectives to the next generation. Their inability to use birth control voluntarily meant that methods other than voluntary contraception were required. These "other methods" were sterilization, mostly voluntary but perhaps in special cases enforced by law. It was this alternative that Huxley embraced in the 1930s and 1940s.

In the early 1930s Huxley was a strong partisan of eugenical sterilization on as large a scale as necessary to ensure that defectives did not propagate. Wersky claims that Huxley subscribed, in part at least, to Sir Richard Gregory's ideas about extensive sterilization, though, and I quote: "he did not want to sterilize en masse various sections of the Welsh and the Irish, or most coal miners and dockers," as was part of Gregory's plan.[33] Nonetheless, he was active in the Eugenics Society's campaign for sterilization from the 1930s through the 1960s, and viewed it as an essential means toward an eugenical end. He greatly admired the sterilization effort in the United States in the first several decades of the century. In a talk given before a birth control organization in New York in 1930, he lavished praise on American eugenicists who had "pioneered" in the passage of eugenical sterilization laws in a number of states. He was particularly impressed with California, which by that date had performed over 12,000, or virtually half, of the eugenical sterilization operations in the United States:

You in this country have been great pioneers of sterilization. In California you have carried that out as part of your public health program and we in England are very confident that if we could get such operations legalized we would follow California's lead, feeling that it would result in nothing but good.[34]

Huxley himself worked for legalization of voluntary eugenical sterilization in England and in the early and mid-1930s supported the move to bring the issue up for legislation. He enthusiastically commended the report of the Departmental Committee on Sterilization in 1934, which supported legalized sterilization of mental defectives. By publicly recommending sterilization as a valid preventive approach, Huxley pointed out, the report "follows a scientific approach, not one dictated by religious, nationalist, political or class motives; and for that reason its findings. . . may well be the harbingers of a new era in social legislation."[35] That new era was to see the application of scientific methods to social problems. As Huxley saw it:

> The next great step, if human progress is to continue, must be man's control of his social environment . . . . [O]ur destiny is in our own hands and . . . no pretended absolute authority should prevent us from tackling our human problems, including those difficult ones of sex and reproduction, in what I would call the humanistic spirit. And by that I mean one which recognizes as its only goal the physical and mental well-being of human beings, and sees in the scientific method its only effective instrument.[36]

In summary, to Huxley eugenics imposed an immeasurably larger and more complex task than the older eugenicists had envisioned. Lacking the more sophisticated mathematics developed later by Fisher, Haldane, and Wright, and largely ignorant of newer findings in genetics, they had naively concluded that mental defectives could be eliminated from the population in just a few generations. The older generation of eugenicists greatly underestimated the difficulty in passing legislation on sterilization and birth control. And they mistakenly saw eugenics and environmental reform as opposed, and so, like the Mendelians and biometricians of an earlier generation, had drawn an artificial battle line. Just as biometrics and Mendelian theory are complementary approaches to the study of heredity, so sociology and human genetics were complementary approaches to the control of human evolution. Huxley emphasized throughout his writings that eugenics is more than merely the study of what and how this or that trait is inherited. Eugenics must also involve action—the setting of social policy. Since "the aim of eugenics is to control the evolution of the human species and guide it in a desirable direction," it must unite theory and practice, heredity and environment, genetics and social theory.[37]

Planning and Control of Human Evolution: Huxley and Biological Engineering

What can be gained by an analysis of Huxley's ideas on eugenics? First, any study of an individual in the history of science can show

how, in a detailed way, various ideas interact in a person's thinking. In Huxley's case, it is interesting to see how he managed to mesh ideas of individualism, collectivity, social planning, liberal politics, Darwinian evolution, and eugenics into some sort of coherent social philosophy.

But such a case study in and of itself would not be particularly important if it did not illuminate broader issues. A second and more important reason to study Huxley's eugenic ideas is that it helps us— at least those who study the history of eugenics—understand the more complex and subtle motivations driving this particular social movement. Among other things, it takes us beyond the stereotypes that eugenical thinking is intrinsically conservative or reactionary or always based on simplistic genetic formulas. A study of Huxley, for example, suggests that while the social environment in which eugenics flourished may have been conservative, the makeup of the movement itself was politically much more heterogeneous. Moreover, the movement was not the same in 1935 as it was in 1915—that is, the movement evolved. Nor was it the same in England and the United States at any one time. This conclusion is not new; it has been made by a number of other writers, including Daniel Kevles, Diane Paul, Donald Mackenzie, and Michael Freeden. But it is demonstrated amply, and with some interesting additional twists, in a study of Huxley's eugenical thinking.

From Eugenics to Population Control

Of particular importance in Huxley's eugenic philosophy was the ease with which, from the 1930s onward, he brought together ideas of eugenics and those of population control. As Allan Chase, Daniel Kevles, Barry Mehler, and others have noted, many of the younger eugenicists saw a clear, logical step from the ideology of eugenics to that of birth, and later population control.[38] Some, such as Raymond Pearl, were closely involved with the old eugenics movement before their disenchantment led them to become prominent advocates of population control.[39] Others, such as Muller and English radicals Huxley, Haldane, and Hogben, were never involved with old-style or mainline eugenics per se, but espoused the newer position. In either case, interest in population control was a natural for eugenicists on several grounds. Birth control for primarily eugenical purposes led to methods of limiting family size that would play a major role in large-scale population control programs after World War II.[40] Concern over differential fertility between classes or racial/ethnic groups was easily broadened into a differential between the "have" and "have-not" nations—or, as Huxley put it, the white nations of western Europe and the black, brown, or yellow nations of the rest of the world.[41] Finally, eugenic improvements in western capitalist countries could be swamped if the population of the rest of the world continued to increase at its current rate of growth. Huxley, like most of those who changed focus from eugenics to population control, carried with him a distinct eugenical ideology expanded to a global level. It was the problem of the differential fertility of lower social and biological stock viewed no longer on a class or national, but on an international, global scale. As Huxley stated as late as 1963:

> The population explosion is making us ask . . . What are people for? Whatever the answer . . . it is clear that the general quality of the world's population is not very high, is beginning to deteriorate, and should and could be improved. It is deteriorating, thanks to genetic defectives who would otherwise have died being kept alive, and thanks to the crop of new mutations due to fallout.[42]

Whereas mainline eugenicists had been mostly concerned with quality, population control ideologues were concerned with quality *and* quantity. A small island of quality (e.g., England) amidst a sea of degeneracy (much of the Third World, which Huxley had visited at various points throughout his life) would be a futile position.[43] Like his American counterpart Raymond Pearl, by the mid-1920s Huxley had begun to see eugenics in populational rather than individual or family terms. No doubt a product, at least in part, of his evolutionary interests, specifically deriving from the populational thinking of the architects of the evolutionary synthesis, Huxley's expanded views of heredity and evolution as properties of populations led him easily from advocacy of eugenics to advocacy of population control.

Influence of Left-Liberal Politics

Another feature of Huxley's approach to eugenics was the left-liberal political background that he brought to his bio-social concerns. While not as left-wing or radical as someone like Haldane or Needham,

Huxley was, nonetheless, much influenced by his left-leaning colleagues. One gets the impression that Huxley was, politically at least, "betwixt-and-between" much of the time. He never committed himself to a consistent radical program, such as Haldane did by joining the British Communist Party. Nor did he even espouse a consistent socialist line at any time. Yet he recognized and overtly pointed to dysgenic features of capitalism and nationalism, and he seemed willing to listen to and absorb particular radical ideas when they were brought to his attention.[44]

Huxley's association with left-liberal politics influenced his eugenical thinking in several ways. First and foremost, it overtly brought to his attention the highly reactionary, racist, and imperialist ideology behind much of the old-style, mainline eugenics. His Marxist and socialist friends raised questions about eugenics from an overtly political perspective. Equally important, Marxism was the only social theory that could claim to be rigorously scientific, and this proved highly amenable to Huxley's thinking. To Huxley, social problems, including those of reproduction, should be approached from a scientific rather than a superstitious, religious, or moralistic viewpoint. He seems to have appreciated in Marxism the scientific outlook as a means of understanding the past in order to change the future.

Eugenics and Scientific Humanism

Of particular importance in Huxley's eugenical thinking was his espousal of "scientific" or "evolutionary humanism"[45] and his simultaneous rejection of formal religion. Evolutionary humanism was a sort of philosophical belief system based upon the application of scientific methods to understanding human social, moral, and ethical life. It was an entirely naturalistic philosophy with no recourse to mystical powers, deities, or salvation. It was not merely a reductionistic approach claiming to solve human problems by the application of existing scientific principles (that is, it did not try simply to biologize human behavior). Rather, as Huxley put it:

> The basic postulate of evolutionary humanism is that mental and spiritual forces—using the term force in a loose and general sense—do have operative effect, and are indeed of decisive importance in the highly practical business of working out human destiny; and that they are not

supernatural, not outside man but within him. Regarded as an evolutionary agency, the human species is a psycho-social mechanism which must
operate by utilizing those forces. We have to understand the nature of
those forces; where, within the psycho-social mechanism, they reside;
and where their points of application are.[46]

The emphasis in evolutionary humanism is on human beings taking
charge of their own futures, of molding and controlling their own
destinies. This was to be accomplished on both the individual and the
collective social levels. The important point of evolutionary humanism was that it was of this world, focused on concrete human problems solved by human beings using rational, scientific methods.
Huxley emphasized the importance of understanding, and then controlling social processes in the same way we have learned to control
natural processes:

> Man has learnt in large measure to understand, control and utilize the
> forces of external nature; he must now learn to understand, control and
> utilize the forces of his own nature. This applies as much to the blind urge
> to reproduction as to personal greed or desire for power, as much to
> arrogance and fanaticism, whether nationalist or religious, as to sadism or
> self-indulgence.[47]

This conception was new, Huxley felt, because it involved working
within the framework of nature, not against it. Evolutionary humanism was thus a departure from the ethos that prevailed in his grandfather's day, which was one of "combating the cosmic process"; today,
Huxley felt, it was a matter or "wrestling with it."[48] By this phrase I
take Huxley to mean that human beings must take an interactionist
rather than adversarial role in dealing with natural and social forces.

The final two features of Huxley's reform eugenic outlook are in
some ways the most central to his own expressed views. The first is
his evolutionary outlook, and the second his commitment to the idea
of scientific management and social planning.

Huxley's Evolutionary Perspective

While all the older (including new-style) eugenicists were evolutionists and Darwinians, few (except perhaps Haldane and maybe Fisher)
were thoroughly imbued with the new evolutionary views emerging
in the 1920s and 1930s. Huxley's evolutionary views were much

influenced by his field experience on bird behavior in the 1910s and 1920s, and by his natural history background as secretary of the Zoological Society. By merging a populational approach to evolution with a synthetic view of Darwinism and Mendelism, Huxley avoided the simplistic view that Ernst Mayr was to call later "bean bag genetics." It is no accident that, among his many other accomplishments, Huxley coined the term "evolutionary synthesis" in the early 1940s.

Huxley's thoroughgoing evolutionary perspective impacted his eugenical thinking in several ways that distinguished him from virtually all other reform eugenicists. The first, as mentioned above, was his populational perspective. As part of this perspective Huxley was able to see the human population as a wide-ranging mixture of genotypes and phenotypes characterized by enormous variability. In no other species, he stated, was variability so wide-ranging (later studies from the 1940s onward were to suggest that *Homo sapiens* is not unique in this regard). Variability was the source not only of certain genetic defects, but also of beneficial variations that could become the raw material for future evolution. From an evolutionist's viewpoint, eugenics could move the human gene pool forward by selecting for desirable genes, as well as prevent degeneration by selecting against undesirable ones. One thing was clear: the human population was an evolving one—it was not static; change was inevitable. From the eugenicists' viewpoint that change could be directed toward desired ends, or allowed to proceed randomly toward undesirable ends.

Another aspect of Huxley's evolutionary perspective was his broad understanding of the complex problems associated with species and species-definitions. His own early work on "clines" dealt with the problem of variation within and between subgroups of a natural species group. As a result of his familiarity with the subtleties of species and subspecies definition, Huxley was able to avoid the typological conception of human beings—specifically the typology of race or ethnic group—that characterized the work of most of the older eugenicists. Huxley saw in the history of the human species the effects of repeated migrations and cross-breeding, suggesting that *Homo sapiens* was a panmictic, global population experiencing gene flow for hundreds of thousands or even a million years. There was thus no racial or ethnic "type," although there were recognizable genotypes and phenotypes associated with certain geographic regions. Still, his evolutionary and systematic perspective made him one of the first to suggest that in many cases variation within local geographic populations could be as large or larger than variation between populations. As a result, Huxley could dismiss the racial theories of earlier eugenicists on scientific as well as social and political grounds.

Huxley and the Movement for Scientific Management

The final aspect of Huxley's world view crucial to understanding his eugenical thinking is his commitment to the notion of social planning, or "scientific management." The concern with social control that motivated many eugenicists from the 1910s onward has been discussed by a number of historians in recent years.[49] In various forms this spirit pervaded the view of many Anglo-American eugenicists of the older as well as the younger generation. But nowhere does it emerge more overtly and consciously than in Huxley's thinking. The concept or "spirit" motivating the new group of social planners emerging in the 1910s and 1920s (especially in England) involved the use of rational, scientific methods to understand and thereby control both the natural and the social worlds. Central to this ideology was the belief that the planners had to be trained experts— as Huxley and many others saw it, middle-class professionals—who would advise lay groups (Parliament, social workers, medical professionals) with informed, scientific judgments. The planners were to be in some sense not just advisers, but integrally involved in the planning process.

As we have seen, Huxley continually spoke of eugenics in terms of planning and controlling human evolution. The human germ plasm had to be managed lest the forces of chaos and deterioration take over. That planning had to be done from a knowledgeable point of view with the facts of modern science in hand. Huxley, like many of his American counterparts, complained that no one doubted the wisdom of managing and controlling the germ plasm of agricultural stocks; yet the same concept was not applied to a much more important commodity, human stocks. Human reproduction had to be managed with the same knowledge and care that we lavish on our domesticated animals and plants. The agricultural analogy appears over and over in Huxley's writings on eugenics from the 1920s through the 1940s, as it did in the writings of many contemporary American eugenicists. [50]

Huxley did more than theorize. He was actively involved in various social planning groups in the 1920s and 1930s. He was one of the core dozen or so members involved in running the Political and Economic Planning committee (PEP), a self-constituted group of progressives who wanted to institute more expert planning at every level of society. He was simultaneously a member of the Next Five Year Group (NFYG), a similar association for which he served in part as a liaison with PEP. In the early 1930s, Huxley admired the Soviet Union's commitment to planning (with its five-year plans) and to the use of trained experts, or technocrats, in industry and agriculture. The next stage in human evolution, he stated, would emerge as human beings recognized that they could and must control their social, as well as their natural environment. A step toward realizing the full potential of such social change would be the ability to control the human genotype in a meaningful way. Whereas the social Darwinists of his grandfather's day were preeminently supporters of the philosophy of laissez-faire, Julian Huxley saw eugenics as a model of the new economic and social spirit of planning and control.

Understanding the strong commitment of many eugenicists to social planning, helps us see how the movement could draw diverse individuals ranging from right-wing reactionaries to communists. The period from the 1890s through the 1930s had seen considerable economic and political chaos in both England (Europe also) and the United States. Periodic recessions, increasingly militant and radical labor union struggles, high unemployment, especially following World War I, and the worldwide capitalist depression from 1929 to the mid-1930s led to constant social turmoil and political instability. Even within the wealthy class, by tradition more attuned to an attitude of laissez-faire in economic and social matters than the working class, increasing numbers of business and political leaders were calling for government planning and control in such areas as labor, import and export, health care (including the training of doctors and teachers), and drugs and food. The concept of planning attracted not only increasing numbers of the more conservative wealthy class, but also portions of the forward-looking, even radical, professional class. Aside from the new and important social/economic role they saw for themselves in a planned society, highly educated professionals such as Huxley sincerely believed that with rational, scientific management many social problems could be solved at their root. The idea of a planned economy was not a purview of either the political right or the left. It was on the common ground of rational, scientific control that progressive—even radical—and conservative eugenicists met and worked to extend the application of management to human reproduction and evolution. Among those who sought this goal, no one was more visionary, more influential, or exemplary than Julian Huxley.

DIANE B. PAUL

The Value of Diversity in Huxley's Eugenics

Garland Allen's thoughtful paper reminds us that eugenics served a wide variety of social ends. Militarism, class privilege, opposition to birth control, and female suffrage were defended and also (though less often) denounced "in the name of eugenics." All eugenicists did agree that differences in mentality and temperament are strongly influenced by differences in genes—hence that intelligence and many personality traits are potentially selectable. For the good of future generations, we should therefore "breed from the best." But here agreement ends. Who are the best? Is social success a reliable measure of genetic worth? What measures of selection are efficacious—and moral?

On these and other issues, eugenicists divide roughly into two groups, which Daniel Kevles has labeled "mainline" and "reform."[1] Mainline eugenics was associated with scientific naiveté and reactionary politics. In the mainline view, the quality of a person's genes is the most important determinant of social success and failure. Those with good heredity will not be thwarted by adverse environments. Conversely, social failure is generally the result of bad heredity, especially "feeblemindedness." The mentally defective are rapidly outbreeding their betters. Thus drastic action is needed. Since feeblemindedness results from a single gene, at high frequency in the population, a policy of sterilizing or segregating the affected would rapidly reduce its incidence.

Reformers, on the other hand, viewed most traits differently: as the product of many genes in complex relations with the environment. They also recognized that no policy could possibly prevent all the affected from breeding. Thus they argued that mainline eugenicists exaggerated the potential efficacy of their proposals. While reformers generally agreed that those at the very bottom, or "social problem group," were biologically inferior, they denied the blanket equation of social success with genetic worth. The effects of nature and nurture could be disentangled only in a society that offered equal opportunities to all its members.

Not all eugenicists fit easily into these broad categories. Was the scientifically sophisticated but politically reactionary R. A. Fisher a reform or a mainline eugenicist? But the categories do reflect, if only imperfectly, real divisions. The boundary between mainline and reform eugenics may be ill-defined, but something important surely separates Julian Huxley from those whom he condemned in his Galton lecture of 1936 for having "converted the distinction between nature and nurture into a hard antithesis, and deliberately or perhaps subconsciously belittled or neglected the effects of the environment and the efforts of social reformers."[2] As Allen notes, Huxley always stressed that the effects of nature and nurture could not be distinguished in a class-based society. He was among the most influential of those who deployed eugenic arguments in the service of social reform.

Allen also suggests that Huxley's thoroughgoing evolutionism produced a perspective on eugenics that was unique even among the reformers. He places particular emphasis on Huxley's celebration of genetic diversity. In Allen's view, Huxley's "populational" thinking led him to emphasize the wide range of genotypes and phenotypes in the human species and the importance of this variability for future evolution. Variability is the source of defects. But it also provides the "raw material" of evolutionary change.

There is no doubt that Huxley always stressed the extent and value of genetic diversity. From his perspective, there was no such thing as a "best type." He thought the analogy between artificial and natural selection to have been seriously misleading in this respect. The stock-breeder aims to produce breeds highly specialized for specific traits, such as milk yield in cattle or speed in race horses. Each breed thus has a much lower variance than the parent species. But such methods applied to human beings would only bring disaster. Human societies benefit from many different skills and qualities, such as physical beauty (which Huxley strongly valued), health and energy, aesthetic and moral sensitivity, manual dexterity, leadership, and scientific genius, and these are not always, or even usually, linked. Huxley's position is succinctly stated in his second Galton lecture of 1962: "Man owes much of his evolutionary success to his unique variability," he wrote. "Any attempt to improve the human species must aim at retaining this useful diversity, while at the same time raising the level of excellence in all its desirable components, and remembering that the selectively evolved characters of organisms are always the results of compromise between different types of advantage, or between advantage and disadvantage."[3]

How exceptional were Huxley's views on diversity? Did they really distinguish his thinking from that of other reform eugenicists? Huxley himself thought not. Indeed, he consistently stressed the ordinariness of his position. To a point, he was right. No eugenicist could deny the fact of genetic variability; without substantial selective variance for important traits, eugenics is pointless. And no eugenicist—reform or otherwise—denied that genetic variation provides the raw material for evolution, and is thus sometimes advantageous. The disputed questions concern its extent and social implications. How much of the standing variation is favorable, or might be in the future? Those who answered "very little" tended to have a different perspective on social policy from those who answered "a lot." Ironically, it was one of Huxley's closest confidants, the geneticist H. J. Muller, whose work was most associated with a variation-reducing view of selection, and hence of eugenics. (Huxley had brought Muller to what was then the Rice Institute in 1915.) In the 1950s and 60s, Muller engaged in a bitter polemic with Theodosius Dobzhansky over the value of genetic diversity. It was a dispute that generated considerable tension for Huxley, whose substantive views were much closer to those of Dobzhansky than to those of his friend, Muller.

In brief, Muller stressed the "precision of adaptation." Since organisms are well-adapted to their environments, nearly all mutations are unreservedly bad, and are removed by selection. Of course favorable mutants sometimes appear, and these provide the raw material for evolution. But they are extremely rare and rapidly become the new normal or "wild type." Most genetic variation is thus transitory. Or at least it would be in nature. But humans have both increased the rate of mutation and decreased that of selection (primarily though improvements in medicine and public health). As a result, the species is genetically deteriorating. A variation-reducing eugenics program is thus urgently needed.

Muller's severest critic was Theodosius Dobzhansky. In Dobzhansky's view also, some variation was unreservedly bad. But he stressed the heterogeneous and changing character of environments—hence the need for a store of genetic variability. Given this need, selection would generally act to preserve variation. As he wrote in a 1953 letter to Huxley: "It does look that balanced polymorphism is of greater importance in adaptive evolution of sexual cross-fertilizing species than we have imagined. . . . This may mean that what we regarded as lethals and hereditary diseases are in reality the raw materials from which the species constructs the co-adapted gene combinations. It will be very useful to consider from this standpoint some of the old problems of human genetics—and eugenics, of course."[4] In other words, disability and disease may be the price a species pays for evolutionary flexibility.

By the late 1950s, Dobzhansky had come to focus almost exclusively on one form of balancing selection: heterozygote advantage or "overdominance." If heterozygotes are generally fitter than homozygotes, then genetic variability is good for individuals as well as species. Dobzhansky had been greatly influenced by the experimental results of his student, Bruce Wallace, who irradiated fruit flies and found that the treated group, with their induced heterozygosity, had a greater viability than the controls. His experiments were seized on as evidence for the virtue of heterozygosity per se.[5] Heterozygote advantage explained why some deleterious genes were maintained at high frequency in the population—for example, the allele that in double dose produces the serious disease, sickle-cell anemia, but when paired with a normal allele only mild symptoms and a more than compensating protection against malaria. One cannot—and would not want to—select against genes of this type. If overdominance were common, eugenics would thus be pointless. (Muller conceded the sickle-cell example, but denied its generality.)

In Dobzhansky's view, Muller seriously underestimated the value of diversity, both genetic and social. Nowhere was this more evident than in Muller's eugenics, which (according to Dobzhansky) aimed at an evolutionarily disastrous uniform type. In his oft-quoted 1962 book, *Mankind Evolving*, Dobzhansky charged that the logical extension of Muller's philosophy would be selection of "the ideal man, or the ideal woman, and to have the entire population of the world, the whole of mankind, carry this ideal genotype."[6]

One of those asked to review the book was Julian Huxley. As early as 1932, he had strenuously denied that any eugenicist held such a view. "No eugenist in his senses ever has suggested, or ever would suggest, that one particular type or standard should be picked out as desirable, and all other types discouraged or prevented from having children," he wrote. "Here biology joins hands with common sense. The dictum of common sense, crystallized into a proverb, is that it takes all kinds to make a world."[7] Now his friend Muller was accused of espousing the very view that Huxley had deemed absurd. He sent Muller a letter, asking for assurance that their views did not conflict. "Surely this is a serious misrepresentation of your (and [Herbert] Brewer's) views?" he wrote, and noted that in his Galton lecture he had pleaded for "varied excellence" to be achieved through married couples' free choice of donors. He asked: "Isn't this your view too?"[8]

Muller quickly replied that Dobzhansky's assertion and similar remarks by L. C. Dunn were "entirely slanderous." He continued: "These are really vicious and unfair attacks by people who do not want to see their own do-nothing stand superseded."[9] A few days later, he wrote again, clarifying his position. Of course he did not believe that there is a single ideal genotype for man. But this is not to deny real differences between him and the Dobzhansky school. "I should not want to hide the fact," he wrote to Huxley, "that I do not share the fantastic view of Dobzhansky and Bruce Wallace that the most advantageous condition for an organism is to have a state of balanced multiple allelism at the great majority of loci, with the further principle acting that the more multiple alleles there are at a locus the better. That is a purely ad hoc construction to bolster their old-fashioned and reactionary view of heterosis. . . ."[10]

Muller never advocated a single human genotype, even in his 1935 eugenic tract *Out of the Night*, where he proposed a program of mass artificial insemination of women with the sperm of particularly estimable men. (Dobzhansky's characterization of Muller's view was based on this proposal.) But as Muller notes, their differences were real. Muller may not have wished for absolute uniformity, but neither did he place the same value on diversity as Dobzhansky—or Huxley—did. A commitment to diversity is today thought to be a Very Good Thing. As a result, Dobzhansky has received much better press than Muller. For Allen, it is clearly one of Huxley's saving graces. But I would like to suggest that the correlates of a commitment to diversity are not always progressive.

We have already seen that, in Dobzhansky's view, suffering and death is the price paid for evolutionary flexibility. Reflecting on this dilemma, Bruce Wallace once remarked that the concept of overdominance was "morally deficient."[11] Overdominance is not really moral or immoral, but it easy to see why even Wallace found it unappealing. Equally unattractive (from at least some perspectives) is the association of diversity with an efficient division of labor. When the reform eugenicists promoted equality of opportunity, it was as a means of separating the genetic sheep from goats—hence, of fitting people into their "natural" slots. As Huxley remarked, it takes all kinds to do the world's work. We might recall that there is an important subgenre of eugenic literature consisting of articles that explain "Why the World Needs More Morons." H. H. Goddard's famous (but apparently rarely read) 1917 essay, "Mental Tests and the Immigrant," is a case in point. That article is usually characterized as a plea for immigration restriction. It is not. For one thing, Goddard believed that the feeblemindedness of immigrants (unlike native WASPs, such as the Kallikaks) was mostly environmental in origin. But he also believed that, in any case, "there is an immense amount of drudgery to be done, an immense amount of work for which we do not wish to pay enough to secure more intelligent workers."[12]

In 1975, the geneticist Jack King reviewed a book by Richard Lewontin on the history of the Muller-Dobzhansky dispute.[13] In his view, Lewontin was wrong to equate Dobzhansky's position with a commitment to social change and Muller's with a defense of the status quo. On the contrary, Muller was a (sometimes) Marxist, committed to the perfectability of human beings. "In fact," writes King, "he was something of a nut on the subject, being justly ridiculed for honestly believing that any sensible woman would prefer the semen of Great Men to that of her own inferior husband." He continues:

> Dobzhansky prefers the human species as it is found, warts and all; low I.Q.'s, dyslexia, schizophrenia, bad backs, obesity, myopia and all ('differences are not deficits'), because human variability is beneficial in that it makes for an efficient division of labour. It is somehow advantageous if labourers are illiterate, if intellectuals have flabby muscles, if baseball players are dull-witted, if artists are tone-deaf and musicians colour blind, if engineers are inarticulate and poets cannot add. If someone is unlucky enough to have all these characteristics, well, homozygotes must perish so that heterozygotes can flourish. If Muller's dissatisfaction with the human species as it is seems overly alarmist, and if his plans for future improvement seem absurdly optimistic, Dobzhansky's satisfaction with the status quo of human biological inequality is depressingly sanguine.[14]

King was not being completely fair to Dobzhansky, who believed that most people were capable of doing most jobs. But he is right to note that Dobzhansky also (and perhaps inconsistently) valued diversity as a means to divide up the world's work efficiently. As he wrote in *Mankind Evolving*, "Equality of opportunity tends to make the occupational differentiation comport with the genetic polymorphism of the population," and went on to note that equality of opportunity "would be meaningless if all people were genetically identical."[15] The latter phrase is a good indication of the strong hereditarian perspective shared by all the reform eugenicists. Given the assumption that differences in mentality, temperament, and character result in large part from differences in genes, different people are necessarily suited to different jobs. Equality of opportunity makes the process of social sorting efficient. For Huxley and the other reformers, science now determines the division of labor . This is perhaps an improvement on tradition and prejudice. But it is not, from every perspective, the ideal.

ELAZAR BARKAN

The Dynamics of Huxley's Views on Race and Eugenics

Garland Allen presents very persuasively the conflicting tendencies in Huxley's attitude to eugenics and race, the tension and ambiguities between his liberalism and elitism, and shows how Huxley integrated these views in the broader evolutionary, as well as political and social, agendas. I especially like the way Allen stressed the priority Huxley gave to his own beliefs when they contradicted scientific knowledge. In my comments, I would like to suggest a different emphasis for a couple of aspects presented by Allen but minimized in his conclusions: first, Huxley's view of race, and second, the question of reformed eugenics.

Allen reminded us of Huxley's book *We Europeans*, which served as a milestone in his opposition to racism, a road upon which he embarked during the thirties as a retaliation against Nazism. Against this recognition of Huxley's commitment to liberalism, I would like to elucidate a different facet of his attitude to the question of race.

It has been pointed out that Huxley's years at the Rice Institute between 1912 and 1916 had a crucial influence upon his intellectual maturity, and this was perhaps most evident in his understanding of the concept of race. During these formative years Huxley first encountered alien people, "others," those whom he knew about, but had never met personally. In my comments I will focus on the years before 1930, a period in which Huxley's ambiguity concerning eugenics and race is most evident.

Huxley's liberalism in these early years was a reflection of elitist English upper-class attitudes toward the others, be they the races of the Empire, the lower classes in England, or blacks in the American South. Huxley's ambiguity represented great trust in the existing social order, of which he was a major beneficiary, combined with pessimism toward the growing numbers of others: non-Europeans and poor alike. This reflected the pessimistic side of his eugenics. His fear of the population explosion was not based on sheer numbers, rather—as Allen justly showed—it stemmed from the composition of those who multiplied so fast. The "economic, technical and cultural progress" of the human species was threatened in Huxley's view "by the high rate of increase of world population," especially "the high differential fertility of various regions, nations and classes." The essence and the target of Huxley's negative eugenics never changed.[1]

Huxley's concepts of race and eugenics developed simultaneously, moving from a focus on a simplistic group dichotomy of superiority and inferiority to a hierarchical system of genetic quality among groups and individuals. In later years, Huxley paid greater attention to the environment and to the importance of social policies of "leveling up" as a preliminary eugenic measure. However, Huxley continued to emphasize the marginal superior genetic endowment that gave the edge to some groups—Europeans over non-Europeans; patricians over plebes—an advantage that was manifested primarily in a small, elite minority.

The influence of Huxley's years at Rice on his racial views is documented in an unpublished and unfinished manuscript from 1918 entitled "The Negro Minds." He referred to his Texas sojourn:

> I had grown up in the transplanted Bostonian conviction, so common in Liberal circles in England, that the negro being a man is therefore a brother, being a brother is therefore an equal. . . . A little life in the Southern States soon taught me more about their minds, more about inborn differences of race.[2]

This presented a conventional misconception about English (and Bostonian) egalitarianism, since many shared Huxley's racial aversions. But the significance of Huxley's confession lies in elucidating the liberal stand on racial differences at the end of World War I.[3] It is important to realize when we speak about a shift from mainline to

reform eugenics that a decade earlier the reform leaders shared the
determinism of the mainliners. Huxley recognized these ambiguities
in his work, and was aware that his liberal humanism had suffered a
humiliating defeat from southern racism. Returning to the United
States in 1924, Huxley revealed glee at the Yankees in his comments
on the blacks' migration to the North:

> The Middle Westerner and to a lesser extent the Yankee are for the first
> time experiencing the negro at baulk and at first hand; and there is a
> certain grim humour in seeing their high moral principles and lovely
> theoretic equalitarianism dissolving under the strain.[4]

Having shared the shame, Huxley was uninhibited in illustrating
his own distaste for blacks:

> You have only to go to a nigger camp-meeting to see the African mind in
> operation—the shrieks, the dancing and yelling and sweating, the sur-
> render to the most violent emotion, the ecstatic blending of the soul of the
> Congo with the practice of the Salvation Army. So far, no very satisfac-
> tory psychological measure has been found for racial differences: that will
> come, but meanwhile the differences are patent.[5]

Huxley's biological determinism encompassed both his eugenic
and racial views. Racial analogies provided an easy literary tool to
introduce the English reader to a similar, but more doubtful, argument for the existence of a comparable genetic gap between classes as
between races, a cleavage that was "inherent and unavoidable."[6]
Environmental advantages notwithstanding, Huxley wrote:

> Baboons or Australian savages can have all these advantages, and will not
> blossom beyond their limits—limits set by their inheritance.[7]

And so he asked, "Why then try to deny the equally obvious fact of
inherited, germinal differences between human beings of a single
nation?"[8] For Huxley, racial and eugenic issues addressed the same
questions.

Among Huxley's most revealing accounts of his views on the
masses, and on alien races, is the chapter titled "Racial Chess" in his
book *Africa View*. Here Huxley recounts how he had first come to
fear blacks:

> I remember once, in central Texas, arriving by car in a little town whose
> streets were crowded . . . almost wholly by negroes; there were hundreds
> of black men to tens of white. I am bound to confess that this first

experience of mine of being in a small minority among human beings of another physical type gave me an emotional jolt; and I began, without any process of ratiocination, to understand why white men living in such circumstances generally took to carrying revolvers and developed a race complex.[9]

This race complex never left Huxley, who admitted that "one could doubtless get over such feelings; but the point is that they arose unbidden."[10] Huxley clearly recognized this tension. His personal aversion to aliens conflicted with his moral commitment to liberalism, but it was only in later years that the conflict inhibited Huxley from pronouncing his bigotry.

By 1930, facing the changed political situation, Huxley distanced himself with the help of Lancelot Hogben from these beliefs, concealing his aversion to aliens and espousing a more sophisticated view of what constitutes inequality among groups. The shifting political scene supplied Huxley with ample emotional support for this growing commitment against racism.

This was first shaped during his journey to East Africa as a delegate of the Colonial Office Advisory Committee on Native Education. His consequent recommendations turned out to be more in line with Hogben's beliefs than with his own of a few years earlier. When the tension between his own political convictions and racial prejudices became a burden for Huxley, he preferred political liberalism and attributed racial prejudice to some distant party, such as "the Dutch at the Cape, or the whites in the Southern States," who "quite sincerely believed that the black men were separated from white by a great gulf which could never be bridged." His egalitarian refuge was that "white and black overlap largely in regard to intelligence, energy, ability and character," leaving the crucial margin in favor of the whites.[11] This remained his view for the next four decades, but while in the thirties it was considered progressive, in the sixties it became conservative.

The more Huxley learned about the "others," the more he came to respect their humanity. His first step was to realize that blacks were not all alike. In January 1930 Huxley published the news in the *Times*: "The Africans, though the ignorant persist in classing them all as merely blacks, natives, or even niggers, show more variety of physical type and way of life than is to be found in all Europe." His second step was to narrow the gap between blacks and whites: "The most important point to realize is that not one of the East or South African tribes is pure Negro; all have a Hamitic admixture, and some are in blood and physical type as closely akin to men of Southern Europe or the Near East as to the negro of West Africa."[12] Here was the transformation to a much more sophisticated view. Huxley recognized the differences in the others and began to treat them as human beings and not intermediate species.

Under these circumstances Huxley began to object to the concept of race: it conveyed more than was justified. "The term [race] is often used as if 'races' were definite biological entities, sharply marked off from each other. This is simply not true." Using scientific terminology, he wrote: "Any given race is characterized by containing within its boundaries a certain assortment of genes. One racial assortment will differ from another in the nature and proportionate abundance of the different kinds of genes of which it is made up, and every race will share some genes with many other races, probably with all." Through migration, slavery and commercialization, "every gradation [exists] between negro and full white." This egalitarian approach still carried with it, however, the old belief that the blacks' mental progress was due to their mixture with more advanced groups: "The Bantu, and still more the Hamitic peoples, have a considerable proportion of more or less white and quite definitely Caucasian blood in their make-up."[13] This duality in language gives the clue to Huxley's double standards: his use of "genes" in scientific discourse when the argument was empirical, rationalist and antiracist; but he yielded to "blood" when racial prejudice gained the upper hand.

Though Huxley continued to view tribes like the Bushman and Pygmy as "descendants of earlier and more primitive buddings of the human stock," as an egalitarian he was concerned with "the average, and the average among the more widespread and successful tribes." As long as Huxley wore the hat of the scientist, the racial differences seemed small and temporary; however, once he employed conjectures, his prejudices crept in. Nowhere was this more evident than in the area of psychology. "Human beings do differ very considerably in inherent mental capacities and potentialities," he wrote, and "there is not the least reason why races should not differ in the average of their inborn mental capacities as they do in their physical traits . . . on biological grounds the[se differences] *should* exist."[14]

Huxley's ambivalent attitude is best illustrated when he argues for European superiority, however small:

> I am quite prepared to *believe* that if we ever devise a really satisfactory method of measuring inborn mental attributes, we shall find the races of Africa slightly below the races of Europe in pure intelligence and probably certain other important qualities. . . . But—and the but is a big one—I am perfectly certain that if this proves to be so, the differences between the racial averages will be small . . . and that the *great majority* will overlap as regards their innate intellectual capacities.[15]

"Great majority" meant no total overlap, which enabled Huxley to concurrently retain his paternalistic views on race and class and be part of the liberal camp, as these views reflected contemporary cultural and scientific values. Human evolution, however, was determined in Huxley's opinion by a margin of excellence that he estimated to be 5 percent, and that determined the European superiority vis-à-vis others. In time, Huxley assigned a greater role to environment and his views on heredity became more sophisticated. Yet while emphasizing general equality, he continued to attribute to genetic endowment a crucial role in elevating the elite above the masses.

This brings me to the second part of my comment on reform eugenics. It seems to me that it is possible to look at the change in the eugenic movement from the late twenties into the forties as a shift of a spectrum rather than as a cleavage or a dichotomy from an old to a new separate entity, from mainline to reform eugenics. Spectrum is a better representation because it emphasizes the continuation within the movement, which was at least as important as the change and innovation. This continuity can be seen on three levels: agenda, institutional, including the main actor, and public reception.

Both the old and young generations in the eugenics movement adhered to what they considered the correct scientific agenda. Even when eugenics was hardly contested, the eugenicists attempted to update their work according to the advances of science. Even Davenport's last major research (on race crossing in Jamaica) was motivated precisely by his attempt to integrate the study of the environment into eugenics and to show the dominance of heredity. His view persisted, but nonetheless he was responding to criticism.

Allen's distinction between good progressive science of population genetics and bad (biometric or Mendelian) of the old generation compares the two generations on an absolute, rather than historical scale. From this perspective one can say that the science of heredity had progressed during those years, and with it eugenics. But this presents eugenics merely as an appendix to biology, lacking independent development. Had that been the situation, one would look for the source of transformation not within the movement, but rather in biology.

Another characteristic claimed in the name of old eugenics is that it was dominated by nonscientists. But this is true only very broadly. There were always nonscientists in the movement, before and after the thirties. The earlier domination of racist ideology was a result of a general consensus accepted by liberals and conservatives alike, as is clear with regard to Huxley. The nonscientists did not differ in this respect from the scientists in the movement.

The institutional frame of the movement remained largely the same until World War II. Clearly there was no revolution, rather continuity with a generational change. The continuity is evident even in the role of the main actors. The new leaders were mainliners and insiders. Both Blacker in England and Osborn in the United States were elected when the movement was dominated by the old generation, and both were perceived in the late twenties as continuing the old tradition. When the movement became defensive against growing academic and public criticism, the shift was motivated by political astuteness together with a growing understanding of the mechanism of heredity. Huxley remained all these years a member of the inner circle of the eugenics movement. Blacker was his former student, and as their correspondence shows, the policies of the movement were determined by a small leadership, of which Huxley was an intimate member. The inside group kept nonmainliners away as a defensive move, and these included radicals and conservatives alike: Lancelot Hogben (despite his close personal friendship with Huxley), Ronald Fisher and his circle ("Rothamsted lobby"), and Ruggles Gates. The leadership tried to censor the radicals in order to preserve minimal public credibility for the movement.

Instead of defining what is mainline eugenics and then evaluating whether Huxley belonged to the mainline or the reform section, we should see Huxley for what he was, namely a liberal and a mainline eugenicist, as was his role in the movement. Then, instead of looking at eugenics as a given entity and classifying the participants into groups, we should look at the participants and evaluate the change in the movement. Then we would talk about the shift of the spectrum, not of a revolution.

The discussion of internal changes in the movement should not overshadow the acknowledged fact that the major change was the movement's decline, not its transformation. A relatively small number remained affiliated with the movement, and few joined it. Clearly whatever internal changes were taking place were initiated by relatively moderate and elitist insiders of whom Huxley was a prominent member.

DANIEL J. KEVLES

Huxley and the Popularization of Science

Julian Huxley's career as a popularizer of science began with a salamander. More precisely, the animal was an axolotl, a Mexican salamander that normally lives an aquatic life as a tadpole with gills and a
dorsal fin but that sometimes metamorphoses into an air-breathing
amphibian. The metamorphosis had been first observed by A. Duméril at the menagerie of Paris' *Muséum d'Histoire Naturelle* in 1865,
but Duméril had succeeded only marginally in attempts at inducing
it. Huxley knew that the administration of mammalian thyroid extract could speed up the metamorphosis of a tadpole into a frog. He
wondered what the application of thyroid extract to the axolotl would
do. At the end of November 1919, Huxley and a colleague began
feeding ox thyroid to two five-inch-long axolotls. Within fifteen
days, the animals began to change color and to absorb their fins and
gills into their bodies. In a few more days, both animals were breath-
ing air and one of them was walking on land.

Huxley published a note about the metamorphosis in *Nature*, on 1
January 1920. The British press took notice, proclaiming, among
other things, that young Huxley had found "the Elixir of Life."
Young Huxley had not found any such thing, and he went to the
trouble to explain that fact to the press, stressing in a long letter that
his experiments implied nothing about the chemical transformation
of human beings. The letter, which offset damage to his reputation
among scientists, was, in effect, his first attempt to write about

science for a general audience—and, as such, it launched his career as a popularizer.[1]

There was, one might easily say, a Huxley family predisposition to interpret science to laypeople. Huxley's grandfather Thomas Henry Huxley was one of the most famous evangels of science in the English-speaking world of the late nineteenth century. Thomas Henry Huxley was not only Darwin's bulldog but an educator of science across the social structure, renowned for pioneering regular scientific lectures to groups of working people. He was a presence in the Huxley family even after his death, and Julian, who as a young child knew him briefly, felt his grandfather's intellectual and social legacy. While on the Rice Institute faculty, Julian Huxley was frequently invited to lecture, not only in Houston but throughout Texas, partly because of the name he bore.[2]

Julian Huxley happened to come along as a popularizer of science when interest in science and technology was mushrooming, and when new means—the broadcast medium especially—to respond to that interest were at hand. The hunger for technical information stemmed to a considerable extent from the triumphs of twentieth-century science—the advent of relativity, quantum mechanics, and genetics, the marvels of chemical products, the powers of scientific medicine, the fascinations of astrophysics. Yet the interest was tempered with ambivalence. Many of the issues raised by late-nineteenth-century science—especially the implications of biology for religion—remained unsettling. Added to them were worries about whether science and technology were outrunning the capacity of their creators to control them, and whether—in the wake of a brutal world war—humankind was capable of moderating the behavior of its bestial self.[3]

Under the circumstances, Huxley felt no need to apologize for his popularizing. In the mid-1920s, he declared, "One of the duties of scientific men—not necessarily of all of them, but certainly of some of them taken as a group—is to make available to the lay public the facts and theories of their science. . . ." He saw "a danger, in these days of manifolded information and broadcast amusement, that the world will become divided into those who have to think for their living and those who never think at all," warning that the "possible isolation of science. . . is a real danger," and that "the more democratic the general civilization . . . the greater is the danger."[4]

Huxley was astonishingly prolific in the traditional medium of popular print. He published numerous articles in general-circulation magazines and he wrote some twenty books of popular science. The first—*Essays of a Biologist*—appeared in 1923; the last, a book on Charles Darwin, was published in 1965. One of them was monumental—*The Science of Life*, which first appeared in thirty-one fortnightly parts, beginning in 1929, and which was brought out between hardcovers in 1931. Written with H. G. Wells and his son, G. P. Wells, it was conceived as a sequel to *The Outline of History*, aimed at the ordinary man, and scored a great success. Huxley's popular writings, always illuminating, ranged from the intellectually demanding—for example, *Evolution in Action*—to instructive entertainments such as *Animal Language*, or *Ants*, which was published in 1930 at the price of sixpence.[5]

Some of the books originated in magazine articles; others in lectures (for which Huxley was much in demand); still others as radio talks. Huxley was one of the first popularizers of science to exploit the broadcast medium. Beginning this phase of his career in the 1920s, he was highly successful at it—perhaps, Ronald Clark has suggested, because of his experience as a university lecturer. In a series on birdwatching and bird behavior, he taught thousands of listeners what they might see among the birds in the neighborhood of the front doorstep. During World War II, he was heard on the radio as the biology expert on the so-called "Brains Trust," whose members held forth during prime time answering abstruse questions.[6]

Huxley infused his secretaryship of the London zoo, from 1935 to 1942, with his commitment to popularization. He wanted, he said, to make the zoo "more than a menagerie," and proposed that "it might become the centre and focus of popular interest in every aspect of animals and animal-life." Huxley was especially concerned to reach children. He appointed a new cadre of assistant curators and encouraged them to give lectures to young people. He fenced off the Fellows' Lawn to establish Pets' Corner, where in good weather children could stroke and be photographed with a lion cub, Shetland pony, small python, and young chimpanzee, among other animals. During the 1935 Christmas holidays, thousands of school children visited special zoo exhibitions that Huxley put on, one designed to illustrate evolution in animals—his *At the Zoo* was suffused with a similar theme—

and another to demonstrate Mendelian heredity. He established a studio of animal art—a building large enough to accommodate lions or tigers and also up to twenty-five students who could come to sketch or paint the beasts. News from the zoo often appeared in the papers, not least because during Huxley's regime a daily press conference was established. And Huxley helped initiate the publication of *Zoo Magazine*, which combined instruction with entertainment and quickly reached a monthly sale of 100,000 copies.[7]

By 1930 Huxley was famous. Readers of *The Spectator* ranked him ahead of James Jeans, Ernest Rutherford, and Bertrand Russell while including him in the lists of Britain's five best brains. One or another Huxley popularization always seemed to be in print. His prowess as a scientific popularizer remained high through World War II and afterward. In 1953, for his work in popularizing science, Huxley was awarded the Kalinga Prize, one thousand pounds, which had been established under the auspices of UNESCO by an Indian who believed that it might encourage the spread of scientific knowledge more widely among the people of his country.[8]

Huxley's powers as a popularizer of science derived in no small part from a keen appreciation of the English language, an ability to use it well, and a sensitivity to the implications of science for fundamental human—and humane—concerns. A product of English public schools and Oxford, he was well and liberally educated. He read Homer and Horace and Catullus for pleasure, sometimes wrote verse in Latin and Greek. While an undergraduate at Oxford, he won the Newdigate Prize for English verse. (It must be added that he spent the prize sum of fifty pounds on a binocular microscope.) One of Huxley's books, popular for more than twenty-five years, was a small volume of poems entitled *The Captive Shrew*. He was a competent if undistinguished versifier, whose poetry was remarkable rather more for its ideas than for its music, reflecting, to quote his rhyme, man's capacity to "sit and look / In quiet, privileged like Divinity / To read the roaring world as in a book."[9]

Huxley's popular scientific prose was enriched here and there by his devotion to poetry. For example, he discussed a male hunting spider that offers the female a nice fly neatly wrapped in silk—adding that if the male found himself in a box from which the female had recently been removed, he would still wrap the fly rather than eat it, and search, "like Shelley with his bouquet, 'That he might there present it!—Oh, to whom?' "[10] Yet the principal contribution of poetry to Huxley's prose, no doubt, was that it helped to make him a master of exposition, a writer of supple and limber style.

To be sure, at times he wrote about nature in a conventionally romantic manner, celebrating, for example, how evolution had realized wonderful possibilities such as "the flowers carpeting the soil, the great trees with the singing birds in their branches, the glistening fish among the reefs of coral. . ." and so on.[11] Yet even his romantic passages often involved unconventional, specific images, a number of them drawn from his own research. Among many examples that one might cite is his description of the recurrent behavior of a male and a female heron in each other's company:

> Generally the hen sits on a lower branch, resting her head against the cock bird's flanks; they look for all the world like one of those inarticulate but happy couples upon a bench in the park in spring. Now and again, however, this passivity of sentiment gives place to wild excitement . . . the two birds raise their necks and wings, and, with loud cries, intertwine their necks. . . . . The long necks are so flexible that they can and do make a complete single turn around each other—a real true-lover's-knot! This once accomplished, each bird then—most wonderful of all—runs its beak quickly and amorously through the just raised aigrettes of the other, again and again, nibbling and clapping them from base to tip. Of this I can only say that it seemed to bring such a pitch of emotion that I could have wished to be a Heron that I might experience it.[12]

Part of what makes Huxley's nature writing compelling is its graphic punch. He explained that crustacea are limited in size by their habit of molting, pointing out that "a crab as big as a cow would have to spend most of its life in retirement growing new armour-plate." He could render the momentary apprehension of a young bird on the verge of its first flight by reporting the "fearful shrill chirping" that the fledgling gave out as it tottered on the edge of its nest, working up courage.[13] He could make the reader not only see but hear the vital activities of nature. He wrote of his approach to a great cliff bastion of birds in Norway that the din of the birds could be detected two miles away and that a half mile away the noise they made "was like the parrot-house at the Zoo heard from just outside," adding, "The chattering and screaming of hundreds of birds blended into one continuous roar."[14] He capsuled the self-transforming powers of developing organisms in a few sentences that reminded human beings that they were biological organisms, too:

> You, like me and every other human being, were once a microscopic spherical ovum, then in turn a double sheet of undifferentiated cells, an embryo with enormous outgrowths enabling you to obtain food and oxygen parasitically from your mother, a creature with an unjointed rod—what biologists call the notochord—in place of a jointed backbone; you once had gill clefts like a fish, you once had a tail, and once were covered with dense hair like a monkey; you were once a helpless infant which had to learn to distinguish objects and to talk; you underwent the transformation of your body and mind that we call puberty. . . . "[15]

Huxley obviously delighted in the stuff of biological science and much of his writings are devoted to sharing with the public his wonder at the marvelous variety and comprehensible logic of life. Yet there was another—and equally important—side to Huxley's efforts at popularization: the implications of biological science for human affairs. He made no sharp distinction between the two categories. They were certainly joined in his mind in his well-known commitment to wildlife preservation, a subject that, not surprisingly, long occupied part of his writings. In a 1930s essay, he warned:

> the world's rhinos are being slaughtered because of the belief of Indians and Chinese in the aphrodisiac qualities of their horns; the whales are being dangerously reduced to make big profits for their slaughterers. . . . fashionable women are still responsible for the death of some of the most beautiful winged creatures in the world . . . ; lizards and snakes are being killed out for shoes.[16]

To Huxley, such depredations not only diminished the animal kingdom but diminished human beings, too, by involving them in the destruction of their natural environment. "Must we confine our knowledge of animals," he asked, "to dead specimens in museums instead of making the world a living museum?"[17]

Huxley often combined between the same hard covers his essays on biology with others on human affairs. For example, the table of contents of his *Man Stands Alone*, which was published in 1941 but contains pieces written between 1926 and 1941, includes several essays on biology as such—notably, "The Size of Living Things," "The Courtship of Animals," "The Intelligence of Birds," "The Way of the Dodo"—and several other essays on biology and man—for instance,

"Eugenics and Society," "Climate and Human History," "The Concept of Race," and "Scientific Humanism." His wonderful book *Ants* is not only a primer on the lives and habits of these remarkable creatures but also an insistent instruction that analogies made between them and man are false and misleading—that there are many "radical differences between social insects and social man."[18]

As the inclusion of the essays on human affairs suggests, his popular scientific writings contained considerable social point. He was a well-connected man of the world, knowledgeable enough about its prominent inhabitants to write an appreciative if somewhat tongue-in-cheek review of *Who's Who, 1935*. ("Ernest Hemingway," he noted, "has had the courage to include *drinking*, [but] nowhere can I find either *gambling* or *women* as recreation.")[19] He was also much involved in the salient social issues of his day, especially those that were in some way connected to biology. Not surprisingly, a good deal of his writing on science and human affairs reflected his social and political commitments.

A feminist of sorts, Huxley discounted sweeping assertions that were often made regarding allegedly innate differences in the aptitudes of men and women. He insisted that what was observed about these differences was a product of upbringing, citing in ironic authority the exclamation of the third-century Greek gossip writer Athenaeus, "Who ever heard of a woman cook?" He found in the courtship, ritual displays, and mating habits of birds not only affirmations of the family but demonstrations that "the family life of birds attains its highest development in these forms which have, we may say, equal sex rights and duties."[20]

Huxley held that sexual compatibility was essential to the happy marriage, that women deserved sexual satisfaction as much as men, that there was nothing wrong or degrading about sexual pleasure dissociated from procreation. He endorsed divorce and birth control. Huxley actively campaigned for contraception, earning the condemnation of Lord Reith for sullying his BBC ether by discussing the subject on the airwaves. However, like his friend J. B. S. Haldane, Huxley was caught between the internalized morality of his Victorian upbringing and his rebellious codes of reasoned belief. He suffered repeated nervous breakdowns, some of which he attributed to "my unresolved conflicts about sex." In a courageous essay on

"Sex Biology and Sex Psychology," he argued that "the bulk of men and women cannot treat sexual problems in a scientific spirit, because of the store of bottled up emotion in the wrong place that they have laid up for themselves by their failure to come to proper terms with their sexual instincts."[21]

Huxley's politics tended to a tepid middle-of-the-roadism until he was jolted to the left by the Depression and the threat of fascism. He was, of course, a eugenicist who proposed at the beginning of the Depression that unemployment relief be made contingent upon the male recipient's agreeing not to father any more children. At times he reached into animal biology to demonstrate the force of genetics in behavior, noting the power of, for example, the brooding instinct. He wrote, "Crows have been known to brood golf-balls, gulls to sit on tobacco-tins substituted for their eggs; and the majestic emperor penguin, if it loses its egg or chick, will even brood lumps of ice in its inhospitable Antarctic home."[22]

In his view, instinct could well overpower environmentally induced instruction. He called to mind Dr. Johnson's comment on the theory that the attraction that men found in the human female's breast derived from the pleasure they had taken in suckling at it in infancy. Dr. Johnson had not noticed, Huxley remarked, that "those who had been hand-fed when babies evinced any passionate fondness for bottles." (Dr. Johnson might have added, one is moved to say, that the grown human female, who also had suckled in infancy, did not show any passionate fondness for the breast either.)[23]

However, Huxley never held to the position that nature invariably overwhelmed nurture, especially not in most human behavior. In *Essays of a Biologist* he said, "Environment plays not merely a large part, but a preponderating one, in [man's] development after the first year or so of life." Though he never really overcame the belief that, on average, members of the very lowest income groups were genetically less well-endowed than members of the upper—especially professional—classes, in the 1930s, he was soon convinced that his views on compulsory sterilization of the unemployed merely aided and abetted Nazism. He came to stress the importance to both eugenics and society of adequate diet, health care, housing, and education. "Don't let's go on pretending it's all the dear old Edwardian Age!" he exclaimed to a friend. In his celebrated 1936 lecture to the

Eugenics Society, Huxley said flatly that a system based on private capitalism and public nationalism was ipso facto dysgenic: it failed to utilize existing reservoirs of valuable genes, and it led to the ultimate dysgenics—war. He declared, "We can't do much practical eugenics until we have more or less equalized the environmental opportunities of all classes and types—and this must be by levelling up."[24]

Huxley had first been sensitized to issues of race during his years at the Rice Institute, when he was exposed in a way that he would not have been in the England of that day to relations between blacks and whites. He later said that his American sojourn had taught him that differences in cultural and social environment have a good deal to do with differences in group behavior. He became passionately engaged with racial issues during the 1930s, in the face of Nazi theory and practice. Huxley did believe, as he explained to the readers of *Harper's* magazine, in 1935, that though it had not been proved, different human groups must possess "innate genetic differences in regard to intelligence, temperament, and other psychological traits." However, he added, "this need not mean that the mental differences are highly correlated with the physical—that a dark skin, for instance, automatically connotes a tendency toward low intelligence or irresponsible temperament."[25]

Huxley may have sometimes been patronizing in his attitudes toward blacks, but he was unalloyed in his rebuttal to claims of racial inferiority among different groups of Caucasians. In 1935, together with the anthropologist A. C. Haddon, he published *We Europeans: A Survey of 'Racial' Problems*. Huxley and Haddon advanced the emerging genetic and anthropological consensus that the concept of "race" made no biological sense. What seemed like a racial group actually consisted of the intermixture of many biological types, the product of successive migrations and intermarriages. The Nazis might claim that Jews constituted a racial type, but in fact in every country Jews overlapped with Gentiles in every conceivable physical characteristic. Jews of one area differed genetically from those of another; they were biologically no more uniform than any people of Europe—including so-called pure Germans. The Nazis might celebrate a Teutonic type—fair, long-headed, tall, and virile; Huxley and Haddon wondered how close a composite of the black-haired Hitler, the broad-faced Rosenberg, the slight Goebbels, and the rotund Goering would come to the Teutonic ideal. Populations differed from each other, Huxley and Haddon stressed, only in the relative proportions of genes for given characters that they possessed. "For existing populations," they maintained, "the word *race* should be banished, and the descriptive and noncommittal term *ethnic groups* should be substituted."[26]

Huxley was a man of the twentieth century, but in a sense he felt the hold of his grandfather's era. While his popular writings reveal his engagement in the immediate issues of his time, they also provide ample evidence that he was absorbed from the days of his young manhood but with increasing intensity in his later years with the long-term questions raised for human beings by Darwin's theory of evolution. Certain of the questions of concern to him—notably about morals and religion—had been spelled out in the nineteenth century, particularly by grandfather Thomas Henry Huxley in 1893, in his Romanes Lecture. Others were implicit in the experience of the twentieth century, notably the doubts raised about the idea of prog-ress and the capacity of man for virtuous behavior. While Huxley dealt with one or more of these matters in a number of his essays and books, his most fully developed reflections upon them are contained in two volumes, *Touchstone for Ethics*, which was published in 1947, and *Evolution in Action*, which appeared in 1954.

In the 1920s, Huxley had posed the question: "Is it possible to speak of progress when at this present moment there are vast poverty-stricken and slum populations with all the great nations, and when these same great nations have just been engaged in the most appalling war in history?"[27] Huxley had argued then that it was possible, indeed, and he kept insisting upon that possibility, even after another world war had produced not only greater carnage but also the means of humanity's own self-destruction. By "progress," he meant the tendency of biological organisms to grow better equipped over evolutionary time to carry on the business of existence and survival. Evolution had taken the simplest organisms and out of them created organisms that were much more complex—and, as such, ever more capable of dealing with the challenges found in their environment. "Biology," he wrote, "presents us with the spectacle of an evolution in which the main direction is the raising of the maximum level of certain qualities of living beings, such as efficiency of organs, coordination, size, accuracy and range of senses, capacity for knowledge, memory and educability, emotional intensity—qualities which in one way or another lead to a more efficient control by the organism over the external world, and to its greater independence."[28]

By Huxley's measures, the "furthest step yet taken in evolutionary progress" was the human species. One might think this view mere anthropomorphic wish fulfillment. Not so, Huxley insisted. Human beings had gone beyond all other forms of organic life when they acquired the capacity for language and cognition, for the organization, analysis, and recording of their experience, and for the transmittal to future generations of what they might learn. Furthermore, unlike ducks, or dogs, or ants, which possessed admirable tools for their particular jobs, human beings were endowed with infinitely variable capabilities. They could specialize, even though they were not born for one or another specialty, and if they came to specialize at one function, they could—unlike, for example, the worker ant— change to another. In Huxley's summary of this argument, "Animal types have limited possibilities, and sooner or later exhaust them: man has an unlimited field of possibilities, and he can never realize all of them."[29]

In these features of man, Huxley found profound long-range consequences. Human beings, diverse in their capacities and self-awareness, were not compelled to pursue solely their individual self-interest, as crude Social Darwinists might have it. They could also cooperate to achieve the common needs of society. More important, man's self-consciousness made possible "not only innumerable single changes, but a change in the very method of change itself"—a transition from evolution by blind processes operating on the opportunities provided by blind chance to evolution by mankind's deliberate choices.[30] Huxley conceded that human beings had so far not used their capacities very wisely to shape the world; and he allowed that savage qualities were to be found in a deplorably large number of them. "Our feet still drag in the biological mud," he wrote, "even when we lift our heads into the conscious air." Still, he found a certain comfort in the belief that evolution had continually raised the upper levels of biological organisms; and further comfort in the recognition that human beings, who had existed for only a moment in evolutionary time, still had generations more before them to work out their problems and realize their possibilities. Huxley summarily declared, "In the light of evolutionary biology man can now see himself as the sole agent of further evolutionary advance on this planet, and one of the few possible instruments of progress in the universe at large. He finds himself in the unexpected position of business manager for the cosmic process of evolution."[31]

Yet by what moral or ethical principles were people to be guided in this cosmic task? Huxley's answer: By the principles implicit in the process of evolution itself. It was here that he took issue with—and departed from—the position advanced by grandfather Huxley. Thomas Henry Huxley had argued in his Romanes Lecture that he could detect no moral purpose in nature. Moral purpose was exclusively a product of human fabrication. And he had declared, "Let us understand, once for all, that the ethical progress of society depends, not on imitating the cosmic process, still less in running away from it, but in combating it." Thus, moral man, although the product of the evolutionary process, had to take arms against it.[32] To Julian Huxley, however, it was incumbent upon mankind to fight *for* it—to struggle to see that evolution continued to flourish. Embracing that aim, human beings could forge an evolutionary ethics, which would start with the principle that it was right to realize ever-new possibilities in evolution and which would consist of further principles extracted from what was necessary for the evolutionary process to proceed. "Anything which permits or promotes open development is right," Huxley held, "anything which restricts or frustrates development is wrong. It is a morality of evolutionary direction."[33]

With evolutionary ethics, conventional religion was no longer necessary. Huxley was a self-proclaimed atheist. He took God to be "a product of biological evolution," a yearning built into the human psyche. But as a result of the advance of psychological and natural science, God, Huxley averred, in one of his delicious contentions, "was no longer a useful hypothesis."[34] He had become a mere vague first cause, not an immanent presence in human affairs. If human beings were to have a religion, it was to be a scientific humanism, a merging of humane values with the evolutionary process. The emergent religion of the near future, he predicted, would not worship supernatural rulers. It would "sanctify the higher manifestations of human nature, in art and love, in intellectual comprehension and aspiring adoration, and will emphasize the fuller realization of life's possibilities as a sacred trust." Here Huxley found the resolution of his grandfather's antithesis between the ethical and the cosmic process. If man was to take charge of evolution, he could, Huxley said, "impose moral principles upon ever-widening areas of the cosmic process, in whose further slow unfolding he is now the protagonist. He can inject his ethics into the heart of evolution."[35]

Huxley's ethics called for diversity rather than uniformity in human society, respect for human differences, social organization that fostered individual development. It was thus the function of the State to enhance the self-realizing opportunities of every individual in society. He condemned "Nazi ethics"—if the term is not an oxymoron—because it exalted a tribal group over universal mankind, the State over the individual. He rejected Marxist materialism because it wrongly reduced mind to mere matter. He warned against trusting the state to achieve a eugenics program by deciding what were good and bad hereditary qualities.[36]

Yet he also recognized that a problem lay at the heart of his ethics—recognized it enough to raise the question: Was the evolutionist not "merely dressing up in new terminology" what moral man had accepted for centuries? He answered the question emphatically in the negative. He claimed that the primacy of human personality was merely a "postulate" of Christianity and liberal democracy, but it was a "fact of evolution." Traditional ethics were grounded in religion, and thus in authority or revelation. Evolutionary ethics was drawn from discoverable scientific principles.[37] Huxley here appeared to be deluding himself. The seemingly cosmic objectivity of his system was to a considerable extent shaped by the very values that he discounted as postulates—and also by his resistance to that scourge of the twentieth century, totalitarianism. He may not actually have resolved the profound issue raised by his grandfather. He did address it with ability and imagination, kept it before the public during some of the cruelest decades of recent history, and did so with morally purposeful intent mixed with a rueful understanding of the human condition.

Julian Huxley was a visionary, but he was no fool. He recognized the enduring force of his grandfather's perception—and in several poignant sentences, he provided a self-revealing coda to his struggle to find for himself and for mankind moral values in an amoral universe. A clearer ethical outlook would not, he cautioned, "prevent us from suffering what we feel as injustice at the hands of the cosmos. . . . Man is the heir of evolution: but he is also its martyr. All living species provide their evolutionary sacrifice: only man knows that he is a victim."[38]

D . L . L E M A H I E U

The Ambiguity of Popularization

In his excellent paper, Daniel Kevles provides both an overview of Julian Huxley's work as a popularizer of science and a detailed analysis of Huxley's more significant, often controversial opinions. Throughout his paper, Kevles accepts as relatively unproblematic the notion that Huxley was a successful popularizer. Huxley, it is stated, published numerous articles in "general circulation magazines" and "some twenty books of popular science." His *Science of Life* was "a great success" and he was a "highly successful" broadcaster. I would like to first discuss the ambiguities of the term "popularization" and then, in an abbreviated fashion, attempt to place Huxley within the cultural context of his day.

The problems of definition in the field of cultural studies are notorious and need not be rehearsed at length. Virtually every major descriptive term, including most notably the word "culture" itself, raises difficulties not easily mitigated by more elaborate attempts at precise definition, or by the substitution of idiosyncratic terminology. The terms "popular" and "popularization" mean different things, of course, to different commentators. Thus, for example, to various elements within the now rapidly aging New Left, the term "popular" was often equated with a more authentic, politically correct "folk culture," which was created apart from and usually in opposition to a despised, politically suspect "mass culture." To others, the term "popular" corresponded more closely to the standard definitions in most dictionaries. It meant to be regarded with favor by most persons or to appeal to the tastes of ordinary individuals. Popular culture and mass culture were terms that, by and large, could be used interchangeably.[1]

Less noticed perhaps, the terms "popularize" and "popularization" have also had both broad and narrow definitions. When used broadly, these terms usually refer to the process whereby a complex work or an idea becomes known in a more accessible fashion to a substantial proportion of the population. In the early twentieth century, for example, Caruso's gramophone recordings popularized a number of operatic themes, just as during the interwar period, the cinema was responsible for popularizing short, simplified versions of some symphonic music. On the other hand, the terms "popularize" and "popularization" have a more restricted meaning in which works or ideas become more widely known among the educated middle classes, or narrower still, the cultivated elite. The Public Broadcasting System popularized Vera Brittain's *Testament of Youth* to the roughly 6 percent of the American population who watch Masterpiece Theatre; and the magazine *Scientific American* makes known recent discoveries in academic science to around 600,000 subscribers.[2]

Clearly, Julian Huxley was more a popularizer of science in this second, more restricted, sense. Although, as Kevles notes, he wrote for "general circulation magazines," most of these journals in Britain appealed to limited audiences, mainly from the educated upper middle classes. During the interwar period, *The Cornhill Magazine*, *The Week-end Review*, *Strand Magazine*, *The Spectator*, *The New Statesman*, and even *The Listener* drew relatively small readerships. In the 1930s, for example, the *Listener* peaked at around 50,000 average weekly net sales, while *The New Statesman* enjoyed a circulation of only 14,000 in 1933.[3] Even allowing for multiple readers of copies delivered to libraries and clubs, these statistics represent only a tiny fraction of the literate adult population in England.

Huxley's "twenty books of popular science" present more of a problem since accurate sales statistics prove difficult to obtain. Still, between the wars most English publishers would break even if a book sold around 1000 copies. In America, where Huxley's books also appeared, any book that sold more than 50,000 copies was considered a nationwide best-seller. Most books, however, sold fewer than 5000 copies.[4] Huxley's *Science of Life* may have been, as Kevles claims, "highly successful," but in the absence of actual sales statistics, that success remains hard to measure.

Huxley's role as a broadcast lecturer also needs to be examined carefully. Few things invigorated intellectuals more during the interwar period than the bracing feeling that if they spoke on the BBC, they addressed a mass audience. In the years before Listener Research, it was easy to assume that if your voice was carried by one of only two available channels, your audience must be vast. Listener Research, which began in 1936, revealed, however, that broadcast talks, particularly by upper-class intellectuals, attracted a minuscule audience, indeed often too small to measure statistically. After 1946, the Third Programme, which broadcast the type of fare that dominated the National network during the Reithian era, rarely attracted more than 1 or 2 percent of listeners, and often even less.[5] Still, even one-half of 1 percent of a broadcast audience at a reasonable hour was probably larger than the weekly circulation of, say, *The Spectator*.

What accounts for Huxley's relatively limited appeal as a popularizer of science? First, during the interwar period, between 85 and 90 percent of Britain's children never went beyond primary school. Indeed, in 1938, only four or five of every *thousand* students who left primary school ever made it to university.[6] Even what Kevles describes as the "graphic punch" of Huxley's prose could not entirely offset the brute fact that Huxley's vocabulary and oftentimes his subject matter, simply went beyond the educational capacity of most working-class men and women. Moreover, neither personal respect for our subject nor the balm of historical understanding will ever completely sanitize those neo-Malthusian opinions of Huxley that would have been insufferable to most British citizens. "The lowest strata," he wrote in 1941, "are reproducing too fast. Therefore . . . they must not have too easy access to relief or hospital treatment lest the removal of the last check on natural selection should make it too easy for children to be produced or to survive; long unemployment should be a ground for sterilization."[7] As an intellectual heir of the crowd psychologists of the late nineteenth and early twentieth centuries, Huxley could hardly have sought the approbation of such a multitude. Indeed, working-class resistance to the patronizing attitudes of a self-regarding intelligentsia remains one of the great untold stories of twentieth-century British cultural history. Finally, anyone who as late as the 1960s considered the world's three greatest threats to be "atomic destruction, over-population, and total vulgarization"[8] cannot be said to have embraced wholeheartedly the mission of popular education.

Yet, if Huxley was a successful popularizer only within the narrow sense of that term, he was not without general renown. During World War II, he became a favorite participant in one of the most popular programs on the BBC. *Brains Trust* began in 1941 and, to the surprise of its originators, became an instant success. In 1942 and 1943, about 21 percent of the listening public tuned in; by 1944, this figure climbed to 25 percent; and at the end of the war, the program reached a peak of almost 30 percent of the listening public. In an internal memorandum, a member of Listener Research reported that the audience for the show was "easily the greatest for any regular spoken word programme, other than the News, and the envy of many a programme of pure entertainment."[9] Questions were asked about the show in Parliament; and the term "brains trust" passed into general circulation.

The success of the program largely depended upon the entertaining verbal exchanges of its varied participants. Though initially conceived as a conventional quiz show, not unlike the more recent *MasterMind*, the format quickly evolved away from factual questions and answers. *Brains Trust* involved brief, clever, usually glib responses to more general, often speculative inquiries. Critics accused the program of trivializing the pursuit of knowledge and caricaturing intellectuals. Nevertheless, Julian Huxley proved himself to be a remarkably capable radio personality. Listener Research discovered that of the original participants in 1941 and 1942, Huxley ranked as most popular, a position he retained in 1943 and 1944 as the number of participants increased substantially. Only in 1945 did Huxley lose his top ranking among the audience for the show.[10]

In conclusion, Huxley was, perhaps, more popular than he was a popularizer. As a prolific advocate of science, he reached only a limited public who welcomed his efforts to improve their understanding. Like John Reith and others during the interwar era, he was a cultural paternalist and uplifter, part of a long tradition in British cultural life. As a radio personality, however, he became, at least briefly, something related but quite different. In a medium that, like television later on, stressed personality over substance, Huxley represented the intellectual. For millions of British people in the early 1940s, he embodied the notion of high intelligence. He was the intellectual as celebrity.

R O B E R T   L .   P A T T E N

The British Context of
Huxley's Popularization

I want to put Daniel Kevles's lucid survey of Julian Huxley's popular writings in the larger context of British culture, and at the same time indicate ways in which, despite Dan LeMahieu's strictures, Huxley fit into the mainstream of Britain's bourgeois cultural life. The comparative ease with which Huxley composed his works, and the widespread acceptance they received—as Kevles mentions, in 1930 *Spectator* readers voted him one of Britain's best brains—owe something to the flourishing of six complexly related traditions of popular communication in the British Isles.

The first, and the earliest of Huxley's efforts, is the legacy of marvelous diaries and journals by amateur naturalists recording their observations of the world around them. Gilbert White's *Natural History . . . of Selborne* (1789) set Darwin to "wondering why every gentleman did not become an ornithologist."[1] British naturalists from White through the Romantic poets to Charles Kingsley and Francis Kilvert have wedded keen observations of natural phenomena to fine writing, frequently raiding English literature—Shakespeare, Wordsworth, Shelley, Keats, Tennyson—for sympathetic language.[2] Julian Huxley absorbed this tradition early, at Eton and Balliol; he admired White,[3] who quoted English literature extensively in the service of his precise descriptions of ornithological and botanical events; and those literate naturalist forebears contribute to what Kevles has so rightly characterized as Huxley's "supple and limber style."

Closely allied to the diaries and journals, especially from the Romantic period forward, is the legacy of poetry about nature. Huxley seems to have been acutely responsive to poetry—everything from biblical songs to the writings of his contemporaries—and to have shared with his audience not just particular lines and verses but also cadences, phrases, and attitudes infused into the English language through the centuries. He gained much from Matthew Arnold's poetry, including at moments that view of humankind as evolutionary martyr which informs Arnold's *Empedocles on Etna* and the Marguerite poems. George Meredith seems to be another significant model for Huxley: his sonnet on "Evolution: At the Mind's Cinema"[4] reads like one of Meredith's lyrics on the evolution of life from blood to brain to spirit. And Huxley used a poem on the nightingale by Meredith's distinguished literary descendant, the Poet Laureate Robert Bridges, to close one of his BBC talks on bird behavior.[5]

A third shaping influence, notably in the African journals, is the tradition of British travel writing from Mungo Park, whom Huxley does not quote, to Darwin's *Journal of Researches* (in Huxley's estimation "one of the best travel books ever written"[6]) and Winston Churchill's *African Journey*. British travel accounts have also been characterized by close attention to nature, coupled with wonder at strangeness and diversity (another Romantic legacy, confounding the early eighteenth-century uniformitarian premise that mankind is everywhere, and at all times, alike). Julian derived from his foreign trips, especially to the "dark continent," not only much biological and anthropological material, but also, as travel writers before and since have done, covert and sometimes overt reflections on Anglo-Saxon culture. He went abroad, in part, to discover home:

> We who belong to an old civilization have many things to be proud of; but we cannot be very proud of the state, or, to put it bluntly, the mess, at which our civilization as a whole is now arrived. We can be proud of our scientific and technical achievements, our knowledge and art, our organization and our wealth; but we cannot, I hope, be proud enough of them to wish to give them to another continent if this also involves the gift of the other concomitants of our civilization, including slums and overgrown cities, gross inequalities of wealth and opportunity, callow discontent and chauvinist nationalism, the over-multiplication of the unfit and the horrors of modern war.[7]

These three traditions converged notably in Darwin's contribution (Vol. III) to the *Narrative of the Surveying Voyages of H.M.S. "Adventure" and "Beagle"* (1839). And what came out of that expedition, twenty years afterward, was of course *The Origin of Species*, a seminal work intended to offer an alternative set of laws governing natural life. The relationship between nature and cosmic order was an especially charged field of study in the nineteenth century; Augustan assumptions about the perfection and moral rule governing all forms of life gave way grudgingly to de-theologized laws, in thermodynamics as well as biology. That scientific inquiry might "look through Nature up to Nature's God"[8] was an assumption Huxley shared with a distinguished host of predecessors: Robert Chambers, Charles Lyell, George Henry Lewes, T. H. Huxley. It is not surprising in this perspective that Julian Huxley would publish his testament, *Religion Without Revelation*, nor that he should "work furiously" to articulate exactly what he could, on the basis of scientific evidence, "believe."[9]

Indeed, this point requires elaboration. That scientific inquiry was not disjunct from religious and moral inquiry was a nineteenth-century conviction vigorously argued at Balliol and in Oxbridge more generally. Julian Huxley wrote within a tradition of discourse that framed issues from a humanistic predetermination. Such early observational treatises as the study of grebe courtship and the organization of ant society presuppose a human analogy to the behavior exhibited. Huxley's religious writings accept certain ontological, epistemological, and hermeneutic premises challenged, *inter alia*, by German metaphysics and its British redactors and more recently by continental anthropology and structuralism and by American positivism. Huxley's theology and methodology remain homocentered in the tradition of nineteenth-century Feuerbachian humanism. In some ways it was a much more comfortable perspective than the radical de-centerings of Schopenhauer's metaphysics or Heisenberg's and Einstein's relativity; but the theological and ethical speculations it generates may be less sophisticated and less applicable to contemporary paradigms.

A fifth tradition flourishing vigorously from the mid-nineteenth century was popular journalism addressing scientific subjects. Dickens's *Household Words* retold Priestley's lectures on combustion for a lay readership; Dickens then used Priestley's explanation of oxidation as one of the paradigms underlying physical and spiritual events in *Bleak House*. Countless other widely circulated periodicals thenceforth offered comprehensible instruction on a variety of scientific and technological subjects. Biology may have a more immediate appeal to a mass audience than physics or inorganic chemistry: the poets and novelists from midcentury regularly employ figures and models from the popular understanding of the natural sciences—Tennyson's famous "nature red in tooth and claw," for instance—though George Eliot by the time of *Daniel Deronda* also embeds more recondite concepts first introduced in algebra and astronomy. There is, in Great Britain, a nourishing circulation of information about natural science that shows up today in *Nature* magazine and the superb films about wildlife made for television. (In a recent survey, British students ranked above Japanese and Americans in their comprehension of biology.) Huxley's periodical essays and films are closely connected to this rich tradition, and his lifetime of commitment to popular education at home and abroad disseminated his scientific and humanistic convictions widely.

Finally, naturalists were not behindhand in mounting the podium to lecture about their field, and about its implications for an understanding of both the material and immaterial worlds. T. H. Huxley is perhaps the most famous Victorian instance, and his talks were grounded in the cultural legacies just enumerated. He was, as Kevles notes, a pioneer in lecturing to working men. Significantly, the American tradition has tended to be otherwise: the pulpit rather than the podium has been the more usual origin of lectures on nature, and nature has from that source often been attacked as an inadequate guide to human behavior. Even today the Stephen Goulds are outnumbered by television evangelists fulminating against the "concept" or "doctrine" of evolution. By contrast, though the BBC might be censorious about airing one or another of Huxley's scientific theories or social policies, it repeatedly invited him to deliver wireless talks on birds and other subjects, and it produced a greatly inspiriting program during the bleakest days of the blitz with the "Brains Trust."

What I am suggesting is that Julian Huxley's popularizations were shaped by aspects of British culture that provided him and his audience common ground, common language, and common interests.

An unfairly negative way of putting it would be to say that Huxley's effectiveness as a popularizer depended on his being *behind* the times, not *before* them. But it is true that his philosophical, scientific, and economic paradigms are in significant ways pre-Heideggerian, pre-Einsteinian, and pre-Keynesian. He was at one with his culture, in some ways distinct from both continental and American developments, in the embedded unresolved conflicts of his thought. He shifts between valorizing multeity in the free marketplace of genetic variation and advocating eugenics and order, between employing scientific rationalism as an explanatory system and a means of control and relying instinctively on his passionate intuitionism, and between arguing for a global concept of process, telic only insofar as life rather than stasis or death is its goal, and repeatedly articulating a biological concept of progress wherein evolution moves toward higher states of adaptability and organization. That shared legacy, product of British empiricism and nineteenth-century scientific humanism, which binds scientist and humanist, specialist and amateur, poet and philosopher, and pedagogue and popularizer, may no longer govern our discourses.

Nearly thirty years ago another British scientist, C. P. Snow, delivered the Rede Lecture on "The Two Cultures and the Scientific Revolution" (1959), to which F. R. Leavis replied in the 1962 Richmond Lecture, "Two Cultures? The Significance of Lord Snow." In hindsight, what appears significant about that acrimonious and uncivil exchange is that it marks a division between science and culture, Snow, a physicist, speaking in optimistic terms about the promise of scientific inquiry, Leavis deploring the absence in science of a moral view of life. As George Steiner observes, the debate between Leavis and Snow articulated a fissure between science and letters that is based on a temporal divide between forward-looking scientific optimists and backward-looking humanistic pessimists.[10]

That is a radical fissure Julian Huxley's writings tried to bridge. More than his brother Aldous, who could posit dystopias emerging from a scientific culture (*Brave New World*), Julian inherited, shared, and transmitted a belief in the benefactions of a scientific engagement with the natural world.[11] Whether in writing about or in advocating the creation of national parks and zoos and bodies like the Nature Conservancy and the IUCN to safeguard precious resources, he believed that science contributed to the preservation of life in its myriad varieties and wondrous adaptations.[12] He insisted on inserting science into UNESCO, but equally he resisted taking "culture" out.

Had Julian Huxley been a physicist at Los Alamos in 1945, a biologist in South Carolina in the 1980s, or director-general of UN-ESCO today, his combination of scientific optimism and cultural humanism might not have found the same public endorsement. At present large segments of Western populations fear that science itself creates dangerous substances (nuclear power, nonbiodegradable substances, atmospheric pollutants, mutant genetic and viral strains), that scientific rationalism may not be the best guide to public policy, and that "value free" science is both a delusive ideal and a disastrous practice. The long twilight of Comtean positivism and Edwardian rationalism by which, as Colin Divall demonstrates, Julian Huxley's formidable intellect was shaped, may at last have come to an end.

Appendix

NANCY L. BOOTHE

The Julian Sorell Huxley Papers
Rice University Library

Rice University first learned of the availability of the papers of Sir Julian Huxley in 1978. This collection seemed to be an appropriate acquisition, not only because of Huxley's distinguished career as a scientist, educator, and influential synthesizer of thought in many areas, but also because he was one of the Rice Institute's original faculty members. As first chairman of the Biology Department, he planned its laboratories and curriculum and chose its original faculty and staff. Alumni and friends, through their diligence and generosity, made the purchase possible in 1980. A Higher Education Act Title II-C grant from the U. S. Department of Education funded the processing of the papers in 1983–84.

The Collection

Ninety-one linear feet in size, the Huxley papers span the years 1899–1980. The major groups of material in the collection are correspondence; manuscripts and typescripts of published and unpublished writings of Huxley's; early family and school records; journals, diaries, and notebooks; materials collected by Huxley on his travels; records related to organizations in which he was active; papers and addresses by others; and photographs, drawings, clippings, and memorabilia.

About one-third of the collection is correspondence, which, because Huxley was in contact with a great many shapers of modern thought, touches on many of the major issues and controversies of this century. His correspondents included scientists, historians, writers, artists, philosophers, musicians, philanthropists, politicians, and other public figures. Approximately eleven hundred of his correspondents were considered of sufficient importance that their letters to and from Huxley were indexed in the guide.

Of particular interest to historians of science are those correspondents who were Huxley's students or scientific colleagues; letters to and from these scientists make up the largest group of correspondence. Examples of these correspondents (some with indications of the numbers of letters to and from Huxley) are J. R. Baker (109), Gavin de Beer (65), C. P. Blacker (62), Jacob Bronowski, Alex Carr-Saunders (67), Wilfrid L. Clark, F. A. E. Crew (70), Theodosius Dobzhansky, L. C. Dunn, Albert Einstein, Charles Elton, E. S. Goodrich, James M. Fisher (56), E. B. Ford (213), J. B. S. Haldane, A. C. Hardy (138), Lancelot Hogben (106), H. B. Kettlewell, Konrad Lorenz (87), Ernst Mayr (113), Peter Medawar, H. J. Muller (107), Joseph Needham, Max Nicholson (over 250), Henry F. Osborn, Bernard Rensch, Peter M. Scott, E. W. Sexton, George G. Simpson, C. P. Snow, Albert Szent-Gyorgyi, Niko Tinbergen (86), W. H. Thorpe (91), and Solly Zuckerman (approximately 350).

Potential researchers in the Julian Huxley papers will be interested to learn that the Rice University Library continues to develop the collection through appropriate additions, by purchase and gift, of manuscripts, correspondence, photographs, books, and other formats. Additions since the date of the Huxley Symposium (1987) include correspondence with Solly Zuckerman, Max Nicholson, Corliss Lamont, and Kenneth Clark, as well as material related to the Idea Systems Group and manuscripts, typescripts, and page proofs of *The Humanist Frame*.

The Library

In 1981, Rice University purchased Huxley's working library as it existed at the time of his death. It consists of approximately 1,200 volumes and is largely, though not exclusively, scientific in nature. In it are presentation copies of major works, many of Huxley's own publications, reprints of his and his colleagues' work, pamphlets, and journals. Because Huxley was a prodigious annotator, the library complements the papers to a significant degree.

The Guide

The 164-page *Guide to the Papers of Julian Sorell Huxley*[1] contains a biographical sketch of Huxley and an eight-page "biographical chronology," a description of the collection as a whole, an explanation of its arrangement, an inventory of the thirteen series into which the papers are divided, and five appendices: Rice-related material in the papers, the Huxley family genealogy, a selected bibliography, information on related manuscript collections in other repositories, and the index to selected correspondents.

Opportunities for Research

The possibilities for research in the Huxley papers are many and varied. There are rich resources in those scientific fields to which he made major contributions: birdwatching, field studies of animal behavior, laboratory research in ontogeny, the study of evolution, integration of the antagonistic schools of Darwinian selectionists and Mendelian geneticists, his stress on experimental methods and concepts in biological research and teaching (e.g., genetics, embryonic development, and cancer research), and the popularization of science.

Outside the boundaries of strictly scientific thought, Huxley's interest in and contributions to such diverse fields as the ethical and moral implications of science, philosophy, religion, humanism, politics, education, anthropology, population studies, family planning, and economics are represented in the correspondence, lecture notes, manuscripts, publications, conference/travel/organizational materials, notebooks, clippings, photographs, journals, and marginalia in books from his library.

Much of Huxley's life and work is documented in this massive collection: his education at Eton and Oxford and subsequent study at the Naples Zoological Station; his early axolotl experiments; his teaching at the Rice Institute and research at Woods Hole Marine Biological Laboratory in the United States; his World War I army service; his Oxford appointment in 1919; his involvement in the 1921 Oxford Spitsbergen Expedition; his marriage to Juliette Baillot; the 1925 appointment to Kings College, London; his tenure as secretary of the London zoo (1935–42); his work as first director-general of UNESCO; his tours to deliver the Conway, Patten, and Romanes lectures; his winning of the Kalinga Prize for the popularization of science in 1953; his "Brains Trust" radio broadcasts and television appearances; his writings and correspondence with publishers and co-authors; his trips to Africa, India, the Near East, and Iceland to study wildlife conservation; and his worldwide travel on behalf of UNESCO.

There is source material on organizations in which Huxley had a founding or significant role, such as the Association of Scientific Workers, the British Trust for Ornithology, the Charles Darwin Foundation for the Galapagos Islands, the Idea Systems Group, the International Committee for Bird Preservation, and the International Union for Conservation of Nature and Natural Resources, the Linnean Society, the Nature Conservancy, the Royal Society, the Society for Experimental Biology, the Wildlife Trust, the World Wildlife Fund, and the Zoological Society of London.

In the "Conference Materials" section of the Huxley papers there are documents, correspondence, minutes, and other material relating to meetings in which he was significantly involved: the bicentenary of the Russian Academy of Sciences in 1945, the Wroclaw Conference in 1948, the 1955–62 Pugwash Conferences, the Darwin Centennial (1959), the Arusha Conference (1961), the CIBA Conference (1963), the Mendel Symposium on Mutational Process (1965), and the Ritualization Symposium (1965).

Examples of Available Information

HUXLEY'S PERSONAL DECISIONS. Edgar Odell Lovett had persuaded the twenty-five-year-old Oxford graduate to become the first faculty member of the Biology Department in the new school in Houston, the Rice Institute. Although Huxley attended Rice's opening in October 1912, which he vividly described in *The Cornhill Magazine*,[2] he delayed taking up his post for a year, which he spent studying in Germany. When Huxley accepted the appointment, as he later wrote to Lovett, "I looked forward to 8 or 10 years in Texas; to setting my Department full on its feet; & seeing the Institute large & stable & settled down before I left."[3]

However, by the summer of 1916, Huxley felt impelled to return to England to serve in his country's armed forces in World War I, intending to return to Rice after the war. The following spring, he wrote to Lovett, resigning his position. One reason for his decision was the likelihood of his getting an appointment at Oxford. "Now there are only five responsible positions in the Zoology Dept. at Oxford. Five years ago they were all filled by good men: one was drowned before the war; two have been killed in France; & one of the few possible Oxford men at other universities died of Fever in camp last year. To be plain, it will be hard to fill up the gaps from Oxford men, and I am the first choice—I think by a pretty large margin."[4]

Another reason for his decision was that he had "come more & more to feel an Englishman . . . less an internationalist. . . . If I am to make any contribution to thought, I must be in touch with the currents of thought, & the only current into which I can freely throw myself will be that of England."[5] In the same letter, he expressed poignantly what his sojourn in Texas had meant to him personally: "I enjoyed my life at Houston—& often indeed feel real pangs of 'homesickness' for Texas & America. Now & again in London, with its restraint, its class-distinctions, its high pressure, its fog, its artificiality, I see mentally the prairie & feel the soft winds from the Gulf. I remember the open warm-heartedness & the freedom & easiness of the Texans,—& I miss it all badly. . . . not only did I enjoy it, but I learnt from it—learnt a great deal—of the world at large, of human nature, of self-reliance, of tolerance, of breadth of view—and have a real sense of gratitude towards Rice & Texas & America. . . . Whatever happens, I shall always think of Rice Institute as the place where I really came of age intellectually. . . ."[6]

INTERNATIONAL RESEARCH SCENE. The Naples Zoological Station had been established in 1875 by Anton Dohrn, its director until 1909, when he was succeeded by his son Reinhard Dohrn. It was funded by grants from German and Italian governmental bodies, by private contributions, and by subscriptions to approximately fifty foreign investigators' tables by learned societies in Great Britain, Germany,

Italy, the United States, Switzerland, Austria, Belgium, Russia, Holland, Rumania, Hungary, and Japan. When Italy entered World War I, Dohrn and other foreign members of the staff were relieved of their positions and replaced by a commission appointed by the Italian government; the station's work all but ceased. At the end of the war, the new Italian government, strongly influenced by Minister of Education Benedetto Croce, reinstated Dohrn as station director. But the American and European scientific communities were slow in renewing their support of the station, concerned that there had been undue control by Germany before the war and that Italian control henceforth might negate the sense of a world research institution.

Julian Huxley, among other scientists, tried to reverse this attitude. In a 4 November 1920 letter to John L. Myres, head of the British Association for the Advancement of Science, Huxley said:

> Since the situation as regards the Naples Zoological Station does not seem to be properly understood in this country, I am venturing to write to you and put you very briefly in possession of the chief facts. . . . He [Dohrn] is willing to do almost anything to make it perfectly clear that no Government shall be exerting any form of pressure on the policy of the Station. I think he would consent to the formation of some form of International advisory committee. He himself is trying to have the Station placed under the guardianship of the League of Nations. . . . I . . . know that his most earnest wish . . . [is] to be able to go back to the Station and make it a really International meeting-place for Scientists of every country. In my subject at least I know that we can not possibly get on without the co-operation of the Germans as well as of other races; and also that there is no other institution on this side of the Atlantic which approaches Naples as a place for the study of marine biology. I sincerely hope that the British Association will continue to help the cause of Internationalism in Science by renewing its grant for a Table at the Naples Station.[7]

Support for the station began to return, and in February 1924, Edmund B. Wilson announced in *Science*, "American biologists will rejoice in the good news that the Zoological Station of Naples is in course of reorganization, with Dr. Reinhard Dohrn as its executive head."[8]

EARLY REACTIONS TO DISCOVERIES. In April 1941, H. J. Muller wrote to Huxley from Amherst:

> I have today got some evidence in support of the prediction I made in the "New Systematics" that in species crosses *recessive* genes from one species

giving sterility, etc., in combination with others from the other species are probably very numerous. For it turns out (if the observations I made this morning are not "premature") that flies *homozygous* for the fourth chromosome of simulans & for the others of melanogaster are sterile in the male, & that the locus in question is included in the small region of the 'Minute-4' deficiency, so it's practically narrowed to a gene, the first such definite cross-sterility gene that's been isolated, I believe. Since the region sampled (the 4th chromosome) is so small in proportion to the total, the finding of such a gene in it may be taken as an indication that such genes are probably numerous, even in species as close as simulans & melanogaster.[9]

CONTROVERSIES. Huxley disagreed with Theodosius Dobzhansky over the latter's position that "natural selection tends to maintain or enhance the Darwinian fitness."[10] The story begins with Huxley's review of Dobzhansky's book *Mankind Evolving: The Evolution of the Human Species*[11] in *Perspectives in Biology and Medicine*,[12] in which Huxley disagreed with Dobzhansky's definition of "Darwinian fitness," i.e., that "fitness is measured solely in terms of reproductive proficiency," a definition shared by other population geneticists but not by Huxley. (A copy of the reprint of the review, marked with Huxley's manuscript comments, exists in Series VI of the collection, "Publications by Julian Huxley.") In addition, the copy of *Mankind Evolving* in the Huxley library is very heavily annotated by Huxley on page margins and end papers. The next step is Huxley's 1962 Galton Lecture, delivered in London on 6 June 1962, of which a manuscript-and-typescript copy exists in Series V of the papers, "Manuscripts, Typescripts, Notes." Published versions of the Galton Lecture appear in the Huxley library in periodical form[13] and as a chapter in a collection of Huxley's essays.[14] Series VI of the papers ("Publications of Julian Huxley") contains two reprint versions of the lecture: a brief summary reprinted from *Nature*[15] and the full-length version reprinted from *Perspectives in Biology and Medicine*.[16] In the Galton Lecture, Huxley took issue with Peter Medawar on the same topic: "The biological *avant garde* has chosen to define *fitness* as 'net reproductive advantage,' to use the actual words employed by Professor Medawar in his Reith Lectures on *The Future of Man*. Any strain of animal, plant, or man which leaves slightly more descendants capable of reproducing themselves than another, is then defined as 'fitter.' This I believe to be an unscientific and misleading definition."[17]

"General Correspondence" (Series III of the collection) contains a terse letter from Dobzhansky to Huxley mentioning the latter's criticism in both review and lecture, and enclosing a copy of a letter Dobzhansky was sending to the editor of *Eugenics Review*. Dobzhansky defended his and Medawar's "geneticism" in that they "describe what has been variously referred to as Darwinian fitness, selective value, or adaptive value of a genotype as the contribution which its carriers make to the gene pool of the succeeding generation relative to the contribution of other genotypes."[18] Dobzhansky ended his letter thus: "The concepts of natural selection and of fitness have evolved and changed since Darwin and Spencer; so have other fruitful concepts in science. Our concept of species is not identical with that of Linnaeus, and of gene with that of Mendel or Bateson. Owners of Greek dictionaries may, of course, manufacture new and fresh terms. I do not find this either expedient or attractive."[19]

K. Hodson, then editor of *The Eugenics Review*, wrote Huxley offering to print a rejoinder along with Dobzhansky's letter.[20] Huxley then turned to colleague H. J. Muller for advice: "In my reply, I would like to adopt the same line of argument, and in many cases the same wording as I have in the enclosed passages from the forthcoming Introduction to the new edition of my *Evolution, the Modern Synthesis*.[21] I would be very grateful if you could read this and tell me whether you think this needs modifying?"[22]

The thorough Muller replied in a two-page cable: "DELIGHTED YOUR LETTER BUT DONT UNDER ESTIMATE REPRDUCTIVE [sic] SELECTION MOST VISIBLE MUTATIONS DROSOPHILA REDUCE BOTH FECUNDITY OF SURVIVORS AND SURVIVAL NEVERTHELESS DOBZHANSKY IS WRONG HIS FITNESS CAN HAVE SELECTION PURSUING SHORT TERM ENDS SUBSEQUENTLY INJURIOUS AND CAN MULTIPLY TRAITS WITHIN SPECIES AT EXPENSE OF WHOLE SPECIES A QUANTITIVELY [sic] VALID CRITERION OF IMPROVEMENT IS INCREASED POTENTIALITY FOR SURVIVAL AND MULTIPLICATION ESPECIALLY UNDER DIFFICULT CONDITIONS."[23]

After the appearance of the Huxley-Dobzhansky interchange in *Eugenics Review*, the two apparently did not correspond again until 1965, when Huxley wrote congratulating Dobzhansky on his selection as a foreign member of the Royal Society, adding a mellow final paragraph: "Since our slight disagreement about 'fitness' I have been collecting all the material I can on the subject, and corresponding with a number of people; I hope, when I get the time, to write something which will help to clarify the situation."[24]

CURIOSITIES. Rewards for archivists and researchers alike in this generally serious and scholarly collection are the rare nuggets Huxley labeled "Curiosities." One startling example is the undated postcard that reads: "Dear Sir Julian, You said on T.V. once that you don't believe in God—well I don't believe in you! Yours truly, Faith Hope, St. Saviours Convent, Harrow."[25]

Another is a letter from a Canadian reader of *Look* who wrote to Huxley in care of the magazine (punctuation, grammar, and spelling are the correspondent's own):

Dear Julian:—

Upon looking over, Look ,I see your on, All about love. Inyour discripion of love , I must say I become, intriested in what you say, there-for this letter.

You state men are thriled to fondle the femeline breast, What I would like to know if they, that is the, opesate seck do they like the? Fondling.

Thank you for this faviour.[26]

Notes

WATERS

I thank Al Van Helden for planning the conference and inviting me to help edit this volume. The Philosophy Department at the University of Vermont generously provided office space, library privileges, and good cheer while I worked on this volume and introduction. Later stages were completed while I was a Visiting Assistant Professor of Biochemistry and Cell Biology at Rice University and then Visiting Fellow at the Center for Philosophy of Science at the University of Pittsburgh, during which time I was supported by a grant from the National Science Foundation (Grant No. DIR 89-12221). I would also like to thank Emily Sher and Jane Maienschein for reading and offering helpful comments on an earlier version of this introduction.

1. Julian S. Huxley, *Problems of Relative Growth* (London: Methuen, 1932); Julian Huxley and G. D. de Beer, *Elements of Experimental Embryology* (Cambridge: Cambridge Univ., 1934); J. S. Huxley, *Evolution, the Modern Synthesis* (London: Allen & Unwin, 1942).

2. See Andrew Huxley, "The Galton Lecture for 1987: Julian Huxley—A Family View," in Milo Keynes and G. Ainsworth Harrison, eds., *Evolutionary Studies, A Centenary Celebration of the Life of Julian Huxley*, (Houndmills, Basingstoke, Hampshire and London: Macmillan, 1989) for a fuller account of Huxley's relatives and their intellectual achievements.

3. Leonard Huxley, *Life and Letters of Thomas Henry Huxley* (London: Macmillan, 1900).

4. Aldous Huxley, *Brave New World* (London: Chatto & Windus, 1958).

5. J. S. Huxley, *Memories* (New York: Harper & Row, 1970), p. 25.

6. Ibid., p. 50.

7. John R. Baker, "Julian Sorell Huxley," *Biographical Memoirs of Fellows of the Royal Society* 22 (1976): 207–238.

8. J. S. Huxley, "Some Phenomena of Regeneration in *Sycon*; With a Note on the Structure of its Collar-cells (Plate 8)," *Philosophical Transactions of the Royal Society of London*, ser. B, 202: 165–89.

9. The first of Huxley's publications on this fieldwork was "The Courtship-habits of the Great Crested Grebe (*Podiceptus cristatus*); With an Addition to the Theory of Sexual Selection," in *Proceedings of the General Meetings for Scientific Business of the Zoological Society of London* 35 (1914): 491–562.

10. Letter from J. S. Huxley to E. G. Conklin dated 26 November 1910 (Julian Huxley Papers, Woodson Research Center, Rice University). Conklin, a biologist at Princeton, had told E. O. Lovett, President of Rice Institute, that he would check out Huxley's suitability as founding faculty member for the Institute's department of biology. He discussed the position with Huxley but did not inform Lovett of Huxley's

interest. When Huxley wrote Conklin the follow-up letter (quoted above), Conklin forwarded the letter to Lovett and admitted that part of the reason he did not inform Lovett of Huxley's interest sooner was "because I covet Huxley for Princeton, and have been hoping that we might have some proposition to lay before him at the same time that you do." (Letter from Conklin to Lovett dated 2 December 1910, E. O. Lovett Papers, Woodson Research Center, Rice University.)

11. Ronald C. Clark, *The Huxleys* (New York: McGraw-Hill, 1968), p. 158.

12. An exchange of letters between Huxley and Rice President E. O. Lovett during Huxley's years at Rice reveals that Huxley's illnesses did interfere with his duties but did not seem to compromise his standing with Lovett. (Letters between Huxley and Lovett, 1912–16, E. O. Lovett Papers, Woodson Research Center, Rice University.)

13. Huxley, *Religion Without Revelation* (New York: Harper, 1927).

14. In his *Memories*, Huxley recalled seeing "a private having his face smashed by a kick from a draught-horse" and added the comment that it was "a foretaste of the multiple horrors of actual war." He continued: "I was getting thoroughly bored with the stupid drill of the Army, and pulled various wires which got me transferred to intelligence." Huxley, p. 111.

15. Juliette Huxley, *Leaves of the Tulip Tree* (Topsfield, Mass.: Salem House, 1988), p. 57.

16. Ibid., p. 79.

17. Ibid., pp. 78–84.

18. Huxley, *Memories*, p. 125.

19. Baker, "Julian Sorrel Huxley," p. 210.

20. Huxley showed that the relation between the weight of the large claw of a fiddler crab, which eventually weighs more than the rest of the crab, and its entire body weight can be expressed by the simple formula: $y = bx^k$, where b is a constant, y the weight of the large claw, x the difference between the total body weight and the weight of the claw, and k is the constant differential growth-ratio. Huxley, *Problems of Relative Growth* (London: Methuen, 1932).

21. Huxley, "Metamorphosis of Axolotl Caused by Thyroid-feeding," *Nature* 104: 435.

22. A. Dumeril first reported the metamorphoses of axolotls into adult forms in 1865. A. Dumeril, "Nouvelles observations sur les axolotls, batraciens urodéles de Mexico (Siredon mexicanus vel Humboldtii) nés dans la Ménagerie des Reptiles au Museum d'Histoire Naturelle, et qui y subissent des métamorphoses," *Comptes Rendus hebdomadaires des séances de l'Académie des sciences* 61 (1865):775–78.

23. Clark, *The Huxleys*, pp. 186–87; Huxley, *Memories*, p. 126; Baker, "Julian Sorell Huxley," p. 211.

24. Huxley, *Essays of a Biologist* (New York: Knopf, 1923).

25. Juliette Huxley, *Leaves*, p. 100.

26. Letter from H. G. Wells to J. S. Huxley dated 6 August 1927, Huxley Papers.

27. Herbert G. Wells, George P. Wells, and Julian S. Huxley, *The Science of Life* (New York: Doubleday, 1931).

28. Huxley, *Memories*, pp. 155–70; William B. Provine, "Introduction," in Mayr and Provine, *The Evolutionary Synthesis, Perspectives on the Unification of Biology* (Cambridge and London: Harvard Univ. Press, 1980), p. 332.

29. Huxley, *Problems of Relative Growth* (London: Methuen, 1932); and Huxley and G. D. de Beer, *Elements of Experimental Embryology* (Oxford: Clarendon Press, 1934).

30. His radio talks covered a wide variety of subjects with titles ranging from "Plans for Tomorrow: The Tennessee Valley Authority—The First of a Series of Eye-witness Accounts of Great Social Experiments Abroad" to "The Future Life—X," "What Do We Know about Immortality?" Jens-Peter Green, "Bibliography," in Baker, "Julian Sorell Huxley," p. 146.

31. Ibid., pp. 146–48.

32. Juliette Huxley, *Leaves*, p. 130.

33. Juliette revealed her side of this story, but Julian did not divulge his (there is no mention of it in Julian's two-volume autobiography). The literature contains hints that much of the story remains to be told (e.g., Olby, this volume, mentions that Julian planned to marry Ms. Waldmeier, but Juliette's account seems to give a different impression).

34. Letter from Julian to Juliette, reprinted in Juliette Huxley, *Leaves*, p. 162.

35. Ibid., p. 163.

36. Clark, *The Huxleys*, p. 207.

37. Or so it is claimed by Clark, p. 209.

38. In May 1930, *The Spectator* announced a contest offering five guineas for the competitors whose list of the five best brains in Great Britain most nearly matched with the majority verdict (*The Spectator* 144: 785). The results were published in June of that year (*The Spectator* 144: 979). G. B. Shaw's name appeared on the most lists (214), followed by Sir Oliver Lodge (183), Lord Birkenhead (162), Winston Churchill (95), Dean Inge (91), and H. G. Wells (86). Huxley's name appeared on 20 lists, which tied him for 16th place, Rutherford's was on 12 lists, which put him at 24th, and Russell's was on 10 lists, which placed him tied at 25th. This contest was conducted only once.

39. Clark, *The Huxleys*, p. 278; Huxley, *Memories*, p. 251.

40. Huxley, *Memories*, p. 251.

41. Clark reports that in 1935, Huxley's first year as secretary, 7,000 children visited the zoo; the number reportedly increased to 55,000 in 1936 and in the following year to at least 97,000. Clark, *The Huxleys*, p. 261.

42. Huxley's opponents never told their side of the story and Clark reports that the minutes of the mutiny are not helpful. Clark also suggests that if the Society's official report of the key meeting is to be believed, the Ministry of Supply imposed paper restrictions that prevented them from explaining the reasons for Huxley's resignation. Clark, *The Huxleys*, p. 268.

43. Clark suggests that the zoo's shift toward public education and popularization was too much for the encrusted Fellows of the Society and noted that Huxley's inability to grasp simple mechanical principles might have decreased his efficiency as an administrator (Clark, pp. 257, 262–63, and 267–68). Baker speculated that former Secretary Sir Peter Chalmers-Mitchell's reluctance to give Huxley a free hand might have played a role and commented that the Fellows thought Huxley was devoting too much time to outside interests, and in particular to his book on evolution (Baker, p. 221). Huxley's account of his years at the zoo supports Clark's primary explana-tion (Huxley, *Memories*, pp. 230–63), but I suspect that Huxley's personality also played an important role in his demise.

44. Huxley, *Memories*, p. 234.

45. Ibid.

46. The neo-Darwinian themes were anticipated in *The Stream of Life* (Julian Huxley; London: Watts; New York and London: Harper, 1926) and *The Science of*

Life (Herbert G. Wells, George P. Wells, and Julian S. Huxley; New York: Doubleday Doran, 1931).

47. Huxley, "Natural Selection and Evolutionary Progress," *Proceedings of the British Association for the Advancement of Science* 106: 81–100.

48. M. Keynes and G. A. Harrison, eds., *Evolutionary Studies*, p. xii; Allen, this volume.

49. The terms "old eugenics" and "reform eugenics" were introduced by Daniel J. Kevles in *In the Name of Eugenics, Genetics and the Uses of Human Heredity* (New York: Knopf, 1985).

50. Huxley, *Memories*, p. 280.

51. Huxley, *Soviet Genetics and World Science, Lysenko and the Meaning of Heredity* (London: Chatto & Windus, 1949).

52. Huxley, *Memories II* (New York: Harper, 1973), p. 15.

53. Although Huxley and others have said the constitutional term was five years, Article VI, Paragraph 2 specifies a six-year term. The original constitution and amendments are reprinted by Richard Hoggart, *An Idea and Its Servants, UNESCO From Within* (New York: Oxford Univ. Press, 1978), pp. 201–213. Armytage, in an interesting overview of Huxley's tenure at UNESCO, reports that it was the American delegation that sought the reduction of Huxley's term to two years. W. H. G. Armytage, "The First Director-General of UNESCO," in Keynes and Harrison, *Evolutionary Studies*, p. 188.

54. James P. Sewell, *UNESCO and World Politics* (Princeton: Princeton Univ. Press, 1975), pp. 111–12.

55. Walter Laves, who was the assistant director of UNESCO under Huxley, and Charles Thomson, who was director of the U.S. National Commission for UNESCO during Huxley's tenure, later summarized Huxley's contribution as follows: "Probably no one person more directly influenced the content and direction of UNESCO's program than did Dr. Huxley in the preparatory stage and during his two years as director-general. Indeed, he was largely responsible for charting the broad course to which the organization became committed during its early years." Walter H. C. Laves and Charles A. Thomson, *UNESCO, Purpose, Progress, Prospects* (Bloomington: Indiana Univ. Press, 1957), p. 295.

56. Nearly all of the second volume reads like a travel log. Huxley, *Memories II*.

57. Ibid., p. 189.

58. Huxley, *Memories* and *Memories II*.

59. Huxley and G. D. de Beer, *Elements of Experimental Embryology* (Oxford: Clarendon Press, 1934).

60. William B. Provine, "Progress in Evolution and the Meaning of Life," in *Evolutionary Progress*, ed. M. H. Nitecki (Chicago: Univ. of Chicago Press, 1988).

61. Quoted from Huxley, *The Uniqueness of Man* (London: Chatto & Windus, 1943), p. 78. Allen (this volume) notes that the quoted selection first appeared in Huxley, "Eugenics and Society," *Eugenics Review*, 1936.

62. LeMahieu's discussion of Huxley's performance on the Brains Trust might lead one to wonder whether Huxley was entirely correct when he attributed the program's popularity to the straight-thinking intellectuals playing off the endearing buffoon. The general public might have viewed the show's personalities in terms of straight man and buffoon, but perhaps listeners thought the role of the buffoon was being played by the intellectuals.

63. David Hubback, "Julian Huxley and Eugenics," in Keynes and Harrison, eds.,

Evolutionary Studies, A Centenary Celebration of the Life of Julian Huxley (Basingstoke and London: Macmillan, 1989), pp. 194–206.

64. Clark does not discuss Huxley's involvement with the society or his contributions to the eugenics literature except for remarking in a single sentence that Huxley prepared a movie for the Eugenics Society (Clark, *The Huxleys*, p. 207). The only mention of Huxley's involvement with eugenics in the 565 pages of his autobiography occurs when he includes the 1949 meeting of the Eugenics Society on a list of seven meetings he attended in the spring of that year (Huxley, *Memories II*, p. 79). Clark (*The Huxleys*) and Huxley (*Memories II*) discuss the latter's position on population control, but they don't explain that it was connected with eugenical views. Only Baker admits that Huxley "had always been a keen eugenist" (Baker, "Julian Sorell Huxley," p. 230). But Baker's discussion of his former teacher's interest in the science of improving the human species is confined to a single sentence, which he completes with the following qualification: "[Huxley] admitted the raising of the genetic level as a part, though only a minor one, of the psychosocial process" (Baker, "Julian Sorell Huxley," pp. 230–31). The official Huxley bibliography (by Jens-Peter Green in Baker) is organized so that it is difficult to determine which publications concern eugenics.

It is also of little help in determining how much of Huxley's broadcast time was devoted to eugenics. In fact, Green's list of Huxley's broadcasts is far from complete. It lists only a handful of the many radio addresses that Huxley must have given. Huxley got in trouble with the BBC for discussing birth control in a 1926 radio debate. The debate, which was being broadcast live, was interrupted by a protest over Huxley's mention of birth control. The Julian Huxley Collection contains a number of letters concerning the incident, some of which accuse the BBC of staging the protest for publicity. This curious broadcast is not listed by Green. I suspect many other controversial broadcasts have also been left off the list and that many of them probably included discussions related to eugenics.

65. John Greene ("The Interaction of Science and World View in Sir Julian Huxley's Evolutionary Biology," *Journal of the History of Biology* 23: 39–55) has written an important essay on the connection between Huxley's world view and his work in evolutionary biology. The Eugenics Society recently held a conference celebrating the life of Huxley, papers from which were published in a volume edited by Keynes and Harrison, *Evolutionary Studies*. In addition to the selections by Andrew Huxley, Armytage, and Hubback mentioned above, the volume includes a useful essay by Durant ("Julian Huxley and the Development of Evolutionary Studies"), which describes Huxley's evolutionary studies and explains how Huxley related his allometric work to evolutionary theory. The volume's remaining selections are concerned mostly with current perspectives on evolutionary issues connected to Huxley's research. Discussions of Huxley's writing and activities can be found in recent works on the various developments in twentieth-century biology. Durant discusses Huxley's political role in the establishment and development of the field of ethology in Britain. Provine ("Introduction") and Viktor Hamburger ("Embryology and the Modern Synthesis in Evolutionary Theory," in *The Evolutionary Synthesis*, ed. Mayr and Provine) remark on Huxley's contribution to the evolutionary synthesis, and Churchill ("The Modern Evolutionary Synthesis and the Biogenetic Law") explains the relation between Huxley's views on rate genes and the biogenetic law.

66. Originally reported by Huxley in "Introduction," in "A Discussion on Ritual-

ization of Behaviour in Animals and Man," *Philosophical Transactions of the Royal Society of London*, ser. B, 251 (1963): 249.

67. Mayr, *Systematics and the Origin of Species* (New York: Columbia Univ. Press, 1942), p. 3.

68. His claims have been uncritically adopted in the secondary literature. See, for example, Clark, *The Huxleys*, p. 156; Baker, "Julian Sorell Huxley," p. 213; and Juliette Huxley, *Leaves*, p. 75.

69. Huxley, *Memories*, p. 89.

70. Baker, "Julian Sorell Huxley," also expresses the idea that Huxley's contributions to laboratory research deserve greater attention.

71. Huxley, *Evolution, the Modern Synthesis*, p. 25.

72. Ibid., p. 26.

73. My hunch is that those who conceive a synthesis as a bringing together of theories will emphasize the role of theoretical population genetics and will tend to regard the work in ecology, systematics, and paleontology as peripheral to the actual synthesis. Those who conceive a synthesis as a bringing together of different fields by relating diverse observations and experimental results to a common framework will view the work of geneticists, ecologists, systematists, and paleontologists as all playing vital roles in a genuine synthesis.

DIVALL

1. Julian S. Huxley, *Essays of a Biologist* (1923; reprint Harmondsworth: Penguin, 1939), p. 11.

2. Huxley's first work dealing at any length with nonscientific topics was published in 1912 when he was twenty-five. This adumbrated several themes central to his scientific humanism but did not provide a complete outline: Huxley, *The Individual in the Animal Kingdom* (Cambridge: Cambridge Univ. Press, 1912).

3. W. M. Simon, *European Positivism in the Nineteenth Century* (Ithaca: Cornell Univ. Press, 1963); J. Burrow, *Evolution and Society* (Cambridge: Cambridge Univ. Press, 1966); David Wiltshire, *The Social and Political Thought of Herbert Spencer* (Oxford: Oxford Univ. Press, 1978).

4. Melvin Richter, *The Politics of Conscience* (London: Weidenfeld & Nicolson, 1964).

5. Susan Budd, *Varieties of Unbelief* (London: Heinemann, 1977); Stefan Collini, *Liberalism and Sociology* (Cambridge: Cambridge Univ. Press, 1979); Greta Jones, *Social Darwinism and English Thought* (Brighton: Harvester Press, 1980).

6. B. Semmel, *Imperialism and Social Reform* (London: Allen & Unwin, 1960); G. R. Searle, *The Quest For National Efficiency, 1899–1914* (Oxford: Oxford Univ. Press, 1971).

7. Budd, *Varieties of Unbelief*, pp. 124–49.

8. Martin Wiener, *Between Two Worlds: The Political Thought of Graham Wallas* (Oxford: Oxford Univ. Press, 1971), pp. 128–29.

9. Huxley, *Memories* (London: Allen & Unwin, 1970), p. 73.

10. Peter Collins, "The British Association as Public Apologist for Science, 1919–1946," in *The Parliament of Science*, ed. Roy McLeod and Peter Collins (Northwood: Science Reviews, 1981), pp. 211–36; W. H. G. Armytage, *Sir Richard Gregory: His Life and Work* (London: Macmillan, 1957), *passim*.

11. Gary Werskey, "Nature and Politics Between the Wars," *Nature* 224 (1969): 462–72.

12. See, for instance, Julian S. Huxley, *The Stream of Life* (London: Watts & Co, 1926), p. 41; D. Paul, "Eugenics and the Left," *Journal of the History of Ideas* 45 (1984): 567–90; G. R. Searle, *Eugenics and Politics in Britain 1900–1914* (Leyden: Noordhoff, 1976), p. 2; G. Greer, *Sex and Destiny: The Politics of Human Fertility* (London: Picador, 1984), pp. 270, 318.

13. Huxley, *If I Were Dictator* (London: Methuen, 1934), p. 16.

14. Huxley, *What Dare I Think?* (London: Chatto & Windus, 1931; London: Phoenix Library, 1933), p. 7.

15. Werskey, *The Visible College* (London: Allen Lane, 1978); William Mc-Gucken, *Scientists, Society & State* (Columbus: Ohio State Univ. Press, 1984), esp. pp. 71–91.

16. Huxley was involved (as a founder) with Political and Economic Planning and (as a member of the executive committee) with the Next Five Years Group: T. C. Kennedy, "The Next Five Years Group and the Failure of the Politics of Agreement in Britain," *Canadian Journal of History* 9 (1974): 45–68.

17. Wiener, *English Culture and the Decline of the Industrial Spirit 1850–1980* (Cambridge: Cambridge Univ. Press, 1981), pp. 98–126.

18. Huxley, *Democracy Marches* (London: Chatto & Windus, 1941), esp. pp. 115–26; *The Uniqueness of Man* (London: Chatto & Windus, 1941), p. ix.

19. Huxley, *If I Were Dictator*, p. 50.

20. Huxley's ideas in the 1940s were very similar to Mannheim's; he favorably reviewed the latter's *Man & Society* for *Nature* 146 (1940): 3–4. But this newfound sensitivity to liberal democracy was probably more a reflection of the long-held views of Huxley's more experienced political collaborators from the 1930s. As far as Hayek was concerned, views such as Huxley's in the 1930s were actually more honest in drawing out the implications of centralized planning: see F. A. Hayek, *The Road to Serfdom* (London: Routledge, 1944), pp. 42–53. On Huxley's educational views, see, for instance, "Education as a Social Function," in *On Living In a Revolution* (London: Chatto & Windus, 1944), pp. 181–96. On UNESCO, see *UNESCO: Its Purpose and Philosophy* (Washington, D.C.: Public Affairs Press, 1948).

21. H. Stuart Hughes, *Consciousness and Society* (London: MacGibbon & Kee, 1959), esp. p. 429.

22. See, for instance, Huxley, *The Individual in the Animal Kingdom*; idem, "The Biological Basis of Individuality," *Journal of Philosophical Studies* 1 (1926): 305–319; idem, "What Is Individuality?" *The Realist* 1 (1929): 109–121.

23. See, for instance, Huxley, "Progress, Biological and Other," in *Essays of a Biologist*, pp. 17–60.

24. I have argued elsewhere that Huxley's political commitment to liberal individualism was underpinned by this conception of individual persons as the highest product of evolution. But Huxley also believed in the reality of supra-individual forms, even implying in quasi-Hegelian fashion that the state had the greater potential: Huxley, *The Individual in the Animal Kingdom*, p. 154. See also Colin Divall, "Capitalising on 'Science': Philosophical Ambiguity in Julian Huxley's Politics, 1920–1950," Ph.D. diss., Manchester Univ., 1985.

25. Huxley, *Progress, Biological and Other*, p. 60.

26. Sandy Lindsay, a pupil of Green's, was a don at Balliol during Huxley's

undergraduate years from 1906–1909. The influence of his mother is brought out in Huxley, *Memories*, pp. 17–22, 70. On Huxley's family background, see also Ronald Clark, *The Huxleys* (London: Heinemann, 1968), pp. 7–141; Juliette Huxley, *Leaves of the Tulip Tree* (London: John Murray, 1988), *passim*, esp. p. 239. On Green's influence generally, see Melvin Richter, *The Politics of Conscience*, esp. p. 13; Andrew Vincent and Raymond Plant, *Philosophy, Politics and Citizenship* (Oxford: Blackwell, 1984).

27. Huxley, *Religion Without Revelation* (London: E. Benn, 1927), pp. 137, 338.

28. Christopher Dawson, *Progress and Religion* (London: Longman, 1929), p. 241.

29. Huxley, "Evolutionary Ethics," in T. H. Huxley and J. S. Huxley, *Evolution and Ethics 1893–1943*, (London: Pilot Press, 1947), pp. 103–52; C. H. Waddington, "The Relations Between Science and Ethics," *Nature* 148 (1941): 270–74.

30. See, for instance, Anthony Flew, *Evolutionary Ethics*, 2d ed. (London: Macmillan, 1967), pp. 27–29.

31. See, for instance, C. D. Broad, Review of "Evolutionary Ethics," *Mind* 53 (1944): 344–67; S. Toulmin, "Contemporary Scientific Mythology," in A. MacIntyre, ed., *Metaphysical Beliefs* (London: SCM Press, 1957), pp. 51–75.

32. He certainly was by 1927, as Whitehead appears in the bibliography to *Religion Without Revelation*. Eddington and Jeans were influences in Huxley's "Man and Reality," in *Science in the Changing World*, ed. Mary Adams (1933; Freeport, N.Y.: Books for Libraries Press, 1968), pp. 186-98. But these thinkers are used here principally because they clearly illustrate an intellectual climate that permeated Huxley's writings from a number of sources in the 1920s.

33. A. S. Eddington, *The Nature of the Physical World* (Cambridge: Cambridge Univ. Press, 1928), pp. 240–242.

34. See, for instance, *Man and Reality*, apparently the only occasion when Huxley tried explicitly to draw out philosophical implications from the physical theory of relativity.

35. Huxley, "Religion Meets Science," *Atlantic Monthly* 147 (1931): 381.

36. Huxley, "Progress, Biological and Other," *The Hibbert Journal* 21 (1923): 436–60.

37. On the philosophical weaknesses generally of this school, see L. Susan Stebbing, *Philosophy and the Physicists* (1937; Harmondsworth: Penguin, 1944); A. O. Lovejoy, *The Revolt Against Dualism: An Inquiry Concerning the Existence of Ideas* (La Salle, Ill.: Open Court Publishing, 1929).

38. One can argue that this was yet another example of the drift of post-Enlightenment thought. See Alisdair MacIntyre, *After Virtue: A Study in Moral Theory* (London: Duckworth, 1981); Leo Strauss, "Three Waves of Modernity," in *Political Philosophy: Six Essays by Leo Strauss*, ed., H. Gilden (Indianapolis: Pegasus, 1975), pp. 81–98; Jürgen Habermas, "The Classical Doctrine of Politics in Relation to Social Philosophy," in *Theory and Practice* (Boston: Beacon Press, 1973), pp. 41–81.

39. A. S. Eddington, *The Nature of the Physical World* (Cambridge: Cambridge Univ. Press, 1928), pp. 329–331.

40. Huxley, *What Dare I Think?*, p. 163.

41. Ibid., p. 162.

42. William McDougall, *Modern Materialism and Emergent Evolution* (London: Methuen, 1929), p. 113.

43. M. Capek argues that Bergson's ontology was closer to Liebnizian panpsychism than to dualism, making it similar to the professed ontology of the emergent

evolutionists. M. Capek, *Bergson and Modern Physics* (Dordrecht: D. Reidel, 1971), p. 30.

44. C. Lloyd Morgan, *Emergent Evolution* (London: Williams & Norgate, 1926); "A Philosophy of Evolution," in J. H. Muirhead, *Contemporary British Philosophy* (London: Allen & Unwin, 1924), pp. 275–306; *Life, Mind and Spirit* (London: Williams & Norgate, 1925); *The Emergence of Novelty* (London: Williams & Norgate, 1933).

45. See, for instance, Morgan, *A Philosophy of Evolution*, p. 205.

46. Huxley, *The Individual in the Animal Kingdom*, p. vii.

47. Huxley, "Progress, Biological and Other," p. 68. Similar statements can be found in Huxley's writings well into the 1960s: see, for instance, "Higher and Lower," in *Essays of a Humanist* (1964; Harmondsworth: Penguin, 1966), p. 43.

48. Huxley, *The Individual in the Animal Kingdom*, pp. 28–31.

49. Huxley, *Religion Without Revelation*, p. 329.

50. Ibid., pp. 361–62.

51. Charles M. Holmes, *Aldous Huxley and the Way to Reality* (Bloomington: Indiana Univ. Press, 1970).

52. Sheldon Wolin, *Politics and Vision: Continuity and Innovation in Western Political Thought* (Boston: Little, Brown, 1960), pp. 352–434.

STANSKY

1. As quoted in Ronald C. Clark, *The Huxleys* (New York: McGraw-Hill, 1968), p. 173.

2. Julian S. Huxley, *Religion Without Revelation* (New York: Harper, 1957), pp. 72–73.

3. Aldous Huxley, *Antic Hay* (1923; New York: Modern Library, 1932), p. 7.

4. Julian Huxley, *Memories* (New York: Harper, 1970), p. 89.

WIENER

1. Quoted in Stefan Collini, *Liberalism and Sociology: L. T. Hobhouse and Political Argument in England, 1880–1914* (Cambridge: Cambridge Univ. Press, 1979), p. 225.

2. Ibid., p. 231.

3. See Noel Annan, "The Intellectual Aristocracy," in *Studies in Social History: A Tribute to G. M. Trevelyan*, ed. J. H. Plumb (London: Longman, 1955).

OLBY

1. Julian S. Huxley, *The Captive Shrew and Other Poems of a Biologist* (Oxford: Blackwell, 1932). Cited in Juliette Huxley, *Leaves of the Tulip Tree* (London: John Murray, 1986), p. 138.

2. Huxley, "Correspondence," *The Times*, cited in Ronald W. Clark, *Sir Julian Huxley, F.R.S.* (London: Roy; New York: Phoenix House, 1960), pp. 32–33.

3. Juliette Huxley, *Leaves of the Tulip Tree*, note 1, p. 165.

4. Jens-Peter Green, *Krise und Hoffnung: der Evolutionshumanismus Julian Huxleys* (Heidelberg: Winter, 1981), p. 3.

5. Huxley, *Evolution, the Modern Synthesis*, 1st ed. (London: Allen & Unwin; New York: Harper, 1942), pp. 22–23. Hereafter *Synthesis*.

6. Ibid., pp. 23–25.

7. Ibid., p. 26.

8. Ibid., p. 27.

9. Ibid., p. 28.

10. Huxley, "Natural Selection and Evolutionary Progress," *Report of the British Association for the Advancement of Science* 106 (1936): 81.

11. Huxley, *Memories* (London: Allen & Unwin; New York: Harper, 1970).

12. J. R. Baker, "Julian Sorell Huxley," *Biographical Memoirs of Fellows of the Royal Society* 22 (1976): 211.

13. H. G. Wells, J. S. Huxley, and G. P. Wells, *The Science of Life* (London: Amalgamated Press, 1931), book 4, esp. ch. 8. Hereafter *Science of Life*.

14. Huxley, "Natural Selection."

15. *Synthesis*, p. 7.

16. C. L. Hubbs, "Evolution, the Modern Synthesis," *The American Naturalist* 77 (1943): 365–68.

17. R. F. Kimball, "The Great Biological Generalization," *Quarterly Review of Biology* 18 (1943): 364–67.

18. As reported by D. Shapere, "The Meaning of the Evolutionary Synthesis," in *The Evolutionary Synthesis*, ed. E. Mayr, and W. B. Provine (Cambridge, Mass.: Harvard Univ. Press, 1980), p. 390.

19. M. J. S. Hodge, "The Structure and Strategy of Darwin's Long Argument," *British Journal for the History of Science* 10 (1977): 237–46.

20. A. Weismann, "The Selection Theory," in *Darwin and Modern Science*, ed. A. C. Seward (Cambridge: Cambridge Univ. Press, 1909), p. 61.

21. R. S. Cowan, "Nature and Nurture in the Work of Francis Galton," *Studies in History of Biology* 1 (1977): 133–208.

22. T. Dobzhansky, "Introduction" to Mayr, *Systematics and the Origin of Species from the Viewpoint of a Zoologist* (New York: Columbia Univ. Press, 1942), p. vii.

23. A. F. Shull, *Evolution* (New York and London: McGraw-Hill, 1936), pp. 209–211.

24. E. B. Ford, *Mendelism and Evolution*, 1st ed. (London: Methuen, 1931), p. 43.

25. N. E. Nordenskiöld, *The History of Biology: A Survey* (New York: Knopf, 1928).

26. E. Radl, *The History of Biological Theories* (London: H. Milford, Oxford Univ. Press, 1930).

27. W. Waagen, "Über die Ansatzstelle der Haftmuskein beim Nautilus und den Ammoniten," *Paleontographica* 17 (1867–70): 185–210.

28. *Synthesis*, p. 23.

29. *Science of Life*, p. 353.

30. H. J. Muller, "Genetic Variability, Twin Hybrids and Constant Hybrids, in a Case of Balanced Lethal Factors," *Genetics* 3 (1918): 422–99. Reprinted in Muller, *Studies in Genetics* (Bloomington: Indiana Univ. Press, 1962), pp. 69–94. Hereafter *Studies*.

31. W. Coleman, "Bateson and Chromosomes: Conservative Thought in Science," *Centaurus* 15 (1970): 228–314. See also R. Olby, "Structural and Dynamical Explanations in the World of Neglected Dimensions," in *A History of Embryology*, eds. T. J. Horder, J. A. Witkowski and C. C. Wylie (Cambridge: Cambridge Univ. Press, 1986), pp. 290–91.

32. Coleman, *Biology in the Nineteenth Century: Problems of Form, Function, and*

Transformation (New York: Wiley, 1971); G. E. Allen, *Life Science in the Twentieth Century* (New York: Wiley, 1975).

33. Huxley, "Natural Selection," p. 265.

34. J. B. S. Haldane, "The Theory of Evolution Before and After Bateson," *Journal of Genetics* 56 (1958): 16.

35. W. Bateson, "Presidential Address to Section E," *Report of the British Association for the Advancement of Science* 84 (1914): 11. Reprinted in B. Bateson, *William Bateson, F.R.S., His Essays and Addresses together with a short Account of his Life* (Cambridge: Cambridge Univ. Press, 1928), p. 285.

36. Muller, "Recurrent Mutations of Normal Genes of Drosophila not Caused by Crossing Over," *Anatomical Record* 26 (1923): 397–98. The subject of recurrent mutations is also discussed in T. H. Morgan, C. B. Bridges, and A. H. Sturtevant, "The Genetics of Drosophila," *Bibliographia Genetica* 2 (1925): 32–62. Chapter 2 is on mutation. Table 1 is from p. 23.

37. Muller, "Mutation," *Eugenics, Genetics and the Family: Proceedings of the Second International Congress of Eugenics* (New York, 1923), p. 112; *Studies*, pp. 221–27.

38. Morgan, Bridges, Sturtevant, "The Genetics of Drosophila," p. 22.

39. Muller, "Mutation," p. 106; *Studies*, p. 221.

40. J. R. G. Turner, "Fisher's Evolutionary Faith and the Challenge of Mimicry," *Oxford Surveys in Evolutionary Biology* 2 (1985): 192.

41. L. A. Callander, "Gregor Mendel: An Opponent of Descent with Modification," *History of Science* 26 (1988): 41–75.

42. R. Olby, *Origins of Mendelism*, 2d ed. (Chicago: Univ. of Chicago Press, 1985), pp. 285–99.

43. C. D. Darlington, *Chromosomes and Plant Breeding* (London: Macmillan 1932).

44. Huxley, "Natural Selection," p. 87.

45. Huxley, *Problems of Relative Growth* (London: Methuen, 1932), p. 214.

46. Ford and Huxley, "Mendelian Genes and Rates of Development in *Gammarus Chevreuxi*," *British Journal of Experimental Biology* 5 (1927): 130.

47. Ibid., p. 132.

48. Huxley, and G. R. de Beer, *The Elements of Experimental Embryology* (Cambridge: Cambridge Univ. Press, 1934), p. 413.

49. Ibid., p. 417.

50. Huxley to L. C. Dunn, 12 April 1932. He wrote: "If I came to work, I should like to have Gammarus as chief material. Is there any chance of getting a team to work on it? That is what I have had for some time in mind here, but conditions are against it. There are few geneticists in England I have none among my students we in England have very few research students anyhow and the Colleges of the University of London are scattered units. So Gammarus work proceeds sporadically at Plymouth, Oxford, University College and King's College, with single workers in each case, and there seems no chance of pushing ahead properly with it." (American Philosophical Society, L. C. Dunn Papers)

51. Huxley to Dunn, 21 February 1934. American Philosophical Society, Dunn Papers.

52. Muller, "Why Polyploidy is Rarer in Animals than in Plants," *The American Naturalist* 59 (1925): 346–55; *Studies*, pp. 455–60.

53. Unsigned review in *Quarterly Review of Biology* 18 (1943): 270.

54. Darlington, *Recent Advances in Cytology* (London: Churchill; Philadelphia: Blakiston, 1932), p. 483.

55. Ibid., p. 430.

56. Ford to Huxley, undated Thursday [1941]. Rice University, Huxley Papers.

57. Huxley, *Synthesis*, p. 594.

58. S. J. Gould, "The Hardening of the Modern Synthesis," in *Dimensions of Darwinism: Themes and Counterthemes in Twentieth Century Evolutionary Theory*, ed. M. Grene (Cambridge: Cambridge Univ. Press, 1983), pp. 71–93.

59. Divall, "From a Victorian to a Modern; Julian Huxley and the English Intellectual Climate," this volume.

60. Huxley, "Natural Selection," p. 96.

61. Ibid., p. 98.

62. Ibid., p. 96.

63. Ibid., p. 99.

64. Huxley, *New Bottles for New Wine* (London: Chatto & Windus; New York: Harper, 1957), p. 15.

65. Ibid., p. 21.

66. Ibid., p. 288.

67. Ibid., p. 287.

68. *Science of Life*, p. 878.

69. Mayr, "Typological versus Population Thinking," in his *Evolution and the Diversity of Life: Selected Essays* (Cambridge, Mass.: Harvard Univ. Press, 1976), pp. 26–29. See also his *Growth of Biological Thought* (Cambridge, Mass.: Harvard Univ. Press, 1982), pp. 45–47.

70. H. Judson, *The Eighth Day of Creation: Makers of the Revolution in Biology* (London and New York: Simon & Schuster, 1979).

71. Huxley, *Synthesis*, p. 614.

72. Ibid., p. 617.

WITKOWSKI

1. The first volume of Huxley's autobiography covers the period of his life in which he carried out his laboratory research. Julian Huxley, *Memories* (London: Allen & Unwin, 1970). The principal biography of the Huxley family is that by Clark and it provides details of other aspects of Huxley's activities: R. W. Clark, *The Huxleys* (London: Heinemann, 1968). The main obituary is by J. R. Baker, one of Huxley's Oxford pupils: "Julian Sorell Huxley," *Biographical Memoirs of Fellows of the Royal Society*, 22 (1976): 207–238. This is reprinted together with an extensive bibliography in J. P. Green, *Julian Huxley, Scientist and World Citizen 1887–1975* (Paris: UNESCO, 1978). Other obituaries dealing with his scientific work include J. Needham, *Nature* 254 (1975): 2–3; P. B. Medawar, *Nature* 254 (1975): 4; E. B. Ford, *Nature* 254 (1975): 5. The most extensive collection of Julian Huxley's papers is held in the Woodson Research Center, Rice University. See S. C. Bates, M. G. Winkler, and C. Riquelmi, eds., *A Guide to the Papers of Julian Sorell Huxley* (Houston: Woodson Research Center, Rice University, 1984).

2. Huxley, *Memories*, p. 73.

3. Ibid., p. 22.

4. Baker, "Huxley," p. 215.

5. Leonard Huxley wrote typical "Life and Letters" biographies of his father and

J. D. Hooker, and was also editor of *The Cornhill Magazine*. L. Huxley, *Life and Letters of Thomas Henry Huxley* (London: Macmillan, 1900).

6. Huxley, *Memories*, pp. 23–25.

7. Ibid., p. 50.

8. Ibid., p. 64.

9. Ibid., p. 65.

10. Ibid., p. 63.

11. Baker, "Huxley," p. 208.

12. M. Ridley, "Embryology and Classical Zoology in Great Britain," in *A History of Embryology*, ed. T. J. Horder, J. A. Witkowski, C. C. Wylie (Cambridge: Cambridge Univ. Press, 1986), pp. 35–67.

13. J. W. Jenkinson, *Experimental Embryology* (Oxford: Clarendon Press, 1909); idem, *Vertebrate Embryology* (Oxford: Clarendon Press, 1913); idem, *Three Lectures on Experimental Embryology* (Oxford: Clarendon Press, 1917).

14. Jenkinson, *Experimental Embryology*, p. 1.

15. Ibid., p. 20.

16. See Ridley, *History*, pp. 52–53.

17. Huxley and G. de Beer, *Elements of Experimental Embryology* (Cambridge: Cambridge Univ. Press, 1934).

18. Huxley, *Problems of Relative Growth*, 2d ed. (1932; New York: Dover Publications, 1972).

19. Baker, "Huxley," pp. 214–15.

20. Jenkinson, *Experimental Embryology*, p. 279.

21. Medawar, *Pluto's Republic* (Oxford: Oxford Univ. Press, 1982), p. 294.

22. E. B. Ford, "Some Recollections Pertaining to the Evolutionary Synthesis," in *The Evolutionary Synthesis*, ed. E. Mayr and W. B. Provine (Cambridge, Mass.: Harvard Univ. Press, 1981), pp. 334–42.

23. Leonard Huxley, *Life and Letters*, vol. 1, pp. 399–400.

24. G. E. Allen, *Thomas Hunt Morgan: The Man and His Science* (Princeton: Princeton Univ. Press, 1978), pp. 54–58.

25. Huxley, *Memories*, p. 83.

26. H. V. Wilson, "On Some Phenomena of Coalescence and Regeneration in Sponges," *Journal of Experimental Zoology* 5 (1907): 245–58.

27. Huxley, "Some Phenomena of Regeneration in *Sycon* with a Note on the Structure of Its Collar Cells," *Philosophical Transactions of the Royal Society of London*, ser. B, 202 (1911): 165–89.

28. Ibid., pp. 169–70.

29. Ibid., p. 176.

30. Jenkinson, *Experimental Embryology*, p. 12.

31. Ridley, *A History*, p. 58.

32. Huxley, "Some Phenomena," p. 173.

33. Ibid.

34. Ibid., p. 180.

35. Ibid., p. 181.

36. Ibid.

37. Ibid., p. 184.

38. Huxley, *Memories*, p. 98.

39. Huxley, "Further Studies on Restitution-Bodies and Free Tissue-Culture in *Sycon*," *Quarterly Journal of Microscopical Science* 65 (1921): 293–322.

40. Allen, "Naturalists and Experimentalists: The Genotype and the Phenotype," *Studies in History of Biology* 3 (1979): 179–209.

41. J. Maienschein, R. Rainger, and K. R. Benson, "Were American Morphologists in Revolt?" *Journal of the History of Biology* 14 (1981): 83–87.

42. J. A. Witkowski, "Ross Harrison and the Experimental Analysis of Nerve Growth: The Origins of Tissue Culture," in T. J. Horder et al., *A History of Embryology*, pp. 149–77.

43. R. G. Harrison, "Cultivation of Tissues in Extraneous Media as a Method of Morphogenetic Study," *Anatomical Records* 6 (1912): 183.

44. Huxley, "Further Studies," p. 318.

45. C. M. Child, *Patterns and Problems of Development* (Chicago: Univ. of Chicago Press, 1941), p.7.

46. Huxley, "Studies in Dedifferentiation. II Dedifferentiation and Resorption in *Perophora*," *Quarterly Journal of Microscopical Science* 65 (1921): 643–97.

47. Huxley, "Early Embryonic Differentiation," *Nature* 113 (1924): 276–78.

48. Ibid., p. 278.

49. Huxley and de Beer, *Elements of Experimental Embryology*, p. x.

50. Ibid., p. xi.

51. G. P. Wells, "Morphogenesis in the Animal Embryo," *Nature* 133 (1934): 890–91.

52. J. Needham, "Morphology and Biochemistry," *Nature* 134 (1934): 275–76.

53. Ridley, "Embryology," p. 47.

54. E. W. MacBride to Huxley, 26 May 1937, Huxley Papers, Woodson Research Center, Rice University.

55. Ibid., 8 May 1922.

56. E. W. MacBride, "Embryology and Predestiny," *Discovery* 15 (1934): 218–20.

57. Ibid., p. 218.

58. Ibid., p. 219.

59. Ibid., p. 220.

60. For a general discussion of fields and gradients, see D. Haraway, *Crystals, Fabrics and Fields* (New Haven: Yale Univ. Press, 1976). For contemporary views of the topic, see J. Needham, *Order and Life* (Cambridge: Cambridge Univ. Press, 1936); L. von Bertalanffy, *Modern Theories of Development* (Oxford: Oxford Univ. Press, 1933), especially pp. 112–20.

61. L. Wolpert, "Gradients, Position and Patterns: A History," in T. J. Horder et al., *History*, pp. 347–62.

62. Needham, *Order and Life*, p. 72.

63. C. H. Waddington, "Morphogenetic Fields," *Science Progress* 29 (1934): 336–46.

64. Ibid.

65. Child, *Patterns and Problems*, p. 277.

66. Ibid.

67. Needham, in *Nature*, p. 2.

68. G. W. Smith, "High and Low Dimorphism," *Mittheilungen aus der Zoologischen Station Neapel* 17 (1906): 312.

69. Huxley, "The Variation in the Width of the Abdomen in the Immature Fiddler Crab Considered in Relation to Its Relative Growth Rate," *The American Naturalist* 58 (1924): 469.

70. T. H. Morgan, "Further Evidence on Variation in the Width of the Abdomen in Immature Fiddler Crabs," *The American Naturalist* 57 (1923): 274–83.

71. Huxley, "Variation," pp. 470–72.

72. Huxley, "Constant Differential Growth-ratios and Their Significance," *Nature* 114 (1924): 895–96.

73. Huxley, *Problems of Relative Growth*, fig. 3.

74. Ibid., p. 2.

75. Ibid., p. 81.

76. E. C. Reeve and J. S. Huxley, "Some Problems in the Study of Allometric Growth," in *Essays on Growth and Form*, ed. W. E. LeGros Clark and P. B. Medawar (Oxford: Clarendon Press, 1945), p. 134.

77. S. J. Gould, "Morphological Channeling by Structural Constraint: Convergence in Styles of Dwarfing and Gigantism in *Cerion*, with a Description of Two New Fossil Species and a Report on the Discovery of the Largest *Cerion*," *Paleobiology* 10 (1984): 172–94.

78. J. F. White and S. J. Gould, "Interpretation of the Coefficient in the Allometric Equation," *The American Naturalist* 64 (1965): 5–18.

79. Reeve and Huxley, "Some Problems," pp. 138–46.

80. Gould, "Allometry and Size in Ontogeny and Phylogeny," *Biological Review* 41 (1966): 598–600.

81. A. G. Cock, "Genetical Aspects of Metrical Growth and Form in Animals," *Quarterly Review of Biology* 41 (1966): 181.

82. Huxley, *Elements of Experimental Embryology*, p. 208.

83. Ibid., pp. 208–212.

84. Gould, "Morphological Channeling."

85. Huxley, *Elements of Experimental Embryology*, p. 216.

86. Ibid., p.x.

87. Gould, "Positive Allometry of Antlers in the 'Irish Elk', *Megaloceros giganteus*," *Nature* 244 (1973): 375–76.

88. Huxley, *Elements of Experimental Embryology*, p. 79.

89. J. Needham, "Chemical Heterogony and the Ground-Plan of Animal Growth," *Biological Review* 9 (1933): 81.

90. G. Teissier, "Recherches Morphologiques et Physiologiques sur la Croissance des Insects," *Travaux de la Station biologique de Roscoff* 9 (1931): 27–238.

91. Needham, "Chemical Heterogony," p. 88.

92. Ibid., p. 95.

93. Ibid., p. 94.

94. Ibid., p. 104.

95. Ibid., p. 107.

96. Needham, in *Nature*, p. 2.

97. J. A. Witkowski, "The Elixir of Life," *Trends in Genetics* 2 (1986): 110–113.

98. Huxley, *Memories*, pp. 125–27.

99. E. Witschi, "Hormonal Regulation of Development in Lower Vertebrates," *Cold Spring Harbor Symposia on Quantitative Biology* 10 (1942): 145.

100. B. M. Allen, "The Results of Extirpation of the Anterior Lobe of the Hypophysis and of the Thyroid of *Rana pipiens* larvae," *Science* 44 (1916): 755–58.

101. Huxley, "Metamorphosis of Axolotl Caused by Thyroid Feeding," *Nature* 104 (1920): 435.

102. *Daily Mail*, 17 February 1920.

103. Huxley, *Memories*, p. 126.

104. Huxley, "The Thyroid Gland and the Control of Animal Growth," *Illustrated London News*, 28 February 1920.

105. Huxley, *Memories*, preface.

106. Huxley, "Studies on Amphibian Metamorphosis II," *Proceedings of the Royal Society of London* ser. B, 98 (1925): 113–46; Huxley and L. T. Hogben, "Experiments on Amphibian Metamorphosis and Pigment Responses in Relation to Internal Secretions," *Proceedings of the Royal Society of London*, ser. B, 93 (1922): 36–53.

107. Huxley and E. B. Ford, "Mendelian Genes and Rates of Development," *Nature* 116 (1925): 861–63.

108. Huxley, "Linkage in *Gammarus chevreuxi*," *Journal of Genetics* 11 (1921): 229–33.

109. Huxley and A. M. Carr-Saunders, "Absence of Prenatal Effects of Lens-Antibodies in Rabbits," *British Journal of Experimental Biology* 1 (1924): 215–48.

110. Huxley, "The Tissue-Culture King," *The Cornhill Magazine* 60 (1926): 422–57, especially pp. 432–41.

111. Huxley, *Memories*, preface.

112. Ibid.

113. Ibid., p. 156.

114. Huxley, *Life and Letters*, vol. 2, pp. 436–39. Also reprinted in Huxley (n. 1), pp. 24–25.

CARLSON

1. There is nothing similar in the American undergraduate or graduate experience. Medawar mentions a similar spell cast by his association with Karl Popper. What Americans lose in not having this broad exposure to powerful intellects, they gain by their intensely focused majors and their heady diversity of graduate course work provided by several first-rate scholars of differing views who inhabit a common department or university, such as Morgan and Wilson for the Drosophila group or Muller, Sonneborn, and Luria for Watson.

CHURCHILL

1. Jan Witkowski this volume.

2. Julian S. Huxley and Gavin R. de Beer, *The Elements of Experimental Embryology* (Cambridge: Cambridge Univ. Press, 1934). This text has been reprinted in a facsimile edition (New York: Hafner, 1963).

3. Juliette Huxley, *Leaves of the Tulip Tree* (Topsfield, Mass.: Salem House, 1986), p. 107.

4. For a discussion of their contrasting views regarding the biogenetic law, see Frederick B. Churchill, "The Modern Evolutionary Synthesis and the Biogenetic Law," in *The Evolutionary Synthesis. Perspectives on the Unification of Biology*, ed. Ernst Mayr and William B. Provine (Cambridge, Mass.: Harvard Univ. Press, 1980), pp. 112–22.

5. Julian S. Huxley, "Natural Selection and Evolutionary Progress," *Proceedings of the British Association for the Advancement of Science* 106 (1936): 81-100.

6. Huxley, "Early Embryonic Differentiation," *Nature* 113 (1924): 276–78.

7. Huxley and de Beer, "Studies in Dedifferentiation IV. Resorption and Differential Inhibition in *Obelia* and *Campanularia*," *Quarterly Journal of Microscopical Science* 67 (1923): 473–95; and "Studies in Dedifferentiation V. Dedifferentiation and Reduction in *Aurelia*," ibid., 68 (1924): 471–79.

8. Hans Spemann, "Die Erzeugung tierischer Chimären durch heteroplastische Transplantation zwischen *Triton cristatus* und *taeniatus*," *Archiv fur Entwicklungsmechanik* 48 (1921): 533–70. See also the account of this postscript in Viktor Hamburger, *The Heritage of Experimental Embryology: Hans Spemann and the Organizer* (New York, Oxford: Oxford Univ. Press, 1988), pp. 37–47. Hamburger also includes in this work a sensitive account of Hilde Proefscholdt Mangold's scientific career, pp. 173–80.

9. Hans Spemann and Hilde Mangold, "Uber Induktion von Embryonalenlagen durch Implantation artfremder Organisatoren," *Archiv für mikroscopischen Anatomie und Entwicklungsmechanik* 100 (1924): 599–638. For an English translation of this paper, see Benjamin H. Willier and Jane M. Oppenheimer, *Foundations of Experimental Embryology* (Englewood Cliffs, N.J.: Prentice-Hall, 1964), pp. 144–84.

10. Charles Manning Child, *Individuality in Organisms* (Chicago: Univ. of Chicago Press, 1915); and *The Origin and Development of the Nervous System from a Physiological Viewpoint* (Chicago: Univ. of Chicago Press, 1921). See also Child to Huxley, 2 November 1920, in Julian S. Huxley Papers, Woodson Research Center, Rice University.

11. Herbert George Wells, Julian S. Huxley, and George Philip Wells, *The Science of Life* (London: Amalgamated Press, 1929–30), see book 4, chap. 5.

12. Huxley, "Spemanns 'Organisator' und Childs Theorie der axialen Gradienten," *Die Naturwissenschaften* 18 (1930): 265.

13. Spemann, *Embryonic Development and Induction* (1938; facsimile reprint, New York: Hafner, 1962), pp. 321–45.

14. Spemann to Huxley, 14 January 1925. Julian S. Huxley Papers, Woodson Research Center, Rice University.

15. Spemann to Huxley, 1 November 1927 and 28 December 1927. Ibid.

16. De Beer to Huxley, 5 April 1926. Ibid.

17. De Beer, "The Mechanics of Vertebrate Development," *Biological Reviews and Biological Proceedings of the Cambridge Philosophical Society* 2 (1926–27):137–97.

18. John R. Baker, "Julian Sorell Huxley," *Biographical Memoirs of Fellows of the Royal Society* 22 (1976): 207–238.

19. Huxley and de Beer, *Elements*, p. 441.

20. Ibid., p. 67.

21. Ibid., p. 139.

22. Thomas Hunt Morgan, *Embryology and Genetics* (New York: Columbia Univ. Press, 1934).

23. Huxley and de Beer, *Elements*, see n. 3, p. 403.

24. Ibid., p. 413.

25. Ibid., p. 442.

26. For a withering attack on the cell theory, see Charles M. Child, "The Significance of the Spiral Type of Cleavage and Its Relation to the Process of Differentiation," *Biological Lectures from the Marine Biological Laboratory of Woods Hole* (n.p., 1899), pp. 231–66. Quotation appears on p. 235. For a detailed account of Child's thinking during this period, see Jeffrey Werdinger, "Embryology at Woods Hole: The Emergence of a New American Biology" (Ph.D. diss., Indiana Univ., 1980), pp.

537–55. The standard biography with bibliography of Child is written by his former student, Libbie H. Hyman, "Charles Manning Child," *Biographical Memoirs of the National Academy of Sciences* 30 (1957): 73–103.

27. F. Seidel, [Review of Huxley and de Beer], *Die Naturwissenschaften* 11 (1936): 175–76. Paul Weiss, *Principles of Development. A Text in Experimental Embryology* (New York: Holt, 1939), pp. 186–89, 373–83.

28. Child, *Individuality*, p. 23.

29. Throughout this account of Spemann's work I follow Viktor Hamburger, *The Heritage*.

30. Ibid., p. 20.

31. Hilde Proefscholdt married Otto Mangold after her 1921 implantation experiments; so she went by her married name in her joint publication with Spemann of 1924.

32. Hamburger, *The Heritage*, p. 33.

33. Ibid., p. 86.

34. Ibid., p. 95.

35. Huxley and de Beer, *The Elements*, particularly pp. 153, 497.

36. Spemann to Huxley, *Embryonic Development*, p. 345.

37. Weiss, *Individuality*, pp. 346–47.

38. Huxley and de Beer, *The Elements*, p. 417.

39. Robert Olby, "Structural and Dynamical Explanations in the World of Neglected Dimensions," in *A History of Embryology*, ed. T. J. Horder, Jan A. Witkowski, and C. C. Wylie (Cambridge: Cambridge Univ. Press, 1986), pp. 275–308.

40. See Natasha X. Jacobs, "From Unit to Unity: Protozoology, Cell Theory and the New Concept of Life," *Journal of the History of Biology* 22 (1989): 215–42; and Marsha L. Richmond, "Protozoa as Precursors of Metazoa: German Cell Theory and Its Critics at the Turn of the Century," ibid., pp. 243–76.

41. Waldemar Schleip, *Die Determination der Primitiventwicklung* (Leipzig: Akademische Verlagsgesellschaft M. B. H., 1929).

42. Weiss, *Principles of Development*.

BURKHARDT

Research for this paper was supported in part by grants from the National Endowment for the Humanities, the National Science Foundation, and the Research Board of the University of Illinois, Urbana-Champaign. I am also pleased to acknowledge the assistance of Mary Winkler, Susan Stewart, and Nancy Boothe and her staff at the Woodson Research Center at Rice University, all of whom helped me in my study of the Huxley papers at Rice University. For permission to quote from the Huxley letters and mauscripts in this paper, I am grateful to Rice University and to the Edward Grey Institute of Field Ornithology, Oxford.

1. See Julian S. Huxley, Introduction, in "A Discussion on Ritualization of Behaviour in Animals and Man," *Philosophical Transactions of the Royal Society of London*, ser. B, 251 (1966): 249. I am not certain of the occasion on which Lorenz made this remark, though a likely possibility is a lecture he gave at the British Museum of Natural History, which Huxley chaired. The letter is mentioned in a letter from Lorenz to Huxley, 26 November 1963, Huxley Papers, Woodson Research Center, Rice University. To the best of my knowledge, however, neither in his published papers nor in his letters to Huxley did Lorenz ever pay Huxley the same

compliment. Indeed, while Lorenz characteristically cited Whitman and Heinroth as the two most important early pioneers of ethology, when it came to citing a *third* pioneer, he has most frequently named Wallace Craig. See, for example, Konrad Z. Lorenz, *Studies in Animal and Human Behavior*, vol. 1 (Cambridge: Harvard Univ. Press, 1970), p. xv.

2. Huxley, *Memories* (London: Allen & Unwin, 1970), pp. 79, 83.

3. Huxley's most important reminiscences about his activities as a birdwatcher, in addition to those in *Memories*, are to be found in *Bird-watching and Bird Behaviour* (London: Chatto & Windus, 1930), and a series of four articles published in *The Countryman* in 1943–44: "Memories and Achievements: A Chapter of Natural History," 28 (1943): 175–80; "Natural History Memories—2, 29 (1944): 28–32; "Natural History Memories—3," 29 (1944): 186–91; and "Natural History Memories—4," 30 (1945): 38–43.

Also important is Huxley's Introduction to Edmund Selous, *Realities of Bird Life: Being Extracts from the Diaries of a Life-loving Naturalist* (London: Constable, 1927), pp. xi–xvi.

For a list of Huxley's writings, see the excellent bibliography compiled by Jens-Peter Green, in J. R. Baker, *Julian Huxley: Scientist and World Citizen 1887–1975* (UNESCO, 1978), pp. 55–184.

4. Peter Medawar, *The Art of the Soluble* (London: Methuen, 1967), p. 151.

5. David Lack to James Fisher, 21 January 1969, Huxley Papers, Woodson Research Center, Rice University.

6. This is not to say, however, that Huxley's notebooks do not deserve careful scrutiny. Reconstructing Huxley's practice in the field and comparing it with that of other field naturalists (and with his own published accounts of his observations) promises to be instructive.

7. For insightful comments on Huxley's work as a scientific birdwatcher, see especially the comments in John R. Baker, "Julian Sorell Huxley," *Biographical Memoirs of Fellows of the Royal Society* 22 (1976): 207–238; John Durant, "Innate Character in Animals and Man: A Perspective on the Origins of Ethology," in *Biology, Medicine, and Society, 1840–1940*, ed. Charles Webster (Cambridge: Cambridge Univ. Press, 1981), pp. 157–92; and David Lack, "Some British Pioneers in Ornithological Research, 1859–1939," *Ibis* 101 (1959): 71–81.

On the history of ethology, see Durant's above-mentioned article; Richard W. Burkhardt, Jr., "On the Emergence of Ethology as a Scientific Discipline," *Conspectus of History* 1 (no. 7, 1981): 62–81; idem, "The Development of an Evolutionary Ethology," in *Evolution from Molecules to Men*, ed. D. S. Bendall (Cambridge: Cambridge Univ. Press, 1983), pp. 429–44; idem, "Theory and Practice in Naturalistic Studies of Behavior Prior to Ethology's Establishment as a Scientific Discipline," in *Interpretation and Explanation in the Study of Animal Behavior*, ed. Marc Bekoff and Dale Jamieson (Boulder, CO: Westview Press, 1990), vol. 2, pp. 6–30; and W. H. Thorpe, *The Origins and Rise of Ethology* (London: Heinemann, 1979).

8. When ethologists provide autobiographical sketches of themselves, they almost invariably recount how, from a very early age, they were enchanted by pets or wildlife. On this point see Konrad Lorenz, "Introduction: The Study of Behaviour," in Jurgen Nicolai, *Bird Life* (New York: Putnam; London: Thames & Hudson, 1974), pp. 14–15.

9. For example, Huxley told an audience of radio listeners in 1930 that a good birdwatcher would tend to become a good naturalist "almost by a natural momentum, provided he has enough time and energy to spend on his hobby." Huxley, *Bird-*

watching and Bird Behaviour, pp. 13 and 17. However, Huxley did on several occasions acknowledge the influence that Selous, Howard, and others had upon the study of bird behavior. See especially his *Bird-watching and Bird Behaviour*, p. vii, and his introduction to Selous, *Realities of Bird Life*. For a broad survey of developments in British natural history from the eighteenth to the twentieth centuries, see David Elliston Allen's excellent *The Naturalist in Britain: A Social History* (London: Allen Lane, 1976). For an analysis of the interaction between amateurs and professionals in Britain in the development of plant ecology in the twentieth century, see P. D. Lowe, "Amateurs and Professionals: The Institutional Emergence of British Plant Ecology," *Journal of the Society for the Bibliography of Natural History* 7 (1976): 517–35.

10. Huxley described his encounter with the green woodpecker in "Memorable Incidents with Birds," *The Listener* 3 (1930): 841–42, and *Memories*, p. 36. Huxley's earliest dated bird notebook, preserved at the Edward Grey Institute of Field Ornithology, Oxford, bears the inscription "1901–1903, Surrey and Hertfordshire."

11. W. H. Mullens and H. Kirke Swann, *A Bibliography of British Ornithology* (London: Macmillan, 1917), p. 96.

12. H. Eliot Howard, *The British Warblers: A History with Problems of Their Lives*, pt. 1, "The Grasshopper Warbler" (London: Porter, 1907–1914), pp. 19–20. On Howard see David Lack, "Some British Pioneers in Ornithological Research," *Ibis* 101 (1959): 71–81.

13. Preface, *The Zoologist*, ser. 4, 7 (1903): iv.

14. Anonymous, "The Unknown Bird World," *Saturday Review* 91 (1901): 683. This new interest in observing the behavior of animals in nature had its counterpart in a concern for the happiness of animals kept in captivity. Another anonymous piece in the *Saturday Review* in 1901 castigated the London zoo as "the bastille of the beasts." The "most unmercifully severe confinement" of the animals at the zoo, the reviewer maintained, produced a great deal of "wholly unjustifiable animal misery." See "The Old Zoo and the New," *Saturday Review* 91 (1901): 330–32, 365–66, 397–98, 433–34.

15. W. Warde Fowler, "In the Ways of Birds," *Saturday Review* 92 (1901): 104–105.

16. On Fowler see Percy Ewing Matheson, "William Warde Fowler (1847–1921)," *Dictionary of National Biography, 1912–1921*, pp. 194–95, and Huxley, "Obituary. W. Warde-Fowler," *British Birds* 15 (1921): 143–44. The characterization of Huxley as "a rather timid undergraduate" is Huxley's own (ibid., p. 143). Fowler wrote *A Year with the Birds. By an Oxford Tutor* (Oxford: Blackwell, 1886). See also his *Kingham Old and New* (Oxford: Blackwell, 1913). Like Selous, Fowler wrote animal books for children.

17. In his obituary of Fowler in *British Birds*, Huxley did not mention Fowler's enthusiasm for Selous. It is difficult to believe, however, that Fowler would not have communicated his enthusiasm about Selous to Huxley when they discussed birds and birdwatching. Selous and Fowler were the only two contemporary ornithologists mentioned in Huxley's 1907 "Habits of Birds" paper (see note 18).

18. On Selous see the works on the history of ethology cited in note 7. Thorpe, in his *Origins and Rise of Ethology*, gives several pages to Selous (pp. 30–33) but exaggerates the difficulties of Selous's prose and fails to mention Selous's work on sexual selection. For additional information on Selous, see K. E. L. Simmons, "Edmund Selous (1857–1934): Fragments for a Biography," *Ibis* 126 (1894), 595–96, and Richard W. Burkhardt, Jr., "Edmund Selous," *Dictionary of Scientific Biography*,

Supplement II 18 (1990): 801–803. Simmons has undertaken an extensive study of Selous's work.

19. Huxley, Introduction, *Realities of Bird Life*, p. xi.

20. Selous, "Observations Tending To Throw Light on the Question of Sexual Selection in Birds, Including a Day-to-Day Diary on the Breeding Habits of the Ruff (*Machetes pugnax*)," *The Zoologist* ser. 4, 10 (1906): 201–219, 285–94, 419–28; 11 (1907): 60–65, 161–82, 367–81; "An Observational Diary on the Nuptial Habits of the Blackcock (*Tetrao tetrix*) in Scandinavia and England," *The Zoologist* ser. 4, 13 (1909): 401–413, 14 (1910): 23–29, 51–56, 176–82, 248–65.

21. Selous, "Observations Tending To Throw Light on the Question of Sexual Selection in Birds," *The Zoologist* ser. 4, 11 (1907): 163.

22. Huxley, Introduction, *Realities of Bird Life*, p. xii.

23. Selous, *The Bird Watcher in the Shetlands* (London: 1905), p. 44.

24. Selous, "An Observational Diary of the Habits—Mostly Domestic—of the Great Crested Grebe (*Podicipes cristatus*) and of the Peewit (*Vanellus vulgaris*), with Some General Remarks," *The Zoologist* ser. 4, 5 (1901): 161–83, 339–50, 454–62; 6 (1902): 133–44 (quoted from 6: 144).

25. In *Bird Life Glimpses* (London: George Allen, 1905), Selous wrote: "With ourselves definite ideas have become greatly developed; but animals may live, rather, in a world of emotions, which would then be much more a cause of their actions, and consequently, of the cries which accompanied them . . ." p.38. Huxley's clearest exposition of this idea was in his popular essay "Ils n'ont que de l'ame: An Essay on Bird Mind," *The Cornhill Magazine* 54 (1923): 415–27, reprinted in *Essays of a Biologist* (London: Chatto & Windus, 1923), pp. 103–129. Durant, "Innate Character in Animals and Man," has called attention to this essay.

26. Huxley, *The Individual in the Animal Kingdom* (Cambridge: Cambridge Univ. Press, 1912), p. 154.

27. Huxley, "A 'Disharmony' in the Reproductive Habits of the Wild Duck (*Anas boschas, L.*)," *Biologisches Zentralblatt* 32 (1912): 621–23.

28. Huxley, *Memories*, p. 68.

29. Huxley to Hardy, 12 May 1921, Huxley Papers, Woodson Research Center, Rice University. In *Memories* Huxley says more about the psychological difficulties he experienced from his guilty feelings regarding sex than about his anxieties over his prospects as a scientist.

30. Huxley, "The Courtship of Birds," *The Listener* (1930), p. 935; *Bird-watching and Bird Behaviour*, pp. 61–62; *Memories*, p. 79.

31. Huxley, "A First Account of the Courtship of the Redshank (*Totanus calidris L.*)," *Proceedings of the Zoological Society of London* (1912): 647.

32. Quote from Huxley, *Bird-watching and Bird Behaviour*, p. 63. The importance of focusing on the behavior of individual birds was a point that had been detailed by Selous, and it was a point that characterized two different remarkable publishing ventures that were undertaken in 1907 but not completed for another half-dozen years: H. Eliot Howard, *The British Warblers: A History with Problems of Their Lives*, 2 vols. (London: R. H. Porter, 1907–1914), and *The British Bird Book*, 4 vols. (London: Jack, 1910–1914), edited by F. B. Kirkman, with Selous and W. P. Pycraft among the contributing authors. Though vol. 1 of *The British Bird Book* did not appear until 1910, Kirkman states in vol. 4, p. 461, that the enterprise was begun in 1907.

33. Letter from Huxley dated 10 April 1911, Huxley Papers, Woodson Research

Center, Rice University. In his published paper ("Redshank," p. 648), Huxley reported his practice of note-taking: "I made a number of notes on the spot, and usually within twenty-four hours embodied what I had seen the day before in a letter to an ornithological friend." Huxley's practice evidently was to include his bird observations in his letters to his fianceé, with the understanding that she would later return at least these sections of the letters to him. File Box 162 in the Huxley Papers at Rice University contains sections from four such letters Huxley wrote from Wales in April 1911. The redshank is featured in the last two of them (7 April and 20 April).

34. Huxley, "Redshank," p. 648. See Selous, "Observations," note 20.

The words Huxley used in 1930 to describe his new awareness of bird behavior upon seeing the courtship of the redshank were much the same as those he used to recount his impressions upon seeing the green woodpecker and reading Selous for the first time. With regard to observing the redshank, he wrote: "Here was a whole new world of watching to be undertaken, rich in sights fascinating in themselves, and full also of meanings which had to be puzzled out and unravelled." *Bird-watching and Bird Behaviour*, p. 63.

35. See Kirkman, *The British Bird Book* (quote 1, p. iii), and Howard, *The British Warblers*, as well as the writings of Selous.

36. Huxley, "Redshank," pp. 651, 654. The italics are Huxley's.

In his autobiography (*Memories*, p. 79), Huxley wrote of his redshank paper: "I am not a little proud that I used the word 'formalized' for some of the male's actions, for we now know that much courtship behaviour is indeed stereotyped in a special formalism; and much prouder of having made field natural history scientifically respectable." In fact, however, Huxley did not use the word "formalized" in his redshank paper. He used the word "formal" there, but he did so in crediting *Selous* with the idea that the combats between the males were formal in character (Huxley, "Redshank," p. 652). In his *Bird-watching and Bird Behaviour*, Huxley again credited Selous with having been the first to stress how hostility between males found an outlet in "formal posturing and mock combats instead of in genuine fighting" (p. 96). When it came to writing his autobiography, however, Huxley failed to mention that it was from Selous that he got the idea of the "formal" nature of the male's actions. (Interestingly, Selous, in his "Observations" article of 1906-1907, did not indicate that the combats between the redshank males were solely formal. He described several male-to-male combats that he believed to be genuine.)

37. Huxley, "The Courtship-Habits of the Great Crested Grebe (*Podiceps cristatus*); with an Addition to the Theory of Sexual Selection," *Proceedings of the Zoological Society of London* 35 (1914): 491–562 (quote on p. 501). See Selous, "An Observational Diary," note 24.

It seems that Selous's writings not only "dovetailed" with Huxley's observations but also influenced Huxley's published accounts of what he himself had seen. Selous's article on the great crested grebe described, for example, how the bird in certain of its courtship actions resembled a penguin (see esp. 5 [1901], 344). Huxley's *field notes* on the great crested grebe never refer to the bird looking like a penguin. In his 1914 paper, however, Huxley proceeded to name two of the crested grebe's actions the "ghostly penguin" and the "penguin dance."

On another matter, John R. Baker, "Huxley," p. 214, notes that Selous used the verb "exhaust" in 1905, describing how in the mutual ceremony of the little grebe the two birds broke the ceremony off "as though it exhausted the matter." Huxley in turn wrote of the mutual ceremonies of the great crested grebe being "self-exhausting." In

Baker's words, "That two independent workers should have chosen words derived from the verb 'exhaust' seems remarkable."

38. Huxley, "The Great Crested Grebe and the Idea of Secondary Sexual Characteristics," *Science*, n.s., 36 (1912): 601–602 (quote on p. 602).

39. Huxley, "Grebe" (1912), p. 602. The question of the transmission and development of secondary sexual characters was one with which Darwin had struggled. Huxley, in his "Redshank" paper, stated that it appeared to be "both more primitive and easier for hereditary characters to be transmitted equally to both sexes" (p. 655). E. B. Poulton, however, upon reading Huxley's assertion, told Huxley that he had found the reverse: "in butterflies I find a great tendency for colours & patterns to be associated with one sex, & often some resistance & apparent difficulty in transferring them to the other." E. B. Poulton to Huxley, 28 September 1912, Huxley Papers, Woodson Research Center, Rice University. For Poulton's thoughts on epigamic characters, see his *Essays on Evolution* (Oxford: Clarendon Press 1908), pp. 379–81, and his *Charles Darwin and the Origin of Species* (London: Longmans, Green, and Co. 1909), pp. 132–43.

40. But see Selous's concluding comments to his great crested grebe article (6 [1902], 144), cited in note 24.

41. Another important book of the period was W. P. Pycraft's *The Courtship of Animals* (London: 1913), which was dedicated to H. Eliot Howard, "whose observations of the courtship of birds recorded in his *History of the British Warblers* constitute a beacon for all engaged in the study of animal behavior." Huxley said "accidental circumstances" prevented him from reading Pycraft's book until after he had completed his own paper. Huxley agreed with Pycraft that Darwin's theory of sexual selection needs modification. He also endorsed Pycraft's views on (1) the principle of ornamentation, and (2) "the necessity for a psychological point of view in our interpretation of the courtship-phenomena of animals." [See Huxley, "Grebe" (1914), p. 559.] John R. Baker's memoir of Huxley notes that Huxley's earlier (1912) grebe paper attributed the grebe's ruff and tufts to sexual, not natural selection, but Baker does not address the question of the intellectual influences that may have helped change Huxley's mind on this matter.

42. Howard's *The British Warblers* was published in nine parts. In the earliest (dated February 1907), in his discussion of the courtship of the grasshopper warbler, Howard wrote: "closer study, devoted for some years to this courtship, convinces me that sexual selection as a rational explanation of the phenomena is impossible. . . ." Though the last of the nine parts of Howard's work is dated October 1914, a correction within part nine has the later date of June 1915.

43. In a footnote on the first page of his paper, Huxley acknowledged that the word "courtship" in his title was perhaps misleading, since courtship, strictly speaking, should apply only to antenuptial behavior. "Love-habits," he said, would have been a better term for what he was talking about. Huxley, "Grebe" (1914), p. 491, note.

44. Ibid., p. 509. See also pp. 514–15.

45. Ibid., p. 516. As Huxley described it, this "constancy" in "marriage," as for example when an unpaired female's attention to a paired male roused the "jealousy" of the male's mate. See ibid., pp. 553–54.

46. Ibid., p. 417.

47. Ibid., p. 556. Whether or not "mutual selection" was a new phrase and concept (Darwin had at least raised the possibility of "a double or mutual process of sexual selection" in *The Descent of Man* [London: Murray, 1871], I, 277; and Selous

had talked about "intersexual selection" in *The Bird Watcher in the Shetlands*, [London: J. M. Dent & Co. 1905], chap. 30), the idea that behavioral displays were built upon actions arising from surplus energy on the bird's part was not new. Both Selous and Howard had used this explanation in their writings.

48. Ibid., pp. 516–17.

49. Ibid., p. 496, speaks of the grebes going through "a curious set ritual."

50. Lack to Huxley, 15 March 1953; Lorenz to Huxley, 2 November 1954 and 26 November 1963; Huxley Papers, Woodson Research Center, Rice University.

51. Huxley, "Grebe" (1914), p. 492. Burkhardt, "On the Emergence of Ethology as a Scientific Discipline," p. 69.

52. Huxley, "Bird-watching and Biological Science. Some Observations on the Study of Courtship in Birds," *Auk* 33 (1916): 142–61, 256–70. In 1921 Huxley told Hardy that during his three years in Texas he had "led the life of an intellectual semi-invalid—never daring to work hard, only doing my routine of lectures & laboratory, still feeling that I shd. never be able to concentrate enough to do real research." See Huxley to Hardy, 12 May 1921, Huxley Papers, Woodson Research Center, Rice University. While Huxley was in Texas, his major ornithological expedition was a trip to Louisiana, where he made observations on the Louisiana heron.

53. Huxley to Hardy, 12 May 1921, Huxley Papers, Woodson Research Center, Rice University.

54. Before his work on the red-throated diver Huxley published several short papers, including "Some Points in the Sexual Habits of the Little Grebe, with a Note on the Occurrences of Vocal Duets in Birds," *British Birds* 13 (1919): 155–58; "The Accessory Nature of Many Structures and Habits Associated with Courtship," *Nature* 108 (1921): 565–66; and a review of H. Eliot Howard's *Territory in Bird Life*, in *Discovery* 2 (1921): 135–36. He also published in 1922 a short paper on "Preferential Mating in Birds with Similar Coloration in Both Sexes," *British Birds* 16 (1922): 99–101.

Before his major paper on the red-throated diver appeared in 1923, a shorter version of his observations on the bird, together with those of G. J. van Oordt, appeared as "Some Observations on the Habits of the Red-throated Diver in Spitsbergen," *British Birds* 16 (1922): 34–46.

55. Huxley, "Courtship Activities in the Red-throated Diver (*Colymbus stellatus Pontopp.*); Together with a Discussion of the Evolution of Courtship in Birds," *Journal of the Linnean Society of London, Zoology* 35 (1923): 253–92 (quote on p. 269).

56. Ibid., p. 273. See also p. 268.

57. Ibid., p. 273. See also p. 283: "the form of the courtship, not merely in details but in broad lines as well, will depend in the main upon other general biological factors affecting the species." Similarly, see p. 290: "It becomes increasingly clear that to interpret the behavior and evolution of a bird, even in apparently only one regard, it is necessary to take into account *all* the circumstances of its life."

58. Ibid., p. 278.

59. Ibid., p. 286.

60. Ibid., p. 288.

61. Huxley, "The Absence of 'Courtship' in the Avocet," *British Birds* 19 (1925): 88–94; Huxley, assisted by F. A. Montague, "Studies on the Courtship and Sexual Life of Birds V: The Oyster-catcher," *Ibis*, ser. 12, 1 (1925): 868–97. Huxley's explanation of the absence of "courtship" in the avocet was that the bird possessed a "very placid 'temperament'," and that as a result there was not enough emotional

tension overflowing into action for courtship displays to develop. Here, if not in many other places, Huxley did credit Howard and Selous "and various other writers" with the view "that 'courtship' displays arise immediately from the excited state of the unsatisfied male bird or of both sexes in species with mutual displays . . ." (p. 92). Before his trip to Holland, Huxley seems to have been especially interested in studying the courtship of godwits.

62. Huxley, "Oyster-catcher," p. 895. Later, Huxley identified "the physiological effect of socialized sexual emotion" that provided a "biological value" to these piping ceremonies. See his "The Present Standing of the Theory of Sexual Selection," in *Evolution: Essays on Aspects of Evolutionary Biology presented to Professor E. S. Goodrich on his Seventieth Birthday*, ed. G. R. de Beer (Oxford: Clarendon Press, 1938), pp. 11–41 (quote from pp. 25–26). In "Memories and Achievements—A Chapter of Natural History," *The Countryman* 28 (1943): 175, Huxley reports how he told an ornithological friend regarding the piping ceremony of the oyster-catcher "that of all the odds and ends I had done in natural history I was in some ways proudest of having shed some light on the meaning of this particular feature of bird life." Huxley, "Oyster-catcher," p. 896, note, says of this paper, "Since the above was written, I have come across further references to this species by Selous (1906). Interesting suggestions are here made concerning the passage from jealousy to courtship and the employment of the piping ceremony to express both, though the evidence is scanty. Observations confirmatory to ours are also recorded." This article by Selous proves in fact to have been the *same* article that Huxley in *1912* said he had not read until after completing his paper on the redshank. See note 34.

63. After spending a weekend of his spring vacation in 1922 with Howard, Huxley wrote, "I feel stimulated to take up systematic observation again; & I feel more than ever impelled to write a general book on the subject!" Huxley to Howard, 24 April 1922. See also a previous letter from Huxley to Howard dated 19 February 1922, as well as Huxley to Howard, 6 December 1923, and 3 March 1924. Howard Papers, Edward Grey Institute of Field Ornithology, Oxford.

64. See the Huxley manuscript titled "Notes by JSH 1925 for a book on Bird Courtship that never got written," Huxley Papers, Woodson Research Center, Rice University.

65. "Biology of Bird Courtship," *Proceedings of the Seventh International Ornithological Congress at Amsterdam, 1930* (1931): 107–108; "Threat and Warning Coloration in Birds, with a General Discussion of the Biological Functions of Colour," *Proceedings of the Eighth International Ornithological Congress, Oxford, 1934* (1938): 430–55. "Darwin's Theory of Sexual Selection and the Data Subsumed by it, in the Light of Recent Research," *The American Naturalist* 72 (1938): 416–33; "The Present Standing of the Theory of Sexual Selection," in *Evolution: Essays on Aspects of Evolutionary Biology*, ed. G. R. de Beer (Oxford: Clarendon Press, 1938), pp. 11–42.

Mary Jane West-Eberhard identifies the "Forgotten Era of Sexual Selection Theory" as beginning in the 1930s as the result of population geneticists redefining fitness in terms of change in gene frequencies and Huxley suggesting that the term "sexual selection" be eliminated. See West-Eberhard, "Sexual Selection, Social Competition, and Speciation," *Quarterly Review of Biology* 58 (1983): 155–83 (see p. 156). Huxley's treatment of sexual selection in his writings of the 1930s and 1940s deserves further historical examination.

66. The radio talks, transmitted by the BBC, appeared first in *The Listener* and then, only slightly modified, as *Bird-watching and Bird Behaviour* (1930).

67. "The Private Life of the Gannets," which Huxley and Lockley filmed in 1934, received an Academy Award for 1937 in the "short subjects" category as the best one-reel documentary. The movie was filmed amid the enormous breeding colony of gannets on the tiny island of Grassholme, off the Welsh coast. The birds were filmed in June and again in August. More than a mile of film was shot (including several days shooting from a herring boat, which produced some superb footage of gannets diving for fish).

The title of the firm did not reflect the subject matter of the film quite so much as it reflected the whim of Hungarian-born British film mogul, Alexander Korda, whose company, London Film, produced the picture. Korda had had considerable success with a series of "private lives" films, including "The Private Life of Helen of Troy" (1927), "Her Private Life" (1929), "The Private Life of Henry VIII" (1933), and "The Private Life of Don Juan" (1934). Korda insisted that Huxley's film be called "The Private Life of the Gannets."

For additional information on the making of the film, see Huxley, *Memories*, pp. 210–11; Huxley, "Making and Using Nature Films," *The Listener* 13 (1935): 595–97, 629; and the letters from Ronald Lockley to Huxley in the Huxley Papers, Woodson Research Center, Rice University.

68. See John R. Durant, "The Making of Ethology: The Association for the Study of Animal Behaviour, 1936–1986," *Animal Behavior* 34 (1986): 1601–16.

69. A. F. J. Portielje, "Zur ethologie bezw. Psychologie von *Botaurus stellaris (L),*" *Ardea* 15 (1926): 1–15; "Zur ethologie bezw. Psychologie von *Phalcrocorax carbo subcormoranus* (Brehm)," *Ardea* 16 (1927): 107–123; "Zur ethologie bezw. Psychologie der Silbermöwe, *Larus argentatus argentatus* Pont.," *Ardea* 17 (1928): 112–49.

70. Jan Verwey, "Die Paarungsbiologie des Fischreihers," *Zoologisches Jahrbuch. Physiologie* 48 (1930): 1–120. Niko Tinbergen, personal interview with the author, 30 April 1979. Huxley's contacts with the Dutch were such that he already knew about the results of Verwey's heron work when he wrote his introduction to Selous's *Realities of Bird Life* in 1927 (see p. xiv).

71. Ernst Mayr played a similar, major role in encouraging the work of David Lack, Konrad Lorenz, and Niko Tinbergen. Mayr in addition encouraged the work of Margaret Morse Nice, the "amateur" ornithologist whose "Life History of the Song Sparrow" was a classic contribution to behavioral ecology and who figured prominently in introducing Lorenzian ecology to the United States. On Nice see Gregg Mitman and Richard Burkhardt, Jr., "Struggling for Identity: the Study of Animal Behavior in America, 1930–1945," in *The Expansion of American Biology*, ed. Keith R. Benson, Jane Maienschein, and Ronald Rainger (New Brunswick: Rutgers Univ. Press, 1991), pp. 164–94.

Demonstrating to nonethologists the importance of ethology was one of the main aims of Huxley's Royal Society symposium on ritualization. On this point see his letters to Wolfgang Wickler, 1 December 1964, and to Konrad Lorenz, 2 December and 24 December 1964, Huxley Papers, Woodson Research Center, Rice University.

72. On 17 April 1925, C. Lloyd Morgan wrote to Huxley: "Yes. We must spur the good E. H. to get on even if he has to leave some problems unsolved," Huxley Papers, Woodson Research Center, Rice University.

73. Huxley to Howard, 16 April 1925, Edward Grey Institute of Field Ornithology, Oxford. The influence of Lloyd Morgan on Howard is noted by Lack, "Some British Pioneers in Ornithological Research," *Ibis* 101 (1959): 73, 75. The extent of this influence remains to be explored. There are more than 130 letters and cards from

Morgan to Howard in the Howard Papers, Edward Grey Institute of Field Ornithology, Oxford. Howard's side of the correspondence, however, is much scantier, there being only ten letters from Howard to Morgan in the Morgan Papers at Bristol.

As the result of a visit to Howard's house in 1934, Huxley published a short paper titled "A Natural Experiment on the Territorial Instinct," a discussion of what had happened to the territorial behavior of a group of ducks when the freezing of a pond forced them to move close together. The paper also appeared in *British Birds* 27 (1934): 270–77. Huxley and Howard also published jointly a short letter to *Nature*: "Field Studies and Physiology: a Further Correlation," *Nature* 133 (1934): 688–89. The last letter from Huxley to Howard in the Howard Papers at the Edward Grey Institute of Field Ornithology is dated 11 September 1940. In this letter Howard once again addressed the issue of memory images.

74. Lack to Huxley, 20 August 1952, and November (no day given), 1971, Huxley Papers, Woodson Research Center, Rice University. See also David Lack, "My Life as an Amateur Ornithologist," *Ibis* 115 (1973): 421–31. esp. p. 427, where he indicates that in the mid-1930s Huxley "was the only senior British zoologist who thought ecology and behaviour important. . . ."

75. Letters with respect to all these subjects are to be found in the Lorenz-Huxley correspondence at Rice University. See especially Lorenz to Huxley, 26 November 1963.

76. N. Tinbergen, "Zur Paarungsbiologie der Flussseeschwalbe (*Sterna hirundo hirundo* L.)," *Ardea* 20 (1931): 1–18 (esp. p. 5).

77. Tinbergen to Huxley, 17 February 1940, Huxley Papers, Woodson Research Center, Rice University.

78. "Scientific Affairs in Europe," *Nature* 156 (1945): 574–79.

79. Tinbergen to Huxley, 20 June 1967. There is a typographical error in this letter: "fill" appears where Tinbergen evidently meant to write "full." See also Tinbergen to Huxley, 18 April 1971, Huxley Papers, Woodson Research Center, Rice University.

80. See especially Tinbergen, "An Objectivistic Study of the Innate Behaviour of Animals," *Bibliotheca biotheoretica* 1 (1942): 39–98. On Tinbergen's and Huxley's different views of the possibility of studying the subjective phenomena going on inside animals, see, for example, Tinbergen to Huxley, 21 January 1959; Huxley to Tinbergen, 27 March 1962; Huxley to Tinbergen, 18 March 1965; Tinbergen to Huxley, 20 March 1965; and Huxley to Tinbergen, 29 March 1965; Huxley Papers, Woodson Research Center, Rice University.

81. Tinbergen to Huxley, 20 June 1967, Huxley Papers, Rice University.

82. Selous, *Bird Life Glimpses*, (London: Allen & Unwin, 1905), p. vi.

83. Huxley to Howard, 19 February 1922, H. Eliot Howard Papers, Edward Grey Institute of Field Ornithology, Oxford University.

84. The extent to which Huxley was able to integrate behavioral studies into his teaching or at least his term-time activities at Oxford deserves closer analysis. John R. Baker indicates that upon returning to Oxford after the war, Huxley gave one course of lectures on experimental zoology, another on genetics, and a third on animal behavior (Baker, "Huxley," p. 210). For the contributions from the Oxford Ornithological Society, see Huxley, "Some Further Notes on the Courtship Behaviour of the Great Crested Grebe," *British Birds* 18 (1924): 129–34, and "Some Points in the Breeding Behaviour of the Common Heron," *British Birds* 18 (1924): 155–63. On the Oxford Ornithological Society, see D. E. Allen, *The Naturalist in*

Britain (London: Allen Lane, 1976), pp. 252–62. The prime mover of the Oxford Ornithological Society, according to Allen, was Huxley's student B. W. Tucker.

85. Tinbergen to Huxley, 18 February 1953, Huxley Papers, Woodson Research Center, Rice University.

86. Selous, for refusing to temper his comments on the iniquities of professional zoologists, was dropped as a contributor to Kirkman's *British Bird Book* project. Huxley, in contrast, was a diplomat capable of moving at ease in a variety of circles, so much so that, while a professional zoologist himself, he was able to deal with the prickly Selous and shepherd Selous's unpublished field notes through publication. Selous ultimately named Huxley as literary executor of Selous's estate. I have come across no evidence that suggests, however, that Huxley did anything in this capacity. See the Edmund Selous Papers, Edward Grey Institute of Field Ornithology, Oxford University, especially F. B. Kirkman to Selous, 3 January 1910, and Selous's "Notebook 4." See also the Selous letters in the Huxley Papers, Woodson Research Center, Rice University, especially Selous to Huxley, 14 March 1926 and 25 July 1926, and F. M. Selous (Selous's wife) to Huxley, 26 March 1934.

87. Fowler to Huxley, 23 October 1916, Huxley Papers, Woodson Research Center, Rice University. Fowler said further that he especially liked the analogies Huxley drew between animal behavior and human behavior.

88. Huxley, "Bird-watching and Biological Science" (1916), p. 161. Distinguishing between different kinds of biological causation has proved to be a continuing problem in the life sciences. It is addressed by Ernst Mayr in his classic article, "Cause and Effect in Biology," *Science* 134 (1961): 1501–1506, reprinted in *Evolution and the Diversity of Life: Selected Essays*, ed. Ernst Mayr (Cambridge: Harvard Univ. Press, 1976), pp. 359–71.

89. Huxley, "The Outlook for Biology," *The Rice Institute Pamphlet* 11 (1924): 241–338. Note especially pages 262–63, 272–73.

90. On Whitman and behavior, see especially Richard W. Burkhardt, Jr., "Charles Otis Whitman, Wallace Craig, and the Biological Study of Behavior in the United States, 1898–1925," in *The American Development of Biology*, ed. Ronald Rainger, Keith Benson, and Jane Maienschein (Philadelphia: Univ. of Pennsylvania Press, 1988), pp. 185–218. On Heinroth see especially Katharina Heinroth, *Oskar Heinroth: Vater der Verhaltensforschung, 1871–1945* (Stuttgart: Wissenschaftliche Verlagsgesellschaft, 1971).

91. Huxley, ms. of Bird Courtship book, Huxley Papers, Woodson Research Center, Rice University.

92. Selous, *Bird Life Glimpses*, pp. 49–50.

93. Huxley, *Memories*, p. 5.

94. For Huxley's own assessment of Lorenz's contributions to ethology—and of what Lorenz has contributed to ethology that Huxley had not—see Huxley, "Lorenzian Ethology," *Zeitschrift für Tierpsychologie* 20 (1963): 402–409. Huxley states, "It is fair to call modern ethology Lorenzian, since it was Lorenz who initiated its practice, and did more than any other man to explore its theoretical basis and its implications for other fields of study, notably human psychology," p. 404.

DURANT

1. Julian S. Huxley, "The Courtship-habits of the Great Crested Grebe (*Podiceps cristatus*) with an Addition to the Theory of Sexual Selection," *Proceedings of the Zoological Society of London* 35 (1914): 509–510.

2. Peter Broks, "Popular Science and Popular Culture: Family Magazines in Britain 1890–1914," in *Three Papers on the Popularisation of Science* (Centre for Science Studies and Science Policy, University of Lancaster and Centre de Recherche en Histoire des Sciences et des Techniques, Cité Des Sciences et de L'Industrie, Paris, 1987), pp. 4–28. I am grateful to Peter Broks for making available to me a draft of chapter 5 of his doctoral dissertation, which is currently under preparation for submission to the University of Lancaster. This extremely interesting chapter contains a lengthy analysis of animal biography in Edwardian popular magazines.

3. E. Selous, *Bird Life Glimpses* (London: George Allen, 1905), p. 23.

4. Huxley, "Courtship-habits," p. 521.

5. Transcript of a program on "Birds of Britain," edited and introduced by James Fisher, and broadcast on the BBC Home Service on Sunday, 8 December 1957, box 158, file 1, Huxley Papers, Woodson Research Center, Rice University.

6. John R. Durant, "Innate Character in Animals and Man: A Perspective on the Origins of Ethology," in *Biology, Medicine and Society, 1840–1940*, ed. C. Webster (Cambridge and London: Cambridge Univ. Press, 1981), pp. 157–92.

7. Selous, *Realities of Bird Life: Being Extracts from the Diaries of a Life-Loving Naturalist* (London: Constable, 1927), p. 341.

8. Huxley, *Bird-watching and Bird Behaviour* (London: Chatto & Windus, 1930), pp. 4–5.

9. Huxley, *Essays of a Biologist* (London: Chatto & Windus, 1923), p. 241.

10. Ibid., pp. 107–108.

11. See Colin Divall, "From a Victorian to a Modern: Julian Huxley and the English Intellectual Climate," this volume.

12. This passage is immediately followed by another and even more striking testimony to the purposelessness of the Darwinian process: "Natural selection," Huxley continues, "though like the mills of God in grinding slowly and grinding small, has few other attributes that a civilized religion would call divine." Julian S. Huxley, "Natural Selection and Evolutionary Progress," *Report of the British Association for the Advancement of Science* (Blackpool, 1936), pp. 81–100, at pp. 95–96.

13. Ibid., p. 100.

14. Huxley, *Evolution: The Modern Synthesis* (Allen & Unwin: London, 1942), p. 387.

15. This, of course, was the real issue at stake in Huxley's disagreement with Theodosius Dobzhansky in the early 1960s over the definition of Darwinian fitness. Huxley objected to Dobzhansky's 'geneticist' definition of fitness in terms of relative reproductive success. This definition, he insisted, neglected the absolute phenotypic improvements by virtue of which relative reproductive success was achieved. In reply, Dobzhansky noted that the notion of biological improvement lacked rigor; but he went on to make an even more telling point. While natural selection did sometimes produce biological improvements, he wrote, "I shall nevertheless persist in my 'geneticism,' and maintain that natural selection does not always or necessarily yield these things" (Dobzhansky to Huxley, 14 March 1963, in Box 34, Huxley Papers, Woodson Research Center, Rice University). Dobzhansky had put his finger on the crucial point, namely that Huxley's progressivist criterion of fitness (and with it his entire philosophy of evolutionary humanism) was simply incoherent.

16. S. J. Gould, "The Hardening of the Modern Synthesis," in *Dimensions of Darwinism*, ed. M. Grene (Cambridge and New York: Cambridge Univ. Press, 1983), pp. 71-93.

17. Huxley, *Evolution: The Modern Synthesis*, pp. 478–80.

18. Huxley, *Essays of a Biologist*, p. 48.

19. Huxley, *Evolution: The Modern Synthesis*, p. 484.

20. Selous, "Observations Tending To Throw Light on the Question of Sexual Selection in Birds, Including a Day-to-Day Diary on the Breeding Habits of the Ruff, *Machetes pugnax*," *The Zoologist* ser. 4, 10 (1906): 201–219, 285–94, and 419–28, and ser. 4, 11 (1907): 60–65, 161–82, and 367–81.

21. Huxley, "The Present Standing of the Theory of Sexual Selection," in *Evolution: Essays on Aspects of Evolutionary Theory*, ed. G. R. de Beer (Oxford: Clarendon Press, 1938), pp. 11–42; and "Darwin's Theory of Sexual Selection and the Data Subsumed by It in the Light of Recent Research," *The American Naturalist* 72 (1938): 416–33.

22. Huxley, *Bird-watching and Bird Behaviour*, pp. 22–23.

23. Ibid., pp. 33–34.

ZUCKERMAN

1. We did not hear, at the Huxley symposium, very much of Hogben, who had one of the most austere minds going. He was very interested in scientific method and did not think much of Julian's more diffuse approach to science.

2. This volume.

3. Julian S. Huxley, "The Courtship-habits of the Great Crested Grebe (*Podiceps cristatus*) with an Addition to the Theory of Sexual Selection," *Proceedings of the Zoological Society of London* 35 (1914): 491–562.

4. W. Rowan, *The Riddle of Migration* (Baltimore: Williams & Wilkins, 1931).

5. I thank Dr. Van Helden for inviting me to the Huxley Symposium. I found it remarkably interesting. It evoked any number of memories, and I wish that I had not moved, as it were, out of the field from which those memories derive.

PROVINE

1. Julian S. Huxley, *Evolution: The Modern Synthesis*, 2d ed. (London: Allen & Unwin, 1962).

2. E. Mayr and W. B. Provine, eds., *The Evolutionary Synthesis* (Cambridge, Mass.: Harvard Univ. Press, 1980).

3. P. H. Barrett, P. Gautrey, S. Herbert, D. Kohn, and S. Smith, eds., *Charles Darwin's Notebooks: 1836–1844* (Ithaca: Cornell Univ. Press; British Museum: Cambridge Press, 1987).

4. T. H. Huxley, "Evolution and Ethics," in *Evolution and Ethics and Other Essays* (London: Macmillan, 1894).

5. J. S. Huxley, *The Individual in the Animal Kingdom* (Cambridge: Cambridge Univ. Press, 1912).

6. Ibid., p. 140.

7. Ibid., p. 154.

8. P. Teilhard de Chardin, *The Phenomenon of Man* (New York: Harper, 1959).

9. Huxley, *The Individual*, p. vii.

10. Huxley, "The Courtship-habits of the Great Crested Grebe (*Podiceps cristatus*); with an Addition to the Theory of Sexual Selection," *Proceedings of the Zoological Society of London* 35 (1914): 491–562.

11. Huxley, *Evolution*, p. 13.

12. Ibid., pp. 564–65.

13. Ibid., p. 576.

14. Ibid., pp. 576–77.

15. All quotes in this paragraph from Huxley, *Evolution*, p. 570.

16. Ibid., p. 571.

17. Ibid.

18. Huxley, *The New Systematics* (Oxford: Clarendon Press, 1940).

19. Mayr, "Where Are We?" *Cold Spring Harbor Symposia on Quantitative Biology* 24 (1959): 1–14.

20. S. Wright, "Genetics and Twentieth Century Darwinism: A Review and Discussion," *American Journal of Human Genetics* 12 (1960): 365–72.

21. Provine, *Sewall Wright and Evolutionary Biology* (Chicago: Univ. of Chicago Press, 1986), chap. 12.

22. G. G. Simpson, *Concession to the Improbable: An Unconventional Autobiography* (New Haven: Yale Univ. Press, 1978).

23. Simpson, *Tempo and Mode in Evolution*, 2d ed. (New York: Columbia Univ. Press, 1984).

24. C. H. Waddington, *Evolution of an Evolutionist* (Ithaca: Cornell Univ. Press, 1975).

25. N. Eldredge, *Unfinished Synthesis: Biological Hierarchies and Modern Evolutionary Thought* (New York: Oxford Univ. Press, 1985).

26. S. J. Gould, "The Hardening of the Modern Synthesis," in *Dimensions of Darwinism*, ed. M. Grene (New York: Cambridge Univ. Press, 1983), pp. 71–93.

27. M. Kimura, *The Neutral Theory of Molecular Evolution* (Cambridge: Cambridge Univ. Press, 1983).

28. G. L. Stebbins and F. G. Ayala, "Is a New Evolutionary Synthesis Necessary?" *Science* 213 (1981): 967–71.

29. D. J. Futuyama, Presidential Address, American Society of Naturalists, *The American Naturalist* 130 (1989): 465–73.

30. J. Antonovics, "The Evolutionary Dys-synthesis: Which Bottles for Which Wine?" *The American Naturalist* 129 (1987): 321–31.

31. Mayr, "On the Evolutionary Synthesis and After," in *Toward a New Philosophy of Biology: Observations of an Evolutionist* (Cambridge, Mass.: Harvard Univ. Press, 1988).

32. Mayr and Provine, eds., *The Evolutionary Synthesis* (Cambridge, Mass.: Harvard Univ. Press, 1980).

33. Provine, "The R. A. Fisher—Sewall Wright Controversy," *Oxford Surveys in Evolutionary Biology* 2 (1985): 197–219.

34. E. B. Ford, *Ecological Genetics* (London: Methuen, 1964).

35. Gould, "The Hardening of the Modern Synthesis," in *Dimensions*, pp. 71–93.

36. Provine, "Founder Effects and Genetic Revolutions in Microevolution and Speciation: An Historical Perspective," in *Genetics, Speciation, and the Founder Principle*, ed. L. V. Giddings, K. Kanshireo, and W. W. Anderson (New York: Oxford Univ. Press, 1989).

37. Y. Delage, *L'Hérédité et les Grandes Problèmes de la Biologie Générale*, 2d ed. (Paris: Librairie C. Reinwald, 1903).

38. W. O. Focke, *Die Pflanzen-Mischlinge: Ein Beitrag zur Biologie der Gewachse* (Berlin: Borntraeger, 1882).

39. W. Bateson, *Materials for the Study of Variation* (London: Macmillan, 1894).

40. E. Baur, *Einführung in die experimentelle Vererbungslehre* (Berlin: Borntraeger, 1911); R. B. Goldschmidt, *Einführung in die Vererbungswissenschaft* (Leipzig: Engelmann, 1928); V. Haecker, *Allegemeine Vererbungslehre* (Braunschweig: Vieweg, 1912).

41. R. C. Punnett, *Mendelism* (Cambridge: Bowes & Bowes, 1905).

42. R. H. Lock, *Recent Progress in the Study of Variation, Heredity, and Evolution* (London: John Murray, 1906); W. Bateson, *Materials*; H. E. Walter, *Genetics: An Introduction to the Study of Heredity* (New York: Macmillan, 1913); W. E. Castle, *Genetics and Eugenics* (Cambridge, Mass.: Harvard Univ. Press, 1916).

43. H. De Vries, *Intracellulare Pangenesis* (Jena: Gustav Fischer, 1889).

44. P. J. Bowler, *The Eclipse of Darwinism: The Anti-Darwinian Evolution Theories in the Decades around 1900* (Baltimore: Johns Hopkins Univ. Press, 1983).

45. V. L. Kellogg, *Darwinism Today* (New York: Holt, 1907).

46. H. W. Conn, *Evolution To-day* (New York: Putnam, 1886); idem, *The Method of Evolution* (New York: Putnam, 1900).

47. Provine, "The Role of Mathematical Population Geneticists in the Evolutionary Synthesis of the 1930s and 1940s," *Studies in History of Biology* 2 (1978): 167–92.

48. On these developments, see Provine, *Sewall Wright and Evolutionary Biology* (Chicago: Univ. of Chicago Press, 1986).

49. H. G. Osborn, *The Earth Speaks to Bryan* (New York: Scribner's, 1925).

50. Ibid., pp. 20–21.

51. Huxley, *Evolution*, p. 574; "Genetics, Evolution, and Human Destiny," in *Genetics in the 20th Century*, ed. L. C. Dunn (New York: Macmillan, 1950), pp. 591–621.

52. Teilhard de Chardin, *The Phenomenon of Man*.

53. Charles R. Darwin, *On the Origin of Species* (London: John Murray, 1859).

54. Simpson, *This View of Life* (New York: Harcourt Brace Jovanovich, 1964).

55. P. B. Medawar, Review of *The Phenomenon of Man* by Teilhard de Chardin, *Mind* 70 (1961): 99–106.

56. Teilhard de Chardin, *The Phenomenon of Man*, p. 11.

57. T. Dobzhansky, *The Biology of Ultimate Concern* (New York: New American Library, 1967).

BEATTY

1. S. J. Gould, "The Hardening of the Modern Synthesis," in *Dimensions of Darwinism: Themes and Counterthemes in Twentieth Century Evolutionary Biology*, ed. M. Grene (Cambridge: Cambridge Univ. Press, 1983).

2. W. B. Provine, "The Development of Wright's Theory of Evolution: Systematics, Adaptation, and Drift," in *Dimensions of Darwinism*, ed. M. Grene (Cambridge: Cambridge Univ. Press, 1983).

3. P. J. Bowler, *The Eclipse of Darwinism: The Anti-Darwinian Theories in the Decades Around 1900* (Baltimore: Johns Hopkins Univ. Press, 1983).

4. V. Kellogg, *Darwinism To-Day* (New York: Holt, 1907).

5. J. Beatty, "Dobzhansky and Drift: Facts, Values, and Change in Evolutionary Biology," in *The Probabilistic Revolution*, ed. L. Kruger et al., vol. 2 (Cambridge: MIT Press, 1987), pp. 271–311.

6. J. S. Huxley, *Evolution: The Modern Synthesis* (London: Allen & Unwin, 1942).

7. T. Dobzhansky, *Genetics and the Origin of Species* (New York: Columbia Univ. Press, 1937), p. 186.

8. Huxley, *Evolution*, p. 45.

9. Ibid., p. 46.

10. Ibid., p. 126.

11. Ibid., p. 127.

12. Ibid., chaps. 5–7.

13. Ibid., pp. 128–29.

14. Ibid., p. 28; see also p. 7.

15. Ibid., p. 387.

16. E. B. Ford, *Ecological Genetics* (London: Methuen, 1964).

17. Huxley, *Evolution: The Modern Synthesis*, 2d ed. (London: Allen & Unwin, 1963), pp. xxii–xxiii.

18. See Ford to Huxley, 27 Aug. 1962; Ford to Huxley, 12 Sept. 1962; Ford to Huxley, 13 Dec. 1962; Huxley to Ford, 14 Dec. 1962, in the Huxley Papers, Woodson Research Center, Rice University.

19. E. Mayr, *Animal Species and Evolution* (Cambridge: Harvard Univ. Press, 1963).

20. Huxley, "Units of Evolution," *Nature* 199: 839.

21. Huxley, *Evolution*, 2d ed., pp. xxi–xxii.

22. Huxley, *Evolution: The Modern Synthesis* (London: Allen & Unwin, 1942), pp. 29–30.

23. Ibid., p. 578.

ALLEN

1. Hermann J. Muller, "Sir Julian Huxley, A Biographical Appreciation," *The Humanist* 2–3 (1962): 51.

2. Julian S. Huxley, *Evolution: The Modern Synthesis* (London: Allen & Unwin, 1942).

3. Huxley, "Eugenics and Society," originally published in *Eugenics Review* (1936); republished in *The Uniqueness of Man* (London: Chatto & Windus, 1941), pp. 34–84. All text references are to the latter printing.

4. Daniel J. Kevles, *In the Name of Eugenics* (New York: Knopf, 1985). Garland E. Allen, "The Eugenics Record Office at Cold Spring Harbor, 1910–1940: An Essay in Institutional History," *Osiris* 2d ser. 2 (1986): 225–64. Barry Mehler, *The American Eugenics Society, 1921–1940* (Ph.D. diss., Univ. of Illinois, 1988).

5. Huxley, "Eugenics and Society," pp. 52–53.

6. Huxley, "Eugenic Sterilization," *Nature* 126 (1930): 504.

7. Ibid., p. 43.

8. Huxley, *Memories*, (London: Allen & Unwin, 1970), pp. 168–69; see also Greta Jones, *Social Hygiene in Twentieth Century Britain* (London: Croom Helm, 1986), pp. 105–106.

9. Huxley, "Eugenics and Society," p. 70.

10. Ibid., p. 34.

11. Ibid., p. 69.

12. Huxley to Blacker, 1935, file C185, Eugenics Society Records, quoted from Kevles, p. 174.

13. Huxley, "Eugenics and Society," p. 68.

14. Ibid., p. 45.

15. Ibid., p. 46; A. M. Carr-Saunders, *Eugenics* (London: Williams & Norgate, 1926), pp. 97, 105, 126.

16. Huxley, "Eugenics and Society," pp. 52–53.

17. Ibid., p. 49.

18. Ibid., p. 50.

19. Ibid., p. 51.

20. Huxley and A. C. Haddon, *We Europeans: A Survey of "Racial" Problems* (London: Jonathan Cape, 1935; Toronto: Nelson, 1935).

21. Ibid., pp. 107–108.

22. Huxley, "Eugenics and Society," p. 53.

23. Ibid., p. 59.

24. Ibid.

25. Ibid., p. 75.

26. Ibid., p. 63.

27. Ibid., p. 74.

28. Ibid.

29. Ibid., p. 75.

30. Ibid., p. 78.

31. Ibid.

32. Ibid. p. 344.

33. Gary Wersky, *The Visible College: The Collective Biography of British Scientific Socialists of the 1930's* (London: Allen Lane, 1978; New York: Holt, Rinehart & Winston, 1979; London: Free Association Books, 1988), p. 42.

34. Huxley, "Eugenic Sterilization," pp. 344, 503.

35. Huxley, "Sterilization: A Social Problem," *New Chronicle* 21 January 1934 p. 2.

36. Ibid., p. 2.

37. Huxley, "Eugenics and Society," p. 68.

38. Birth control refers to limiting the number of children in individual families. The term tends to include not only the actual methods employed but also the constellation of ideas and values that make such practices acceptable, or even desirable, to individual families. Population control, on the other hand, refers to limiting the growth rate and ultimate size of whole populations—social classes, ethnic groups, or nations. It includes the methods and ideology of birth control, but applied to a population at large.

39. Garland E. Allen, "From Eugenics to Population Control in the Work of Raymond Pearl," in *The Expansion of American Biology*, ed. Keith R. Benson, Jane Maienschein, and Ronald Rainger (New Brunswick, N.J.: Rutgers Univ. Press, 1991): 231–61.

40. Allen Chase, *The Legacy of Malthus: The Social Costs of the New Scientific Racism* (New York: Knopf, 1977); Linda Gordon, *Woman's Body, Woman's Right* (New York: Grossman, 1976).

41. Huxley, "The Applied Science of the Next Hundred Years: Biological and Social Engineering," *Life and Letters* 19 October 1934: 42.

42. Quoted from Kevles, pp. 260–61.

43. John R. Baker, "Julian Sorell Huxley," *Biographical Memoirs of Fellows of the*

Royal Society 22 (1976): 207–238; Huxley, "Towards a Higher Civilization"; idem, "Sterilization: A Social Problem"; idem, "World Population," *Scientific American* 194, no. 3 (1956): 64–76; idem, "Population and Human Fulfillment," in *New Bottles for New Wine* (New York: Harper, 1957), pp. 168–212.

44. Wersky, *The Visible College*, p. 242.

45. Huxley used both terms in his writings, though in later years he appears to have gradually dropped "scientific humanism" and referred most of the time to "evolutionary humanism." For simplicity I will use only "evolutionary humanism" when speaking of Huxley's views on this subject.

46. Huxley, "Evolutionary Humanism," p. 295.

47. Ibid., pp. 296–97.

48. Ibid., p. 310.

49. Michael Freeden, "Eugenics and Progressive Thought: A Study in Ideological Affinity," *The Historical Journal* 22, no. 3 (1979): 645–71; Greta Jones, "Eugenics and Social Policy Between the Wars," *The Historical Journal* 25, no. 3 (1982): 717–28; Barbara Kimmelman, "Genetics and Eugenics in an Agricultural Context: The American Breeders Association, 1903–1913," *Social Studies of Science* 13 (1983): 163–204; Donald Mackenzie, "Karl Pearson and the Professional Middle Class," *Annals of Science* 36 (1979): 125–43; Diane Paul, "Eugenics and the Left," *Journal of the History of Ideas* 45 (1984): 567–90; Kevles, *In the Name of Eugenics*; Allen, "Naturalists and Experimentalists: The Genotype and the Phenotype," *Studies in History of Biology* 3 (1979): 179–210.

50. Allen, "Chevaux de course et chevaux de trait: Metaphores et analogies agricoles dans l'eugenisme Americaine, 1910–1940," in *Histoire de la Genetique*, ed. Jean-Louis Fischer and William H. Schneider (Paris: Science en Situation, 1990), pp. 83–98.

PAUL

1. Daniel J. Kevles, *In the Name of Eugenics* (New York: Knopf, 1985).

2. Julian S. Huxley, "Eugenics and Society," *Eugenics Review* 28 (1936): 13.

3. Huxley, "Eugenics and Evolutionary Perspective," in *Essays of a Humanist* (New York: Harper, 1964), p. 259.

4. Theodosius Dobzhansky to Julian Huxley, 11 July 1953, Huxley Papers, Woodson Research Center, Rice University.

5. For a more detailed discussion of these events, see John Beatty, "Weighing the Risks: Stalemate in the Classical-Balance Controversy," *Journal of the History of Biology* 20 (1987): 289–319.

6. Theodosius Dobzhansky, *Mankind Evolving* (New Haven: Yale Univ. Press, 1962), p. 329.

7. Huxley, *What Dare I Think?* (London: Chatto & Windus, 1931), pp. 116–17.

8. Ibid.

9. H. J. Muller to Julian Huxley, 9 July 1962, Muller Papers, Manuscripts Department, Lilly Library, Indiana University, Bloomington.

10. Muller to Huxley, 13 July 1962. Muller Papers, Manuscripts Department, Lilly Library, Indiana University, Bloomington.

11. Bruce Wallace, "Some of the Problems Accompanying an Increase in Mutation Rates in Mendelian Populations," in *Effects of Radiation on Human Heredity* (Geneva: World Health Organization, 1957), p. 59.

12. Henry H. Goddard, "Mental Tests and the Immigrant," *The Journal of Delinquency* 2 (1917): 269.

13. R. C. Lewontin, *The Genetic Basis of Evolutionary Change* (New York: Columbia Univ. Press, 1974).

14. Jack C. King, review in *Annals of Human Genetics* (1975), p. 508.

15. Dobzhansky, *Mankind Evolving*, p. 244.

BARKAN

1. Julian S. Huxley, "Eugenics in Evolutionary Perspective," in *Essays of a Humanist* (New York: Harper, 1964), pp. 252, 267. Also "The Vital Importance of Eugenics—Letter to the Editor," *Harpers Monthly*, 163 (1931):325.

2. Huxley Papers, Woodson Research Center, Rice University, Box 58:8.

3. See John W. Cell, *The Highest Stage of White Supremacy: The Origins of Segregation in South Africa and the American South* (Cambridge: Cambridge Univ. Press, 1987).

4. Huxley, "America Revisited. III. The Negro Problem," *The Spectator*, 29 November 1924, pp. 821–22.

5. Ibid.

6. Huxley, "Nature and Nurture," *The New Leader*, 29 February 1924, Huxley Papers, Woodson Research Center, Rice University.

7. Huxley, "Eugenics and Heredity," Letter to the Editor, *The New Statesman*, 1924, Huxley Papers, Woodson Research Center, Rice University.

8. Huxley, "Nature and Nurture," Huxley Papers.

9. Huxley, *Africa View* (New York: Harper, 1931), p. 394.

10. Ibid.

11. Huxley, "Why Is The White Man in Africa?" *Fortnightly Review*, 137 (January 1932) p. 65.

12. *Africa View*, pp. 6–7, 15.

13. Ibid., pp. 395, 396, 400.

14. Ibid., pp. 404–405 (emphasis added).

15. Ibid., pp. 405–406 (emphasis added).

KEVLES

1. Ronald W. Clark, *The Huxleys* (New York: McGraw-Hill, 1968), pp. 186–87; J. R. Baker, "Julian Sorell Huxley," *Biographical Memoirs of Fellows of the Royal Society* 22 (1976): 211, 217; Stephen Jay Gould, *Ontogeny and Phylogeny* (Cambridge: Harvard Univ. Press, 1977), pp. 177–78.

2. Clark, *The Huxleys*, pp. 179, 162.

3. See, for example, Daniel J. Kevles, *The Physicists: The History of a Scientific Community in Modern America* (New York: Knopf, 1978), pp. 170–84.

4. Julian S. Huxley, *Essays in Popular Science* (London: Chatto & Windus, 1926), pp. v, vii.

5. Baker, "Huxley," pp. 211–12, 235–38.

6. Clark, *The Huxleys*, pp. 204, 278–79.

7. Ibid., pp. 256–61; Huxley, *At the Zoo* (London: Allen & Unwin, 1936).

8. Clark, *The Huxleys*, p. 204; Baker, "Huxley," p. 212.

9. Clark, *The Huxleys*, pp. 145, 204–205; Baker, "Huxley," p. 232.

10. Huxley, *Man Stands Alone*, 2d ed. (New York: Harper, 1941), pp. 192–93.

11. Huxley, *Evolution in Action* (New York: Harper, 1953), pp. 86–87.

12. Huxley, *Essays of a Biologist* (New York: Knopf, 1923), pp. 171–72.

13. Huxley, *Man Stands Alone*, p. 148; idem, *Bird-watching and Bird Behaviour*, quoted in Clark, *The Huxleys*, pp. 171–72.

14. Quoted in Clark, *The Huxleys*, p. 189.

15. Huxley, *Evolution in Action*, pp. 11–12.

16. Huxley, *Man Stands Alone*, pp. 187–88.

17. Ibid., pp. 184–85.

18. Ibid., p. v; Huxley, *Ants* (London: Dobson, 1930), p. 3.

19. Huxley, *Man Stands Alone*, pp. 254–55.

20. Ibid., pp. 111, 201–202.

21. Kevles, *In the Name of Eugenics: Genetics and the Uses of Human Heredity* (New York: Knopf, 1985), p. 125; Huxley, *Essays of a Biologist*, pp. 167–68.

22. Kevles, *In the Name of Eugenics*, pp. 123–26; Huxley, *Man Stands Alone*, pp. 211–12; Huxley, *What Dare I Think?: The Challenge of Modern Science to Human Action & Belief* (London: Chatto & Windus, 1931), p. 211.

23. Huxley, *Man Stands Alone*, p. 211.

24. Huxley, *Essays of a Biologist*, p. 86; Kevles, *In the Name of Eugenics*, p. 174.

25. Clark, *The Huxleys*, pp. 172–73; Huxley, "The Concept of Race," *Harper's* 170 (May 1935): 692.

26. Huxley and A. C. Haddon, *We Europeans: A Survey of "Racial" Problems* (London: Jonathan Cape, 1935), pp. 18, 25–26, 68, 91, 96–97, 103, 104, 107, 184, 261, 263, 267–68.

27. Huxley, *Essays of a Biologist*, pp. 55–57.

28. Ibid., pp. 74–85.

29. Huxley, *Evolution in Action*, pp. 126, 147–49.

30. Huxley, *Essays of a Biologist*, pp. x, 45–46.

31. Huxley, ed., *The Humanist Frame* (New York: Harper, 1961), p. 20; Huxley, *Evolution in Action*, pp. 149–50, 153.

32. Clark, *The Huxleys*, p. 118; Huxley, *Essays of a Biologist*, p. 71. T. H. Huxley's Romanes Lecture is reprinted in T. H. Huxley and Julian Huxley, *Touchstone for Ethics* (New York: Harper, 1947), pp. 67–112.

33. Huxley, *Evolution in Action*, p. 167; Huxley and Huxley, *Touchstone for Ethics*, p. 136.

34. Huxley, *Essays of a Biologist*, p. 208; Huxley, *Man Stands Alone*, pp. 277–78.

35. Huxley, *The Humanist Frame*, p. 26; Huxley and Huxley, *Touchstone for Ethics*, pp. 155–56.

36. Huxley and Huxley, *Touchstone for Ethics*, pp. 31–32, 140, 147–48; Huxley, *Evolution in Action*, pp. 36–37, 96–97, 173–75.

37. Huxley and Huxley, *Touchstone for Ethics*, p. 144; Huxley, *Evolution in Action*, pp. 164–68.

38. Huxley and Huxley, *Touchstone for Ethics*, p. 155.

LEMAHIEU

1. For a good discussion of the problems of definition in this field, see, among others, Raymond Williams, *Keywords: A Vocabulary of Culture and Society* (New York: Oxford Univ. Press, 1976); and C. W. E. Bigsby, "The Politics of Popular

Culture," in *Approaches to Popular Culture*, ed. C. W. E. Bigsby (Bowling Green: Bowling Green Univ. Popular Press, 1976), pp. 3–25.

2. On the percentage of total population who watch public television, see Ronald E. Frank and Marshall G. Greenberg, *Audiences for Public Television* (Beverly Hills: Sage Publications, 1982), pp. 80, 225. Statistics on television viewing involve a number of ambiguities and can be calculated in a variety of ways. See also *A Public Trust: The Report of the Carnegie Commission on the Future of Public Broadcasting* (New York: Bantam, 1979), pp. 329–41. On the circulation of *Scientific American*, see *Ulrich's International Periodicals Directory, 1987–88* (New York: Bowker, 1987), p. 1585.

3. On the circulation statistics for *The Listener*, see Mark Pegg, *Broadcasting and Society* (London: Croom Helm, 1983), p. 106. For *The New Statesman*, see Edward Hyams, *The New Statesman: The History of the First Fifty Years, 1913–1963* (London: Longman, 1963), p. 183.

4. Michael Joseph, *The Commercial Side of Literature* (New York: Harper, 1926), p. 245.

5. The BBC did not publish most results of its listening research in the 1940s, but the files can be found in the BBC Written Archives Centre, Reading. See, for example, "Memorandum to Broadcasting Committee, 1949," Listener Research, BBC Written Archives Centre. On the history of Listener Research, see Robert Silvey, *Who's Listening? The Story of the BBC Audience Research* (London: Allen & Unwin, 1974).

6. John Stevenson, *British Society, 1914–45* (Harmondsworth: Penguin, 1984), p. 257. Thus, university education was clearly dominated by the middle and upper classes. See also Brian Simon, *The Politics of Educational Reform, 1920–1940* (London: Lawrence & Wishart, 1974).

7. Huxley, *Man in the Modern World* (London: Chatto & Windus, 1947), p. 42. The essay in which the quotation appears was first published in 1941.

8. Huxley, *Memories* (New York: Harper & Row, 1970), p. 20.

9. Asa Briggs, *The History of Broadcasting in the United Kingdom, Volume Three, The War of Words* (Oxford: Oxford Univ. Press, 1970), p. 561.

10. Ibid., p. 562. On the "Brains Trust," see also H. Thomas, *Britain's Brains Trust* (London: Chapman & Hall, 1944).

PATTEN

1. Quoted by Julian S. Huxley in J. S. Huxley and H. B. D. Kettlewell, *Charles Darwin and His World* (New York: Viking, 1985), p. 10.

2. E. D. H. Johnson, *The Poetry of Earth* (New York: Athenaeum, 1966).

3. Huxley notes in *Memories II* (New York: Harper & Row, 1973), p. 21, that Gilbert White was the first to distinguish by calls and songs among the chiff-chaff, willow-warbler, and wood-warbler.

4. At the beginning of *Essays of a Biologist* (1923; reprint, New York: Knopf, 1929), a book dedicated to Julian's "colleagues and friends at the Rice Institute."

5. "The Everyday Life of a Bird," in *Bird-Watching and Bird Behaviour* (London: Dobson, 1949), originally BBC program, spring 1949.

6. Huxley and Kettlewell, *Darwin*, p. 50.

7. Epilogue to *Africa View* (New York and London: Harper, 1931), p. 452.

8. Quoted in a review of *Elements of Geology* by Charles Lyell, *Quarterly Review* 64, no. 127 (June 1839): 103.

9. Huxley, *Religion Without Revelation* (London: Ernest Benn, 1927); his account of its composition appears on pp. 153–54 of *Memories* (London: Allen & Unwin, 1970), immediately following a paragraph on his aunt Mary Ward's *Robert Elsmere*, which helped to convert him "to what I must call a religious humanism." Further reflections on this testament appear in *Memories II*, pp. 106–107.

10. George Steiner, "A False Quarrel?" pp. 13–14 of "The Two Cultures Revisited," *Cambridge Review* 108 (March 1987).

11. See, for example, the "Conclusion" to *Memories II*, pp. 257–59, or the early statement in the "Preface" to *Essays of a Biologist*, p. viii. Aldous could also praise scientific interventions in human life, especially with regard to mind-altering chemicals that brought release and vision.

12. The orientation of this conference has led to neglecting Huxley's work as a quasi-civil servant, writing reports and chairing committees that instigated major reforms in third-world education and that fostered in manifold ways ecological and environmental conservation efforts.

BOOTHE

1. Sarah C. Bates, Mary G. Winkler, and Christina Riquelmy, *A Guide to the Papers of Julian Sorell Huxley* (Houston: Woodson Research Center, 1984). The guide is available for $15, including tax and postage. Order from the Woodson Research Center, Rice University Library, Rice University, Houston, TX 77251.

2. Julian S. Huxley, "Texas and Academe," *The Cornhill Magazine* 703 (1918): 53–65.

3. Huxley to Edgar Odell Lovett, 15 April 1917, p. 3. President's Papers (Edgar Odell Lovett), Woodson Research Center, Rice University.

4. Ibid., pp. 1–2.

5. Ibid., p. 2.

6. Ibid., pp. 3–4.

7. Huxley to John L. Myres, 4 November 1920, pp. 1–2, Huxley Papers.

8. Edmund B. Wilson, "Reorganization of the Naples Zoological Station," *Science* 59 (February 1924): 182–83.

9. Hermann J. Muller to Huxley, 22 April 1941, p. 2, Huxley Papers. (The author thanks the Lilly Library, Indiana University, for permission to publish quotations from the Muller correspondence.)

10. Theodosius Dobzhansky, "Natural Selection and Fitness," *Eugenics Review* 55 (July 1963): 129.

11. Dobzhansky, *Mankind Evolving: The Evolution of the Human Species* (New Haven and London: Yale Univ. Press, 1962).

12. *Perspectives in Biology and Medicine* 6 (Autumn 1962): 144–48.

13. Huxley, "Eugenics in Evolutionary Perspective," *The Eugenics Review* 54 (October 1962): 123–41.

14. Huxley, *Essays of a Humanist* (New York and Evanston: Harper & Row, 1964), pp. 251–80.

15. Huxley, "Eugenics in Evolutionary Prespective," reprinted from *Nature* 195 (21 July 1962): 227–28.

16. Ibid., reprinted from *Perspectives in Biology and Medicine* 6 (Winter 1963): 155–87.

17. Huxley, *Essays of a Humanist*, p. 261.

18. Dobzhansky, "Natural Selection," p. 129.

19. Ibid.

20. K. Hodson to Huxley, 18 March 1963, Huxley Papers.

21. Huxley, *Evolution, the Modern Synthesis*, 2d ed. (London: Allen & Unwin, 1963; New York: Wiley, 1963).

22. Huxley to Muller, 1 April 1963, Huxley Papers.

23. Muller to Huxley, 7 April 1963, Huxley Papers.

24. Huxley to Dobzhansky, 6 April 1965, Huxley Papers.

25. Faith Hope to Huxley, n.d., Huxley Papers.

26. ——— to Huxley, 1956(?), Huxley Papers.

Sources

Adams, M., ed. 1933. *Science in the changing world*. Freeport, NY: Books for Libraries.

Allen, B. M. 1916. The results of extirpation of the anterior lobe of the hypophysis and of the thyroid of *Rana pipiens* larvae. *Science* 44: 755–58.

Allen, D. E. 1976. *The Naturalist in Britain: a social history*. London: Allen Lane.

Allen, G. E. 1975. *Life science in the twentieth century*. New York: John Wiley & Sons.

———. 1978. *Thomas Hunt Morgan: the man and his science*. Princeton: Princeton Univ. Press.

———. 1979. Naturalists and experimentalists: the genotype and the phenotype. *Studies in History of Biology* 3: 179–209.

———. 1980. The work of Raymond Pearl: from eugenics to population control. *Science for the People* 12, no. 4, pp. 22–28.

———. 1986. The Eugenics Record Office at Cold Spring Harbor, 1910–1940. An essay in institutional history. *Osiris*, 2d ser. 2: 225–64.

———. 1988. Chevaux de course et chevaux de trait: metaphores et analogies agricoles dans l'eugenisme Americaine, 1910–1940. In *Histoire de la Genetique*, edited by Jean-Louis Fischer and William H. Schneider, pp. 83–98. Paris: Science en Situation, 1990.

Annan, N. 1955. The intellectual aristocracy. In *Studies in social history : a tribute to G. M. Trevelyan*, edited by J. H. Plumb. London: Longman.

Antonovics, J. 1987. The evolutionary dys-synthesis: which bottles for which wine? *The American Naturalist* 129: 321–31.

Armytage. W. H. G. 1957. *Sir Richard Gregory: his life and work*. London: Macmillan.

———. 1989. The first director-general of UNESCO. In *Evolutionary studies: a centenary celebration of the life of Julian Huxley*, edited by M.

Keynes and G. A. Harrison, pp. 186–93. See Keynes, M., and Harrison, G. A. 1989.

Ayala, F. G. 1981. See Stebbins, G. L., and Ayala, F. G. 1981.

Baker, J. R. 1976. Julian Sorrell Huxley, 22 June 1887—14 February 1975, elected F. R. S. 1938. *Biographical Memoirs of Fellows of the Royal Society* 22: 207–238.

————. 1978. *Julian Huxley, scientist and world citizen 1887–1975*. Paris: UNESCO.

Barrett, P. H.; Gautry, P. J.; Herbert, S.; and Smith, S., eds. 1987. *Charles Darwin's notebooks: 1836–1844*. Ithaca: Cornell Univ. Press; British Museum: Cambridge Univ. Press.

Bates, S. C.; Winkler, M. G.; and Riquelmi, C. 1984. *A guide to the papers of Julian Sorell Huxley*. Houston: Woodson Research Center, Rice Univ.

Bateson, B. 1928. *William Bateson, F. R. S., naturalist: his essays and addresses together with a short account of his life*. Cambridge: Cambridge Univ. Press.

Bateson, W. 1894. *Materials for the study of variation*. London: Macmillan.

————. 1909. *Mendel's principles of heredity*. Cambridge: Cambridge Univ. Press.

————. 1914. Presidential address to section E. *Report of the British Association for the Advancement of Science 84*. In *William Bateson, F. R. S.*, pp. 275–96. See Bateson, B. 1928.

Baur, E. 1911. *Einführung in die experimentelle Vererbungslehre*. Berlin: Borntraeger.

Beatty, J. 1987a. Dobzhansky and drift: facts, values, and chance in evolutionary biology. In *The probabilistic revolution*, 2 vols., edited by L. Kruger, L. J. Daston, G. Gigerenzer, M. Heidelberger, and M. S. Morgan, vol. 2, pp. 271–311. Cambridge, Mass.: MIT Press.

————. 1987b. Weighing the risks: stalemate in the classical-balance controversy. *Journal of the History of Biology* 20: 289–319.

Bekoff, M., and Jamieson, D., eds. 1990. *Interpretation and explanation in the study of animal behavior*. Boulder, CO.: Westview Press.

Bendall, D. S., ed. 1983. *Evolution from molecules to men*. Cambridge: Cambridge Univ. Press.

Benson, K. R. 1981. See Maienschein, J.; Rainger, R.; and Benson, K. R. 1981.

————. 1988. See Rainger, R.; Benson, K. R.; and Maienschein, J. 1988.

————, Maienschein, J., and Rainger, R. 1991. *The expansion of American biology*. New Brunswick, N. J.: Rutgers Univ. Press.

Bigsby, C. W. E. 1976. The politics of popular culture. In *Approaches to popular culture*, edited by C. W. E. Bigsby, pp. 3–25. Bowling Green: Bowling Green Univ. Popular Press.

Boothe, N. L. 1987. The Julian Sorell Huxley Papers, Rice University Library, Houston, Texas. *The Mendel Newsletter*, no. 27, pp. 1–11.

Bowler, P. J. 1983. *The eclipse of Darwinism: the anti-Darwinian evolution theories in the decades around 1900.* Baltimore: Johns Hopkins Univ. Press.

Bridges, C. B. 1925. See Morgan, T. H.; Bridges, C. B.; and Sturtevant, A. H. 1925.

Briggs, A. 1960–1979. *The history of broadcasting in the United Kingdom,* 4 vols. Oxford, New York: Oxford Univ. Press.

Broad, C. D. 1944. Review of *Evolutionary ethics. Mind* 53: 344–67.

Broks, P. 1987. Popular science and popular culture: family magazines in Britain, 1890–1914. In *Three papers on the popularisation of science.* Lancaster: Centre for Science Studies and Science Policy, Univ. of Lancaster; Paris: Centre de Recherche en Histoire des Sciences et des Techniques, Cité des Sciences et de l'Industrie.

Budd, S. 1977. *Varieties of unbelief.* London: Heinemann.

Burkhardt, R. W., Jr. 1981. On the emergence of ethology as a scientific discipline. *Conspectus of History* 1, no. 7, pp. 62–81.

———. 1983. The development of evolutionary ethology. In *Evolution from molecules to men,* pp. 429–44. See Bendall, D. S. 1983.

———. 1988. Charles Otis Whitman, Wallace Craig, and the biological study of animal behavior in the United States, 1898–1925. In *The American development of biology,* pp. 185–218. See Rainger, R.; Benson, K. R.; and Maienschein, J. 1988.

———. 1990a. Edmund Selous. *Dictionary of scientific biography* 18: 801–803.

———. 1990b. Theory and practice in naturalistic studies of behavior prior to ethology's establishment as a scientific discipline. In *Interpretation and explanation in the study of animal behavior* 2: 6–30. See Bekoff, M., and Jamieson, D. 1990.

Burrow, J. 1966. *Evolution and society.* Cambridge: Cambridge Univ. Press.

Callander, L. A. 1988. Gregor Mendel: an opponent of descent with modification. *History of Science* 26: 41–75.

Capek, M. 1971. *Bergson and modern physics.* Dordrecht: Reidel.

Carr-Saunders, A. M. 1924. See Huxley, J. S., and Carr-Saunders, A. M. 1924.

———. 1926. *Eugenics.* London: Williams and Norgate.

Castle, W. E. 1916. *Genetics and eugenics.* Cambridge, Mass.: Harvard Univ. Press.

Cell, J. W. 1982. *The highest stage of white supremacy: the origins of segregation in South Africa and the American South.* Cambridge: Cambridge Univ. Press.

Chase, A. 1976. *The legacy of Malthus: the social costs of new scientific racism.* New York: Knopf.

Child, C. M. 1899. The significance of the spiral type of cleavage and its relation to the process of differentiation. *Biological Lectures from the Marine Biological Laboratory of Woods Hole,* pp. 231–66.

———. 1915. *Individuality in Organisms.* Chicago: Univ. of Chicago Press.

————. 1921. *The origin and development of the nervous system from a physiological viewpoint*. Chicago: Univ. of Chicago Press.

————. 1941. *Patterns and problems of development*. Chicago: Univ. of Chicago Press.

Churchill, F. B. 1980. The modern evolutionary synthesis and the biogenetic law. In *The evolutionary synthesis: perspectives on the unification of biology*, pp. 112–22. See Mayr, E., and Provine, W. B. 1980.

Clark, R. W. 1960. *Sir Julian Huxley, F. R. S.* London: Roy; New York: Phoenix House.

————. 1968. *The Huxleys*. London: Heinemann; New York: McGraw-Hill.

Clark, W. E. LeG., and Medawar, P. B., eds. 1945. *Essays on growth and form*. Oxford: Clarendon Press.

Cock, A. G. 1966. Genetical aspects of metrical growth and form in animals. *Quarterly Review of Biology* 41: 131–90.

Coleman, W. 1970. Bateson and chromosomes: conservative thought in science. *Centaurus* 15: 228–314.

————. 1971. *Biology in the nineteenth century: problems of form, function, and transformation*. New York: John Wiley & Sons.

Collini, S. 1979. *Liberalism and sociology: L. T. Hobhouse and political argument in England, 1880–1914*. Cambridge: Cambridge Univ. Press.

Collins, P. 1981. The British Association as public apologist for science, 1919–1946. In *The parliament of science*, edited by R. MacLeod and P. Collins, pp. 211–36. Northwood: Science Reviews.

Conn, H. W. 1886. *Evolution today*. New York: Putnam.

————. 1900. *The method of evolution*. New York: Putnam.

Cowan, R. S. 1977. Nature and nurture: the interplay of biology and politics in the work of Francis Galton. *Studies in History of Biology* 1: 133–208.

Darlington, C. D. 1932a. *Chromosomes and plant breeding*. London: Macmillan.

————. 1932b. *Recent advances in cytology*. London: Churchill; Philadelphia: P. Blakiston's Sons.

Darwin, C. R. 1859. *On the origin of species*. London: John Murray.

————. 1871. *The descent of man*. London: John Murray.

Dawson, C. 1929. *Progress and religion*. London: Longman.

de Beer, G. 1923. See Huxley, J. S., and de Beer, G. R. 1923.

————. 1924. See Huxley, J. S., and de Beer, G. R. 1924.

————. 1926–27. The mechanics of vertebrate development. *Biological Reviews and Biological Proceedings of the Cambridge Philosophical Society* 2: 137–97.

————, ed. 1938. *Evolution: essays on aspects of evolutionary biology presented to Professor E. S. Goodrich on his seventieth birthday*. Oxford: Clarendon Press.

De Vries, H. 1889. *Intracellulare pangenesis*. Jena: Gustav Fischer; Chicago: Open Court.

————. 1901–1903. *Die mutationtheorie*. 2 vols. Leipzig: Von Veit.

Delage, Y. 1895. *L'Hérédité et les grandes problèmes de la biologie générale*. 2d ed., 1903. Paris: Librairie C. Reinwald.

————, and Goldsmith, M. 1912. *The theories of evolution*. London: Frank Palmer; New York: B. W. Huebsch.

Divall, C. 1985. Capitalising on 'science': philosophical ambiguity in Julian Huxley's politics, 1920–1950. Ph.D. dissertation, Manchester Univ.

Dobzhansky, T. 1937. *Genetics and the origin of species*. New York: Columbia Univ. Press.

————. 1942. Introduction to *Systematics and the origin of species from the viewpoint of a zoologist*. See Mayr, E. 1942.

————. 1962. *Mankind evolving: the evolution of the human species*. New Haven and London: Yale Univ. Press.

————. 1963. Natural selection and fitness. *Eugenics Review* 55: 129.

————. 1967. *The biology of ultimate concern*. New York: New American Library.

Duméril, A. 1865. Nouvelles observations sur les axolotls, Batraciens urodéles de Mexico (*Siredon mexicanus vel Humboldtii*) nés dans la Ménagerie de Reptiles au Muséum d'Histoire Naturelle, et que y subsissent des métamorphoses. *Comptes rendus hebdomadaires des séances de l'Académie des sciences* 61: 775–78.

Durant, J. R. 1981. Innate character in animals and man: a perspective on the origins of ethology. In *Biology, medicine, and society, 1840–1940*, pp. 157–92. See Webster, C. 1981.

————. 1986. The making of ethology: the Association for the Study of Animal Behaviour, 1936–1986. *Animal Behaviour* 34: 1606–16.

————. 1989. Julian Huxley and the development of evolutionary studies. In *Evolutionary studies: a centenary celebration of the life of Julian Huxley*, pp. 26–40. See Keynes, M., and Harrison, G. A. 1980.

Eddington, A. S. 1928. *The nature of the physical world*. Cambridge: Cambridge Univ. Press.

Eldredge, N. 1985. *Unfinished synthesis: biological hierarchies and modern evolutionary thought*. New York: Oxford Univ. Press.

Fischer, J.-L., and Schneider, W. H., eds. 1988. *Histoire de la genetique*. Paris: Science en Situation.

Flew, A. 1967. *Evolutionary ethics*. 2d ed. London: Macmillan.

Focke, W. O. 1881. *Die Pflanzen-Mischlinge: Ein Beitrag zur Biologie der Gewächse*. Berlin: Borntraeger.

Ford, E. B. 1925. See Huxley, J. S., and Ford, E. B. 1925.

————. 1931. *Mendelism and evolution*. London: Methuen.

————. 1964. *Ecological genetics*. London: Methuen.

————. 1975. Obituary of J. S. Huxley. *Nature* 245: 5.

————. 1980. Some recollections pertaining to the evolutionary synthesis. In *The evolutionary synthesis*, 334–42. See Mayr, E., and Provine, W. B. 1980.

————, and Huxley, J. S. 1927. Mendelian genes and rates of development in *Gammarus Chevreuxi*. *British Journal of Experimental Biology* 5: 112–34.

Fowler, W. W. 1886. *A year with birds. By an Oxford tutor*. Oxford: Blackwell.

————. 1901. In the ways of birds. *Saturday Review* 92: 104–105.

————. 1913. *Kingham old and new*. Oxford: Blackwell.

Frank, R. E., and Greenberg, M. G. 1982. *Audiences for public television*. Beverley Hills: Sage Publications.

Freeden, M. 1979. Eugenics and progressive thought: a study in ideological affinity. *The Historical Journal* 22: 645–71.

Futuyama. D. J. 1989. Presidential address, American Society of Naturalists: Speciational trends and the role of species in macroevolution. *The American Naturalist* 130: 465–73.

Gautrey, P. J. 1987. See Barrett, P. H.; Gautrey, P. J.; Herbert, S.; Kohn, D.; and Smith, S., eds. 1987.

Goddard, H. H. 1917. Mental tests and the immigrant. *Journal of Delinquency* 2: 243–77.

Goldschmidt, R. B. 1928. *Einführung in die Vererbungswissenschaft*. Leipzig: Engelmann; Berlin: Julius Springer.

Gordon, L. 1976. *Woman's body, woman's right*. New York: Grossman.

Gould, S. J. 1965. See White, F. J., and Gould, S. J. 1965.

————. 1966. Allometry and size in ontogeny and phylogeny. *Biological Review* 41: 587–640.

————. 1973. Positive allometry of antlers in the "Irish Elk," *Megaloceros giganteus*. *Nature* 244: 375–76.

————. 1977. *Ontogeny and phylogeny*. Cambridge, Mass.: Harvard Univ. Press.

————. 1983. The hardening of the modern synthesis. In *Dimensions of Darwinism: themes and counterthemes in twentieth-century evolutionary theory*, pp. 71–93. See Grene, M. 1983.

————. 1984. Morphological channeling by structural constraint: convergence in styles of dwarfing and gigantism in *Cerion* with a description of two new fossil species and a report on the discovery of the largest *Cerion*. *Paleobiology* 10: 172–94.

Green, J. P. 1978. Bibliography. In *Julian Huxley, scientist and world citizen*, pp. 57–184. See Baker, J. R. 1978.

————. 1981. *Krise und Hoffnung: der Evolutionshumanismus Julian Huxleys*. Heidelberg: Winter.

Greenberg, M. G. 1982. See Frank, R. E., and Greenberg, M. G. 1982.

Greene, J. C. 1990. The interaction of science and world view in Sir Julian Huxley's evolutionary biology. *Journal of the History of Biology* 23: 39–55.

Greer, G. 1984. *Sex and destiny: the politics of human fertility*. London: Picador; New York: Harper.

Grene, M., ed. 1983. *Dimensions of Darwinism: themes and counterthemes in twentieth-century evolutionary theory.* Cambridge: Cambridge Univ. Press; Paris: La Maison des Sciences de l'Homme.

Habermas, J. 1973. The classical doctrine of politics in relation to social philosophy. In J. Habermas, *Theory and practice,* pp. 41–81. Boston: Beacon Press.

Haddon, A. C., and Huxley, J. S. 1935. Racial myths and ethnic fallacies. *Discovery* 16: 252–57.

Haecker, V. 1912. *Allegemeine Vererbungslehre.* Braunschweig: Vieweg.

Haldane, J. B. S. 1958. The theory of evolution before and after Bateson. *Journal of Genetics* 56: 11–27.

Hamburger, V. 1980. Embryology and the modern synthesis in evolutionary theory. In *The evolutionary synthesis: perspectices on the unification of biology,* pp. 97–111. See Mayr, E., and Provine, W. B. 1980.

———. 1988. *The heritage of experimental embryology: Hans Spemann and the Organizer.* New York: Oxford Univ. Press.

Haraway, D. 1976. *Crystals, fabrics and fields.* New Haven: Yale Univ. Press.

Harrison, R. G. 1912. Cultivation of tissues in extraneous media as a method of morphogenetic study. *Anatomical Records* 6: 181–93.

Hayek, F. A. 1944. *The road to serfdom.* London: Routledge.

Heinroth, K. 1971. *Oskar Heinroth: Vater der Verhaltensforschung, 1871–1945.* Stuttgart: Wissenschaftliche Verlagsgesellschaft.

Herbert, S. 1987. See Barrett, P. H.; Gautrey, P. J.; Herbert, S.; Kohn, D.; and Smith, S., eds. 1987.

Hodge, M. J. S. 1977. The structure and strategy of Darwin's long argument. *British Journal for the History of Science* 10:237–46.

Hogben, L. T. 1922. See Huxley, J. S., and Hogben, L. T. 1922.

Hoggart, R. 1978. *An idea and its servants, UNESCO from within.* New York: Oxford Univ. Press.

Holmes, C. M. 1970. *Aldous Huxley and the way to reality.* Bloomington: Indiana Univ. Press.

Horder, T. J.; Witkowski, J. A.; and Wylie, C. C., eds. 1986. *A history of embryology.* Cambridge: Cambridge Univ. Press.

Howard, H. E. 1907–14. *The British warblers: a history with problems of their lives,* 2 vols. London: R. H. Porter.

Hubback, D. 1989. Julian Huxley and eugenics. In *Evolutionary studies: a centenary celebration of the life of Julian Huxley,* pp. 194–206. See Keynes M., and Harrison, G. H. 1989.

Hubbs, C. L. 1943. Review of *Evolution, the modern synthesis.* *The American Naturalist* 77: 365–68.

Hughes, H. S. 1959. *Consciousness and society.* London: MacGibbon & Kee; New York: Knopf.

Huxley, Aldous. 1923. *Antic Hay.* New York: George H. Doran.

———. 1932. *Antic Hay.* New York: Modern Library.

———. 1958. *Brave new world.* London: Chatto & Windus.

Huxley, Andrew. 1989. The Galton lecture for 1987: Julian Huxley—a family view. In *Evolutionary studies: a centenary celebration of the life of Julian Huxley*, pp. 9–25. See Keynes, M., and Harrison, G. A. 1989.

Huxley, Juliette. 1986. *Leaves of the tulip tree*. London: John Murray; Topsfield, Mass.: Salem House.

Huxley, J. S. 1911. Some phenomena of regeneration in *Sycon*; with a note on the structure of its collar cells. *Philosophical transactions of the Royal Society of London*, ser. B 202: 165–89.

———. 1912a. A disharmony in the reproductive habits of the wild duck (*Anas boschas*, L.). *Biologisches Zentralblatt* 32: 621–23.

———. 1912b. A first account of the courtship of the redshank (*Totanus calidris* L.). *Proceedings of the Zoological Society of London*, pp. 647–56.

———. 1912c. The great crested grebe and the idea of secondary sexual characteristics. *Science* 36: 601–602.

———. 1912d. *The individual in the animal kingdom*. Cambridge: Cambridge Univ. Press; New York: G. P. Putnam's Sons.

———. 1914. The courtship habits of the great crested grebe *(Podiceps cristatus)*; with an addition to the theory of sexual selection. *Proceedings of the Zoological Society of London* 35: 491–562.

———. 1916. Bird-watching and biological science. Some observations on the study of courtship in birds. *Auk* 33: 142–61, 256–70.

———. 1918. Texas and Academe. *The Cornhill Magazine* 45: 53–65.

———. 1919. Some points in the sexual habits of the little grebe, with a note on the occurrences of vocal duets in birds. *British Birds* 13: 155–58.

———. 1920a. Metamorphosis of axolotl caused by thyroid feeding. *Nature* 104: 435.

———. 1920b. The thyroid gland and the control of animal growth. *Illustrated London News*, 28 February.

———. 1921a. The accessory nature of many structures and habits associated with courtship. *Nature* 108: 565–66.

———. 1921b. Further studies in restitution-bodies and free tissue-cultures in *Sycon*. *Quarterly Journal of Microscopical Science* 65: 293–322.

———. 1921c. Linkage in *Gammarus chevreuxi*. *Journal of Genetics* 11: 229–33.

———. 1921d. Obituary of W. W. Fowler. *British Birds* 15: 143–44.

———. 1921e. Review of H. E. Howard, *Territory in Bird Life*. *Discovery* 2: 135–36.

———. 1921f. Studies in dedifferentiation. II Dedifferentiation and resorption in *Perophora*. *Quarterly Journal of Microscopical Science* 65: 643–97.

———. 1922. Preferential mating in birds with similar coloration in both sexes. *British Birds* 16: 99–101.

———. 1923a. Courtship activities in the red-throated diver (*Colymbus stellatus* Pontopp.); together with a discussion of the evolution of courtship in birds. *Journal of the Linnean Society of London, Zoology* 35: 253–92.

———. 1923b. *Essays of a biologist*. London: Chatto & Windus; New York: Knopf, 1929.

———. 1923c. Ils n'ont que de l'âme: An essay on bird mind. *The Cornhill Magazine* 54: 415–27.

———. 1923d. Progress, biological and other. *The Hibbert Journal* 21: 436–60. Reprinted in *Essays of a biologist*, pp. 3–65. See Huxley, J. S. 1923b.

———. 1924a. "America Revisited. III. The Negro Problem," *The Spectator*, 29 November 1924, pp. 821–22.

———. 1924b. Constant differential growth-ratios and their significance. *Nature* 114: 895–96.

———. 1924c. Early embryonic differentiation. *Nature* 113: 276–78.

———. 1924d. Eugenics and heredity. *The New Statesman* 23: 281–82.

———. 1924e. Nature and nurture. *The New Leader*, 29 February.

———. 1924f. The outlook in biology. *Rice Institute Pamphlets* 11: 241–338.

———. 1924g. Some further notes on the courtship behaviour of the great crested grebe. *British Birds* 18: 129–34.

———. 1924h. Some points in the breeding behaviour of the common heron. *British Birds* 18: 155–63.

———. 1924i. The variation in the width of the abdomen in the immature fiddler crab considered in relation to its relative growth rate. *The American Naturalist* 58: 468–75.

———. 1925a. The absence of "courtship" in the avocet. *British Birds* 19: 88–94.

———. 1925b. Studies on amphibian metamorphosis II. *Proceedings of the Royal Society of London* ser. B 98:113–46.

———. 1926a. The biological basis of individuality. *Journal of Philosophical Studies* 1: 305–319.

———. 1926b. *The stream of life*. London: Watts; New York and London: Harper.

———. 1926c. The tissue-culture king. *The Cornhill Magazine* 60: 422–57.

———. 1926d. *Essays in popular science*. London: Chatto & Windus.

———. 1927a. Introduction. In *Realities of bird life: being extracts from the diaries of a life-loving naturalist*, pp. xi–xvi. See Selous, E. 1927.

———. 1927b. *Religion without revelation*. London: Ernest Benn; New York: Harper.

———. 1929. What is individuality? *The Realist* 1: 109–121.

———. 1930a. *Ants*. London: Ernest Benn; New York: J. Cape, H. Smith.

———. 1930b. Bird mind. *Atlantic Monthly* 146: 473–82.

———. 1930c. *Bird-watching and bird behaviour*. London: Chatto & Windus; London: Dobson, 1949.

———. 1930d. The courtship of birds. *The Listener*, pp. 935–37.

————. 1930e. Eugenic sterilisation. *Nature* 126: 503.

————. 1930f. Memorable incidents with birds. *The Listener* 3: 841–42.

————. 1930g. Spemanns "Organisator" und Childs Theorie der axialen Gradienten. Translated by H. Spemann. *Die Naturwissenschaften* 18: 265.

————. 1930h. Towards a higher civilization. *Birth Control Review* (December): 342–45.

————. 1931a. *Africa view.* London: Chatto & Windus; New York and London: Harper.

————. 1931b. Religion meets science. *Atlantic Monthly* 147: 373–383.

————. 1931c. Biology of bird courtship. *Proceedings of the VIIth International Ornithological Congress at Amsterdam, 1930,* pp. 107–108.

————. 1931d. The vital importance of eugenics. *Harper's Monthly Magazine* 163: 324–31.

————. 1931e. *What dare I think? The challenge of modern science to human action and belief.* London: Chatto & Windus; New York: Harper.

————. 1932a. *The captive shrew and other poems of a biologist.* Oxford: Blackwell.

————. 1932b. *Problems of relative growth.* London: Methuen; New York: Harper.

————. 1932c. Why is the white man in Africa? *Fortnightly Review* 137: 60–69.

————. 1933. Man and reality. In *Science in the changing world,* pp. 186–98. See Adams, M. 1933.

————. 1934a. The applied science of the next hundred years: biological and social engineering. *Life and Letters* 19: 38–46.

————. 1934b. *If I were dictator.* London: Methuen.

————. 1934c. A natural experiment on the territorial instinct. *British Birds* 27: 270–77.

————. 1934d. Sterilisation: a social problem. *New Chronicle,* 22 January.

————. 1935a. The concept of race in the light of modern genetics. *Harper's Monthly Magazine* 170: 689–98.

————. 1935b. Making and using nature films. *The Listener* 13: 595–97, 629.

————. 1936a. *At the zoo.* London: Allen & Unwin; Toronto: Nelson.

————. 1936b. Eugenics and society. *Eugenics Review* 28: 11–31. Reprinted in *The uniqueness of man* (American title, *Man stands alone*), pp. 34–84. See Huxley, J. S. 1941.

————. 1936c. Natural selection and evolutionary progress. *Proceedings of the British Association for the Advancement of Science* 106: 81–100.

————. 1938a. Darwin's theory of sexual selection and the data subsumed by it, in the light of recent research. *The American Naturalist* 72: 416–33.

————. 1938b. *The present standing of the theory of sexual selection.* In *Evolution: essays on aspects of evolutionary biology presented to Pro-*

fessor E. S. Goodrich on his seventieth birthday, pp. 11–42. See de Beer, G. R. 1938.

———. 1938c. Threat and warning coloration in birds, with a general discussion of the biological functions of color. *Proceedings of the Eighth International Ornithological Congress, Oxford, 1934*, pp. 430–55.

———. 1939. *Essays of a biologist.* Harmondsworth: Penguin.

———, ed. 1940a. *The new systematics.* Oxford: Clarendon Press.

———. 1940b. Science, natural and social. *The Scientific Monthly* 50: 5–16.

———. 1940c. War and reconstruction. *Nature* 145: 330–34.

———. 1941a. The concept of race in the light of modern genetics. In *The uniqueness of man* (American title *Man stands alone*), pp. 106–126. See Huxley, J. S. 1941c.

———. 1941b. *Democracy marches.* London: Chatto & Windus.

———. 1941c. *The uniqueness of man.* London: Chatto & Windus. American title *Man stands alone.* New York and London: Harper.

———. 1942. *Evolution, the modern synthesis.* London: Allen & Unwin; New York: Harper, 2d ed. 1962.

———. 1943. Memories and achievements: a chapter of natural history. *The Countryman* 28: 175–80.

———. 1944a. Natural history memories—2. *The Countryman* 29: 28–32.

———. 1944b. Natural history memories—3. *The Countryman* 29: 186–91.

———. 1944c. *On living in a revolution.* London: Chatto & Windus; New York: Harper.

———. 1945. Natural history memories—4. *The Countryman* 30: 38–43.

———. 1947a. Evolutionary ethics. In Huxley, T. H., and Huxley, J. S., *Evolution and ethics 1893–1943*, pp. 103–52. London: Pilot Press.

———. 1947b. *Man in the modern world.* London: Chatto & Windus.

———. 1948. *UNESCO: its purpose and philosophy.* Washington, D.C.: Public Affairs Press.

———. 1949. *Soviet genetics and world science, Lysenko and the meaning of heredity.* London: Chatto & Windus.

———. 1950. Genetics, evolution, and human destiny. In *Genetics in the 20th century*, edited by L. C. Dunn, pp. 591–621. New York: Macmillan.

———. 1953. *Evolution in action.* London: Chatto & Windus; New York: Harper.

———. 1956. World population. *Scientific American* 194, no. 3, pp. 64–76.

———. 1957a. Evolutionary humanism. In *New bottles for new wine*, pp. 279–312. See Huxley, J. S. 1957b.

———. 1957b. *New bottles for new wine.* London: Chatto & Windus; New York: Harper.

———. 1957c. Population and human fulfillment. In *New bottles for new wine*, pp. 168–212. See Huxley, J. S. 1957b.

―――. 1957d. *Religion without revelation*. New and rev. ed. London: M. Parrish; New York: Harper.

―――. 1959. Introduction. In *The phenomenon of man*. See Teilhard de Chardin, P. 1959.

―――, ed. 1961. *The humanist frame*. London: Allen & Unwin; New York: Harper.

―――. 1962a. Eugenics in evolutionary perspective. *The Eugenics Review* 54: 123–41. Reprinted in *Perspectives in Biology and Medicine* (1963) 6: 155–87; and *Essays of a humanist* (Huxley, J. S. 1964a), pp. 251–80. Summary in *Nature* (1962) 195: 227–28.

―――. 1962b. Review of Dobzhansky, T. 1962. *Perspectives in Biology and Medicine* 6: 144–48.

―――. 1963a. *Evolution: the modern synthesis*. 2d. ed. London: Allen & Unwin; New York: Wiley.

―――. 1963b. Lorenzian ethology. *Zeitschrift für Tierpsychologie* 20: 402–409.

―――. 1963c. Units of evolution. *Nature* 199: 838–40.

―――. 1964a. *Essays of a humanist*. London: Chatto & Windus; Toronto: Clarke Irwin; New York: Harper.

―――. 1964b. Eugenics and evolutionary perspective. In *Essays of a humanist*, pp. 251–80. See Huxley, J. S. 1964a.

―――. 1966. A discussion of ritualization of behaviour in animals and man—Introduction. *Philosophical Transactions of the Royal Society of London*, ser. B 251: 249–71.

―――. 1970. *Memories*. London: Allen & Unwin; New York: Harper.

―――. 1972. *Problems of relative growth*. 2d ed. New York: Dover.

―――. 1973. *Memories II*. London: Allen & Unwin; New York: Harper.

Huxley, J. S., and Carr-Saunders, A. M. 1924. Absence of prenatal effects of lens-antibodies in rabbits. *British Journal of Experimental Biology* 1: 215–48.

Huxley, J. S., and de Beer, G. R. 1923. Studies in dedifferentiation IV. Resorption and differential inhibition in *Obelia* and *Campanularia*. *Quarterly Journal of Microscopical Science* 67: 473–95.

―――. 1924. Studies in dedifferentiation V. Dedifferentiation and reduction in *Aurelia*. *Quarterly Journal of Microscopical Science* 68: 471–79.

―――. 1934. *Elements of experimental embryology*. Cambridge: Cambridge Univ. Press.

Huxley, J. S., and Ford, E. B. 1925. Mendelian genes and rates of development. *Nature* 116: 861–63.

―――. 1927. See Ford, E. B., and Huxley, J. S. 1927.

Huxley, J. S., and Haddon, A. C. 1935. *We Europeans: a survey of "racial" problems*. London: Jonathan Cape; Toronto: Nelson.

Huxley, J. S., and Hogben, L. T. 1922. Experiments on amphibian metamorphosis and pigment responses in relation to internal secretions. *Proceedings of the Royal Society of London* ser. B 93: 36–53.

Huxley, J. S., and Howard, H. E. 1934. Field studies and physiology: a further correlation. *Nature* 133: 688–89.

Huxley, J. S., and Huxley, T. H. 1947. See Huxley, T. H., and Huxley, J. S. 1947.

Huxley, J. S., and Kettlewell, H. B. D. 1965. *Charles Darwin and his world*. London: Thames & Hudson; New York: Viking Press.

Huxley, J. S., and Montague, F. A. 1925. Studies on the courtship and sexual life of birds. V. The oyster-catcher. *Ibis* 12th ser. 1: 868–97.

Huxley, J. S., and Reeve, E. C. 1945. In *Essays on growth and form*, pp. 121–156. See Clark, W. E. LeG., and Medawar, P. B. 1945.

Huxley, J. S., and van Oordt, G. J. 1922. Some observations on the habits of the red-throated diver in Spitsbergen. *British Birds* 16: 34–46.

Huxley, J. S., Wells, H. G., and Wells, G. P. 1931. See Wells, H. G.; Wells, G. P.; and Huxley, J. S. 1931.

Huxley, L. 1900. *Life and letters of Thomas Henry Huxley*. London: Macmillan.

Huxley, T. H. 1894. Evolution and ethics. In Huxley, T. H., *Evolution and ethics and other essays*, pp. 46–116. London: Macmillan.

———, and Huxley, J. S. 1947. *Evolution and ethics 1893–1943*. London: Pilot Press. American title *Touchstone for ethics 1893–1943*. New York: Harper.

Hyams, E. 1963. *The new statesman: the history of the first fifty years, 1913–1963*. London: Longman.

Hyman, L. H. 1957. Charles Manning Child. *Biographical Memoirs of the National Academy of Sciences* 30: 73–103.

Jacobs, N. X. 1989. From unity to unity: protozoology, cell theory, and the new concept of life. *Journal of the History of Biology* 22: 215–42.

Jamieson, D. 1990. See Bekoff, M., and Jamieson, D. 1990.

Jenkinson, J. W. 1909. *Experimental embryology*. Oxford: Clarendon Press.

———. 1913. *Vertebrate embryology*. Oxford: Clarendon Press.

———. 1917. *Three lectures on experimental embryology*. Oxford: Clarendon Press.

Johnson, E. D. H. 1966. *The poetry of Earth*. New York: Athenaeum.

Jones, G. 1980. *Social Darwinism and English thought*. Brighton: Harvester Press.

———. 1982. Eugenics and social policy between the wars. *The Historical Journal* 25: 717–28.

———. 1986. *Social hygiene in twentieth century Britain*. London: Croom Helm.

Joseph, M. 1926. *The commercial side of literature*. New York: Harper.

Judson, H. 1979. *The eighth day of creation: makers of the revolution in biology*. New York: Simon & Schuster.

Kellogg, V. L. 1907. *Darwinism today*. New York: Holt.

Kennedy, T. C. 1974. The Next Five Years Group and the failure of the politics of agreement in Britain. *Canadian Journal of History* 9: 45–68.

Kevles, D. J. 1978. *The physicists: the history of a scientific community in modern America*. New York: Knopf.

―――. 1985. *In the name of eugenics: genetics and the uses of human heredity*. New York: Knopf.

Keynes, M., and Harrison, G. A., eds. 1989. *Evolutionary studies: a centenary celebration of the life of Julian Huxley*. Basingstoke and London: Macmillan.

Kimball, R. F. 1943. The great biological generalization. *Quarterly Review of Biology* 18: 364–67.

Kimmelman, B. 1983. Genetics and eugenics in an agricultural context: the American Breeders' Association, 1903–1913. *Social Studies of Science* 13: 163–204.

Kimura, M. 1983. *The neutral theory of molecular evolution*. Cambridge: Cambridge Univ. Press.

King, J. L. 1975. Review of Lewontin, R. C. 1974. *Annals of Human Genetics* 38: 507–510.

Kirkman, F. B. 1910–14. *The British bird book*, 4 vols. London: Jack.

Kohn, D. 1987. See Barret, P. H.; Gautrey, P. J.; Herbert, S.; Kohn, D.; and Smith, S., eds. 1987.

Lack, D. 1959. Some British pioneers in ornithological research, 1859–1939. *Ibis* 101: 71–81.

―――. 1973. My life as an amateur ornithologist. *Ibis* 115: 421–31.

Laves, W. H. C., and Thomson, C. A. 1957. *UNESCO: purpose, progress, prospects*. Bloomington: Indiana Univ. Press.

Lewontin, R. C. 1974. *The genetic basis of evolutionary change*. New York: Columbia Univ. Press.

Lock, R. H. 1906. *Recent progress in the study of variation, heredity, and evolution*. London: John Murray.

Lorenz, K. Z. 1970. *Studies in animal and human behavior*, vol. 1. Cambridge: Harvard Univ. Press.

―――. 1974. Introduction: the study of behaviour. In *Bird life*. See Nicolai, J. 1974.

Lovejoy, A. O. 1929. *The revolt against dualism: an inquiry concerning the existence of ideas*. LaSalle, Ill.: Open Court Publishing.

Lowe, P. D. 1976. Amateurs and professionals: the institutional emergence of British plant ecology. *Journal of the Society for the Bibliography of Natural History* 7: 517–35.

MacBride, E. W. 1934. Embryology and predestiny. *Discovery* 15: 218–20.

MacIntyre, A., ed. 1957. *Metaphysical beliefs*. London: SCM Press.

―――. 1981. *After virtue: a study in moral theory*. London: Duckworth.

MacKenzie, D. 1979. Karl Pearson and the professional middle class. *Annals of Science* 36: 125–143.

Maienschein, J. 1988. See Rainger, R.; Benson, K. R.; and Maienschein, J. 1988.

―――. 1991. See Benson, K. R.; Maienschein, J.; and Rainger, R. 1991.

Maienschein, J.; Rainger, R.; and Benson, K. R. 1981. Were American morphologists in revolt? *Journal of the History of Biology* 14: 83–87.

Mangold, H. 1924. See Spemann, H., and Mangold, H. 1924.

Matheson, P. E. 1927. William Warde Fowler (1847–1921). *Dictionary of National Biography 1912–1921*, pp. 194–95.

Mayr, E. 1942. *Systematics and the origin of species from the viewpoint of a zoologist*. New York: Columbia Univ. Press.

———. 1959. Where are we? *Cold Spring Harbor Symposia on Quantitative Biology* 24: 1–14.

———. 1961. Cause and effect in biology. *Science* 134: 1501–1506.

———. 1963. *Animal species and evolution*. Cambridge, Mass.: Harvard Univ. Press.

———. 1976. *Evolution and the diversity of life: selected essays*. Cambridge, Mass.: Harvard Univ. Press.

———. 1982. *Growth of Biological Thought*. Cambridge, Mass.: Harvard Univ. Press.

———. 1988. On the evolutionary synthesis and after. In E. Mayr, *Toward a new philosophy of biology: observations of an evolutionist*. Cambridge, Mass.: Harvard Univ. Press.

Mayr, E., and Provine, W. B., eds. 1980. *The evolutionary synthesis: perspectives on the unification of biology*. Cambridge, Mass.: Harvard Univ. Press.

McDougall, W. 1929. *Modern materialism and emergent evolution*. London: Methuen.

McGucken, W. 1984. *Scientists, society and state*. Columbus: Ohio State Univ. Press.

Medawar, P. B. 1945. See Clark, W. E. LeG., and Medawar, P. B. 1945.

———. 1961. Review of *The phenomenon of man*. *Mind* 70: 99–106.

———. 1967. *The art of the soluble*. London: Methuen.

———. 1975. Obituary of Julian Huxley. *Nature* 254: 4.

———. 1982. *Pluto's republic*. Oxford: Oxford Univ. Press.

Mehler, B. 1988. A history of the American Eugenics Society, 1921–1940. Ph.D. diss., Univ. of Ill.

Mitman, G., and Burkhardt, Jr., R. Struggling for identity: the study of animal behavior in America, 1930–1945. In *The expansion of American biology*, pp. 164–94. See Benson, K. R.; Maienschein, J.; and Rainger, R. 1991.

Montague, F. A. 1925. See Huxley, J. S., and Montague, F. A. 1925.

Morgan, C. L. 1924. A philosophy of evolution. In *Contemporary British philosophy*, pp. 275–306. See Muirhead, J. H. 1924.

———. 1925. *Life, mind and spirit*. London: Williams & Norgate.

———. 1926. *Emergent evolution*. London: Williams & Norgate.

———. 1933. *The emergence of novelty*. London: Williams & Norgate.

Morgan, T. H. 1923. Further evidence on variation in the width of the abdomen in immature fiddler crabs. *The American Naturalist* 57: 274–83.

————. 1934. *Embryology and genetics.* New York: Columbia Univ. Press.

Morgan, T. H.; Bridges, C. B.; and Sturtevant, A. H. 1925. The genetics of drosophila. *Bibliographia Genetica* 2: 3–267.

Muirhead, J. H., ed. 1924. *Contemporary British philosophy.* London: Allen & Unwin.

Mullens, W. H., and Swann, H. K. 1917. *A bibliography of British ornithology.* London: Macmillan.

Muller, H. J. 1918. Genetic variability, twin hybrids and constant hybrids, in a case of balanced lethal factors. *Genetics* 2: 422–99. Reprinted in *Studies in Genetics,* pp. 69–94. See Muller, H. J. 1962.

————. 1923a. Mutation. In *Eugenics, genetics, and the family: Proceedings of the 2nd International Congress of Eugenics* 1: 106–112. Baltimore: Williams and Wilkins.

————. 1923b. Recurrent mutations of normal genes of drosophila not caused by crossing over. *Anatomical Record* 26: 397–98.

————. 1925. Why polyploidy is rarer in animals than in plants. *The American Naturalist* 59: 346–53.

————. 1962a. Sir Julian Huxley: a biographical appreciation. *The Humanist,* no. 2–3, pp. 51–55.

————. 1962b. *Studies in genetics.* Bloomington: Indiana Univ. Press.

Needham, J. 1933. Chemical heterogony and the ground-plan of animal growth. *Biological Reviews* 9: 79–109.

————. 1934. Morphology and biochemistry. *Nature* 134: 275–76.

————. 1936. *Order and life.* Cambridge: Cambridge Univ. Press.

————. 1975. Obituary of Julian Huxley. *Nature:* 254: 2–3.

Nicolai, J. 1974. *Bird life.* New York: Putnam; London: Thames and Hudson.

Nordenskiöld, N. E. 1928. *The history of biology: a survey.* New York: Knopf.

Olby, R. 1985. *Origins of Mendelism.* 2d. ed. Chicago: Univ. of Chicago Press.

————. 1986. Structural and dynamical explanations in the world of neglected dimensions. In *A history of embryology,* pp. 275–308. See Horder, T. J. 1986.

The old zoo and the new. 1901. *Saturday Review* 91: 330–32, 365–66, 397–98, 433–43.

Osborn, H. F. 1925. *The earth speaks to Bryan.* New York: Scribner.

Paul, D. 1984. Eugenics and the left. *Journal of the History of Ideas* 45: 567–90.

Pegg, Mark. 1983. *Broadcasting and society.* London: Croom Helm.

Plant, R. 1984. See Vincent, A., and Plant, R. 1984.

Plumb, J. H., ed. 1955. *Studies in social history: a tribute to G. M. Trevelyan.* London: Longman.

Portielje, A. F. J. 1926. Zur ethologie bezw. Psychologie von *Botaurus stellaris* (L.). *Ardea* 15: 1–15.

————. 1927. Zur ethologie bezw. Psychologie von *Phalcrocorax carbo subcormoranus* (Brehm). *Ardea* 17: 112–49.

Poulton, E. B. 1908. *Essays on Evolution*. Oxford: Clarendon Press.

————. 1909. *Charles Darwin and the Origin of Species*. London: Longman.

Preface. 1903. *The Zoologist* ser. 4 7: iv, edited by W. L Distant.

Provine, W. B. 1978. The role of mathematical population geneticists in the evolutionary synthesis of the 1920s and 1940s. *Studies in History of Biology* 2: 167–92.

————. 1980. Introduction to section on England. In *The evolutionary synthesis: perspectives on the unification of biology*, pp. 329–34. See Mayr, E., and Provine, W. B. 1980.

————. 1983. The development of Wright's theory of evolution: systematics, adaptation, and drift. In *Dimensions of Darwinism: themes and counterthemes in twentieth century evolutionary theory*, pp. 43–70. See Grene, M. 1983.

————. 1985. The R. A. Fisher–Sewall Wright controversy. *Oxford Surveys in Evolutionary Biology* 2:197–219.

————. 1986. *Sewall Wright and evolutionary biology*. Chicago: Univ. of Chicago Press.

————. 1988. Progress in evolution and meaning in life. In *Evolutionary progress*, edited by M. H. Nitecki, pp. 49–74. Chicago: Univ. of Chicago Press.

————. 1989. Founder effects and genetic revolutions in microevolution and speciation: an historical perspective. In *Genetics, speciation, and the founder principle*, edited by Giddings, L. V., Kaneshiro, K. Y., and Anderson, W. W., pp. 43–76. New York: Oxford Univ. Press.

A public trust: the report of the Carnegie Commission on the future of public broadcasting. 1979. New York: Bantam.

Punnett, R. C. 1905. *Mendelism*. Cambridge: Bowes and Bowes.

Pycraft, W. P. 1913. *The courtship of animals*. London: Hutchinson.

Radl, E. 1930. *The history of biological theories*. London: H. Milford, Oxford Univ. Press.

Rainger, R. 1981. See Maienschein, J.; Rainger, R.; and Benson, K. R. 1981.

————; Benson, K. R.; and Maienschein, J., eds. 1988. *The American development of biology*. Philadelphia: Univ. of Pennsylvania Press.

————. See Benson, K. R., Maienschein, J., and Rainger, R., eds. 1991.

Reed, J. 1978. *From private vice to public virtue: the birth control movement and American society since 1830*. New York: Basic Books.

Reeve, E. C., and Huxley, J. S. 1945. Some problems in the study of allometric growth. In *Essays on growth and form*, pp. 121–56. See Clark, W. E. LeG., and Medawar, P. B. 1945.

Review of C. Lyell. *Principles of Geology*. 1839. *Quarterly Review* 64: 103.

Review of Ernst Mayr, *Systematics and the origin of species from the viewpoint of a zoologist*. 1943. *Quarterly Review of Biology* 18: 270.

Richmond, M. L. 1989. Protozoa as precursors of metazoa: German cell theory and its critics at the turn of the century. *Journal of the History of Biology* 22: 243–76.

Richter, M. 1964. *The politics of conscience*. London: Weidenfeld and Nicolson.

Ridley, M. 1986. Embryology and classical zoology in Great Britain. In *A history of embryology*, 35–67. See Horder, T. J.; Witkowski, J. A.; and Wylie, C. C. 1986.

Riquelmi, C. 1984. See Bates, S. C.; Winkler, M. G.; and Riquelmi, C. 1984.

Rowan, W. 1931. *The riddle of migration*. Baltimore: Williams and Wilkins.

Schleip, W. 1929. *Die Determination der Primitiventiwicklung*. Leipzig: Akademische Verlagsgesellschaft M. B. H.

Schneider, W. H. 1988. See Fischer, J.-L., and Schneider, W. H. 1988.

Scientific affairs in Europe. 1945. *Nature* 156: 576–79.

Searle, G. R. 1971. *The quest for national efficiency, 1899–1914*. Oxford: Oxford Univ. Press; Berkeley: Univ. of California Press.

———. 1976. *Eugenics and politics in Britain 1900–1914*. Leyden: Noordhoff International Publishing.

Seidel, F. 1936. Review of Huxley, J. S., and de Beer, G. R., *Elements of experimental embryology*. *Die Naturwissenschaften* 11: 175–76.

Selous, E. 1901–1902. An observational diary of the habits—mostly domestic—of the great crested grebe (*Podicipes cristatus*) and the peewit (*Vanellus vulgaris*), with some general remarks. *The Zoologist* ser. 4, 5: 161–83, 339–50, 454–62; 6: 133–44.

———. 1905a. *Bird life glimpses*. London: Allen & Unwin.

———. 1905b. *The bird watcher in the Shetlands*. New York: Dutton.

———. 1906–1907. Observations tending to throw light on the question of sexual selection in birds, including a day-to-day diary on the breeding habits of the ruff (*Machetes pugnax*). *The Zoologist* ser. 4, 10: 201–219, 285–94, 419–28; 11: 60–65, 161–82, 367–81.

———. 1909–1910. An observational diary on the nuptial habits of the blackcock (*Tetrao tetrix*) in Scandinavia and England. *The Zoologist* ser. 4, 13: 401–413; 14: 23–29, 51–56, 176–82, 248–65.

———. 1927. *Realities of bird life: being extracts from the diaries of a life-loving naturalist*. London: Constable.

Semmel, B. 1960. *Imperialism and social reform*. London: Allen & Unwin.

Seward, A. C., ed. 1909. *Darwin and modern science*. Cambridge: Cambridge Univ. Press.

Sewell, J. P. 1975. *UNESCO and world politics*. Princeton: Princeton Univ. Press.

Shapere, D. 1980. The meaning of the evolutionary synthesis. In *The evolutionary synthesis*, pp. 388–98. See Mayr, E., and Provine, W. B. 1980.

Shull, A. F. 1936. *Evolution*. New York and London: McGraw-Hill.

Silvey, R. 1974. *Who's listening? The story of the BBC Audience Research*. London: Allen & Unwin.

Simmons, K. E. L. 1984. Edmund Selous (1857–1934): fragments for a biography. *Ibis* 125: 595–96.

Simon, B. 1974. *The politics of educational reform, 1920–1940.* London: Lawrence and Wishart.

Simon, W. E. 1963. *European positivism in the nineteenth century.* Ithaca: Cornell Univ. Press.

Simpson, G. G. 1944. *Tempo and mode in evolution.* New York: Columbia Univ. Press. 2d ed. 1984.

———. 1964. *This view of life.* New York: Harcourt, Brace, and World.

———. 1978. *Concession to the improbable: an unconventional autobiography.* New Haven: Yale Univ. Press.

Smith, G. W. 1906. High and low dimorphism. *Mittheilungen aus der Zoologischen Station zu Neapel* 17: 312–37.

Smith, S. 1987. See Barrett, P. H.; Gautrey, P. J.; Herbert, S.; Kohn, D.; and Smith, S., eds. 1987.

Spemann, H. 1921. Die Erzeugung tierischer Chimären durch heteroplastische embryonale Transplantation zwischen *Triton cristatus* und *taeniatus*. *Roux's Archiv* 48: 533–70.

———. 1938. *Embryonic development and induction.* Facsimile reprint. New York: Hafner Publishing Co.

———, and Mangold, H. 1924. Über Induktion von Embryonalen lagen durch Implantation aertfremder Organisatoren. *Roux's Archiv* 100: 599–638.

Stebbing, L. S. 1937. *Philosophy and the physicists.* Harmondsworth: Penguin.

Stebbins, G. L., and Ayala, F. G. 1981. Is a new evolutionary synthesis necessary? *Science* 213: 967–71.

Steiner, G. 1987. The two cultures revisited. *Cambridge Review* 108: 13–14.

Stevenson, J. 1984. *British society, 1914–45.* Harmondsworth: Penguin.

Strauss, L. 1975. Three waves of modernity. In *Political philosophy: six essays by Leo Strauss.* Indianapolis: Pegasus.

Sturtevant, A. H. 1925. See Morgan, T. H.; Bridges, C. B.; and Sturtevant, A. H. 1925.

Swann, H. K. 1917. See Mullens, W. H., and Swann, H. K. 1917.

Teilhard de Chardin, P. 1959. *The phenomenon of man.* New York: Harper.

Teissier, G. 1931. Recherches morphologiques et physiologiques sur la croissance des insectes. *Travaux de la station biologique de Roscoff.* 9: 27–238.

Thomas, H. 1944. *Britain's Brains Trust.* London: Chapman Hall.

Thorpe, W. H. 1979. *The origins and rise of ethology.* London: Heinemann.

Tinbergen, N. 1931. Zur Paarungsbiologie der Flussseeschwalbe (*Sterna hirundo hirundo* L.). *Ardea* 20:1–18.

———. 1942. An objectivistic study of the innate behaviour of animals. *Bibliotheca biotheoretica* 1: 39–98.

————. 1945. Letter to J. S. Huxley. In *Scientific Affairs in Europe*, pp. 577–78. See *Scientific Affairs in Europe*. 1945.

Toulmin, S. 1957. Contemporary scientific mythology. In *Metaphysical beliefs*, pp. 51–75. See MacIntyre, A. 1957.

Turner, J. R. G. 1985. Fisher's evolutionary faith and the challenge of mimicry. *Oxford Surveys in Evolutionary Biology* 2: 159–196.

Ulrich's international periodicals directory. 1987–88. 1987. New York: Bowker.

The unknown bird world. 1901. *Saturday Review* 91: 683.

van Oordt, G. J. 1922. See Huxley, J. S., and van Oordt, G. J. 1922.

Verwey, J. 1930. Die Paarungsbiologie des Fischreihers. *Zoologisches Jahrbuch. Physiologie* 48: 1–120.

Vincent, A., and Plant, R. 1984. *Philosophy, politics and citizenship*. Oxford: Blackwell.

von Bertalanffy, L. 1933. *Modern theories of development*. Oxford: Oxford Univ. Press.

Waagen, W. 1867–70. Über die Ansatzstelle der Haftmuskeln beim Nautilus und den Ammoniten. *Paleontographica* 17: 185–210.

Waddington, C. H. 1934. Morphogenic fields. *Science Progress* 29: 336–46.

————. 1941. The relations between science and ethics. *Nature* 148: 270–74.

————. 1975. *Evolution of an evolutionist*. Ithaca: Cornell Univ. Press.

Wallace, B. 1957. Some of the problems accompanying an increase of mutation rates in Mendelian populations. In *Effects of radiation on human heredity*, pp. 57–62. Geneva: World Health Organization.

Walter, H. E. 1913. *Genetics: an introduction to the study of heredity*. New York: Macmillan.

Webster, C., ed. 1981. *Biology, medicine, and society, 1840–1940*. Cambridge: Cambridge Univ. Press.

Weismann, A. 1909. The selection theory. In *Darwin and modern science*. See Seward, A. C. 1909.

Weiss. P. 1939. *Principles of development. A text in experimental embryology*. New York: Henry Holt.

Wells, G. P. 1931. See Wells, H. G.; Huxley, J. S.; and Wells, G. P. 1931.

————. 1934. Morphogenesis in the animal embryo. *Nature* 133: 890–91.

Wells, H. G.; Huxley, J. S.; and Wells, G. P. 1931. *The science of life*. London: Amalgamated Press; Garden City, N.Y.: Doubleday. (Originally published in 3 vols. 1929–30. London: Amalgamated Press.)

Werdinger, J. 1980. Embryology at Woods Hole: the emergence of a new American biology. Ph.D. diss., Indiana Univ.

Werskey, G. 1969. Nature and politics between the wars. *Nature* 224: 462–72.

————. 1978. *The visible college: The collective biography of British scientific socialists of the 1930s*. London: Allen Lane. (New York: Holt Rinehart and Winston, 1979; London: Free Association Books, 1988.)

West-Eberhard, M. J. 1983. Sexual selection, social competition, and speciation. *Quarterly Review of Biology* 58: 155–83.

White, J. F., and Gould, S. J. 1965. Interpretation of the coefficient in the
allometric equation. *The American Naturalist* 99: 5–18.

Wiener, M. J. 1971. *Between two worlds: the political thought of Graham
Wallas.* Oxford: Oxford Univ. Press.

———. 1981. *English culture and the decline of the industrial spirit 1850–
1980.* Cambridge: Cambridge Univ. Press.

Williams, R. 1976. *Keywords: a vocabulary of culture and society.* New
York: Oxford Univ. Press.

Willier, B. H., and Oppenheimer, J. M. 1964. *Foundations of experimental
embryology.* Englewood Cliffs, N.J.: Prentice-Hall.

Wilson, E. B. 1924. Reorganization of the Naples Zoological Station. *Science*
59: 182–83.

Wilson, H. V. 1907. On some phenomena of coalescence of regeneration in
sponges. *Journal of Experimental Zoology* 5: 245–58.

Wiltshire, D. 1978. *The social and political thought of Herbert Spencer.*
Oxford: Oxford Univ. Press.

Winkler, M. G. 1984. See Bates, S. C.; Winkler, M. G.; and Riquelmi, C.
1984.

Witkowski, J. A. 1986a. See Horder, T. J.; Witkowski, J. A.; and Wylie,
C. C. 1986.

———. 1986b. The elixir of life. *Trends in Genetics* 2: 110–13.

———. 1986c. Ross Harrison and the experimental analysis of nerve
growth: the origins of tissue culture. In *A history of embryology,* pp.
149–77. See Horder, T. J.; Witkowski, J. A.; and Wylie, C. C. 1986.

Witschi, E. 1942. Hormonal regulation of development in lower vertebrates. *Cold Spring Harbor Symposia on Quantitative Biology* 10: 145–
51.

Wolin, S. 1960. *Politics and vision: continuity and innovation in western
political thought.* Boston: Little Brown.

Wolpert, L. 1986. Gradients, positions and pattern: a history. In *A history of
embryology,* pp. 347–62. See Horder, T. J.; Witkowski, J. A.; and
Wylie, C. C. 1986.

Wright, S. 1960. Genetics and twentieth century Darwinism: a review and
discussion. *American Journal of Human Genetics* 12: 365–72.

Wylie, C. C. 1986. See Horder, T. J.; Witkowski, J. A.; and Wylie, C. C.
1986.

Films and Radio Programs
Huxley, J. S., and Lockley, R. 1934. *Private life of the Gannets.*

Index